T0139831

Lecture Notes in Networks and Systems   **989**

Series Editor

Janusz Kacprzyk ⓘ, *Systems Research Institute, Polish Academy of Sciences, Warsaw, Poland*

Advisory Editors

Fernando Gomide, *Department of Computer Engineering and Automation—DCA, School of Electrical and Computer Engineering—FEEC, University of Campinas— UNICAMP, São Paulo, Brazil*

Okyay Kaynak, *Department of Electrical and Electronic Engineering, Bogazici University, Istanbul, Türkiye*

Derong Liu, *Department of Electrical and Computer Engineering, University of Illinois at Chicago, Chicago, USA*

    Institute of Automation, Chinese Academy of Sciences, Beijing, China

Witold Pedrycz, *Department of Electrical and Computer Engineering, University of Alberta, Alberta, Canada*

    Systems Research Institute, Polish Academy of Sciences, Warsaw, Poland

Marios M. Polycarpou, *Department of Electrical and Computer Engineering, KIOS Research Center for Intelligent Systems and Networks, University of Cyprus, Nicosia, Cyprus*

Imre J. Rudas, *Óbuda University, Budapest, Hungary*

Jun Wang, *Department of Computer Science, City University of Hong Kong, Kowloon, Hong Kong*

The series "Lecture Notes in Networks and Systems" publishes the latest developments in Networks and Systems—quickly, informally and with high quality. Original research reported in proceedings and post-proceedings represents the core of LNNS.

Volumes published in LNNS embrace all aspects and subfields of, as well as new challenges in, Networks and Systems.

The series contains proceedings and edited volumes in systems and networks, spanning the areas of Cyber-Physical Systems, Autonomous Systems, Sensor Networks, Control Systems, Energy Systems, Automotive Systems, Biological Systems, Vehicular Networking and Connected Vehicles, Aerospace Systems, Automation, Manufacturing, Smart Grids, Nonlinear Systems, Power Systems, Robotics, Social Systems, Economic Systems and other. Of particular value to both the contributors and the readership are the short publication timeframe and the worldwide distribution and exposure which enable both a wide and rapid dissemination of research output.

The series covers the theory, applications, and perspectives on the state of the art and future developments relevant to systems and networks, decision making, control, complex processes and related areas, as embedded in the fields of interdisciplinary and applied sciences, engineering, computer science, physics, economics, social, and life sciences, as well as the paradigms and methodologies behind them.

Indexed by SCOPUS, INSPEC, WTI Frankfurt eG, zbMATH, SCImago.

All books published in the series are submitted for consideration in Web of Science.

For proposals from Asia please contact Aninda Bose (aninda.bose@springer.com).

Álvaro Rocha · Hojjat Adeli ·
Gintautas Dzemyda · Fernando Moreira ·
Aneta Poniszewska-Marańda
Editors

Good Practices and New Perspectives in Information Systems and Technologies

WorldCIST 2024, Volume 5

 Springer

Editors
Álvaro Rocha
ISEG
Universidade de Lisboa
Lisbon, Portugal

Hojjat Adeli
College of Engineering
The Ohio State University
Columbus, OH, USA

Gintautas Dzemyda
Institute of Data Science and Digital
Technologies
Vilnius University
Vilnius, Lithuania

Fernando Moreira
DCT
Universidade Portucalense
Porto, Portugal

Aneta Poniszewska-Marańda
Institute of Information Technology
Lodz University of Technology
Łódz, Poland

ISSN 2367-3370   ISSN 2367-3389 (electronic)
Lecture Notes in Networks and Systems
ISBN 978-3-031-60226-9   ISBN 978-3-031-60227-6 (eBook)
https://doi.org/10.1007/978-3-031-60227-6

This Springer imprint is published by the registered company Springer Nature Switzerland AG
The registered company address is: Gewerbestrasse 11, 6330 Cham, Switzerland

If disposing of this product, please recycle the paper.

Preface

This book contains a selection of papers accepted for presentation and discussion at the 2024 World Conference on Information Systems and Technologies (WorldCIST'24). This conference had the scientific support of the Lodz University of Technology, Information and Technology Management Association (ITMA), IEEE Systems, Man, and Cybernetics Society (IEEE SMC), Iberian Association for Information Systems and Technologies (AISTI), and Global Institute for IT Management (GIIM). It took place in Lodz city, Poland, 26–28 March 2024.

The World Conference on Information Systems and Technologies (WorldCIST) is a global forum for researchers and practitioners to present and discuss recent results and innovations, current trends, professional experiences, and challenges of modern Information Systems and Technologies research, technological development, and applications. One of its main aims is to strengthen the drive toward a holistic symbiosis between academy, society, and industry. WorldCIST'24 is built on the successes of: WorldCIST'13 held at Olhão, Algarve, Portugal; WorldCIST'14 held at Funchal, Madeira, Portugal; WorldCIST'15 held at São Miguel, Azores, Portugal; WorldCIST'16 held at Recife, Pernambuco, Brazil; WorldCIST'17 held at Porto Santo, Madeira, Portugal; WorldCIST'18 held at Naples, Italy; WorldCIST'19 held at La Toja, Spain; WorldCIST'20 held at Budva, Montenegro; WorldCIST'21 held at Terceira Island, Portugal; WorldCIST'22 held at Budva, Montenegro; and WorldCIST'23, which took place at Pisa, Italy.

The Program Committee of WorldCIST'24 was composed of a multidisciplinary group of 328 experts and those who are intimately concerned with Information Systems and Technologies. They have had the responsibility for evaluating, in a 'blind review' process, the papers received for each of the main themes proposed for the conference: A) Information and Knowledge Management; B) Organizational Models and Information Systems; C) Software and Systems Modeling; D) Software Systems, Architectures, Applications and Tools; E) Multimedia Systems and Applications; F) Computer Networks, Mobility and Pervasive Systems; G) Intelligent and Decision Support Systems; H) Big Data Analytics and Applications; I) Human-Computer Interaction; J) Ethics, Computers & Security; K) Health Informatics; L) Information Technologies in Education; M) Information Technologies in Radiocommunications; and N) Technologies for Biomedical Applications.

The conference also included workshop sessions taking place in parallel with the conference ones. Workshop sessions covered themes such as: ICT for Auditing & Accounting; Open Learning and Inclusive Education Through Information and Communication Technology; Digital Marketing and Communication, Technologies, and Applications; Advances in Deep Learning Methods and Evolutionary Computing for Health Care; Data Mining and Machine Learning in Smart Cities: The role of the technologies in the research of the migrations; Artificial Intelligence Models and Artifacts for Business Intelligence Applications; AI in Education; Environmental data analytics; Forest-Inspired

Computational Intelligence Methods and Applications; Railway Operations, Modeling and Safety; Technology Management in the Electrical Generation Industry: Capacity Building through Knowledge, Resources and Networks; Data Privacy and Protection in Modern Technologies; Strategies and Challenges in Modern NLP: From Argumentation to Ethical Deployment; and Enabling Software Engineering Practices Via Last Development Trends.

WorldCIST'24 and its workshops received about 400 contributions from 47 countries around the world. The papers accepted for oral presentation and discussion at the conference are published by Springer (this book) in six volumes and will be submitted for indexing by WoS, Scopus, EI-Compendex, DBLP, and/or Google Scholar, among others. Extended versions of selected best papers will be published in special or regular issues of leading and relevant journals, mainly JCR/SCI/SSCI and Scopus/EI-Compendex indexed journals.

We acknowledge all of those that contributed to the staging of WorldCIST'24 (authors, committees, workshop organizers, and sponsors). We deeply appreciate their involvement and support that was crucial for the success of WorldCIST'24.

March 2024

Álvaro Rocha
Hojjat Adeli
Gintautas Dzemyda
Fernando Moreira
Aneta Poniszewska-Marańda

Organization

Conference

Honorary Chair

Hojjat Adeli   The Ohio State University, USA

General Chair

Álvaro Rocha   ISEG, University of Lisbon, Portugal

Co-chairs

Gintautas Dzemyda   Vilnius University, Lithuania
Sandra Costanzo   University of Calabria, Italy

Workshops Chair

Fernando Moreira   Portucalense University, Portugal

Local Organizing Committee

Bożena Borowska   Lodz University of Technology, Poland
Łukasz Chomątek   Lodz University of Technology, Poland
Joanna Ochelska-Mierzejewska   Lodz University of Technology, Poland
Aneta Poniszewska-Marańda   Lodz University of Technology, Poland

Advisory Committee

Ana Maria Correia (Chair)   University of Sheffield, UK
Brandon Randolph-Seng   Texas A&M University, USA

Chris Kimble	KEDGE Business School & MRM, UM2, Montpellier, France
Damian Niwiński	University of Warsaw, Poland
Eugene Spafford	Purdue University, USA
Florin Gheorghe Filip	Romanian Academy, Romania
Janusz Kacprzyk	Polish Academy of Sciences, Poland
João Tavares	University of Porto, Portugal
Jon Hall	The Open University, UK
John MacIntyre	University of Sunderland, UK
Karl Stroetmann	Empirica Communication & Technology Research, Germany
Marjan Mernik	University of Maribor, Slovenia
Miguel-Angel Sicilia	University of Alcalá, Spain
Mirjana Ivanovic	University of Novi Sad, Serbia
Paulo Novais	University of Minho, Portugal
Sami Habib	Kuwait University, Kuwait
Wim Van Grembergen	University of Antwerp, Belgium

Program Committee Co-chairs

| Adam Wojciechowski | Lodz University of Technology, Poland |
| Aneta Poniszewska-Marańda | Lodz University of Technology, Poland |

Program Committee

Abderrahmane Ez-zahout	Mohammed V University, Morocco
Adriana Peña Pérez Negrón	Universidad de Guadalajara, Mexico
Adriani Besimi	South East European University, North Macedonia
Agostinho Sousa Pinto	Polytechnic of Porto, Portugal
Ahmed El Oualkadi	Abdelmalek Essaadi University, Morocco
Akex Rabasa	University Miguel Hernandez, Spain
Alanio de Lima	UFC, Brazil
Alba Córdoba-Cabús	University of Malaga, Spain
Alberto Freitas	FMUP, University of Porto, Portugal
Aleksandra Labus	University of Belgrade, Serbia
Alessio De Santo	HE-ARC, Switzerland
Alexandru Vulpe	University Politechnica of Bucharest, Romania
Ali Idri	ENSIAS, University Mohamed V, Morocco
Alicia García-Holgado	University of Salamanca, Spain

Almir Souza Silva Neto	IFMA, Brazil
Álvaro López-Martín	University of Malaga, Spain
Amélia Badica	Universiti of Craiova, Romania
Amélia Cristina Ferreira Silva	Polytechnic of Porto, Portugal
Amit Shelef	Sapir Academic College, Israel
Ana Carla Amaro	Universidade de Aveiro, Portugal
Ana Dinis	Polytechnic of Cávado and Ave, Portugal
Ana Isabel Martins	University of Aveiro, Portugal
Anabela Gomes	University of Coimbra, Portugal
Anacleto Correia	CINAV, Portugal
Andrew Brosnan	University College Cork, Ireland
Andjela Draganic	University of Montenegro, Montenegro
Aneta Polewko-Klim	University of Białystok, Institute of Informatics, Poland
Aneta Poniszewska-Maranda	Lodz University of Technology, Poland
Angeles Quezada	Instituto Tecnologico de Tijuana, Mexico
Anis Tissaoui	University of Jendouba, Tunisia
Ankur Singh Bist	KIET, India
Ann Svensson	University West, Sweden
Anna Gawrońska	Poznański Instytut Technologiczny, Poland
Antoni Oliver	University of the Balearic Islands, Spain
Antonio Jiménez-Martín	Universidad Politécnica de Madrid, Spain
Aroon Abbu	Bell and Howell, USA
Arslan Enikeev	Kazan Federal University, Russia
Beatriz Berrios Aguayo	University of Jaen, Spain
Benedita Malheiro	Polytechnic of Porto, ISEP, Portugal
Bertil Marques	Polytechnic of Porto, ISEP, Portugal
Boris Shishkov	ULSIT/IMI - BAS/IICREST, Bulgaria
Borja Bordel	Universidad Politécnica de Madrid, Spain
Branko Perisic	Faculty of Technical Sciences, Serbia
Bruno F. Gonçalves	Polytechnic of Bragança, Portugal
Carla Pinto	Polytechnic of Porto, ISEP, Portugal
Carlos Balsa	Polytechnic of Bragança, Portugal
Carlos Rompante Cunha	Polytechnic of Bragança, Portugal
Catarina Reis	Polytechnic of Leiria, Portugal
Célio Gonçalo Marques	Polytenic of Tomar, Portugal
Cengiz Acarturk	Middle East Technical University, Turkey
Cesar Collazos	Universidad del Cauca, Colombia
Cristina Gois	Polytechnic University of Coimbra, Portugal
Christophe Guyeux	Universite de Bourgogne Franche Comté, France
Christophe Soares	University Fernando Pessoa, Portugal
Christos Bouras	University of Patras, Greece

Christos Chrysoulas	London South Bank University, UK
Christos Chrysoulas	Edinburgh Napier University, UK
Ciro Martins	University of Aveiro, Portugal
Claudio Sapateiro	Polytechnic of Setúbal, Portugal
Cosmin Striletchi	Technical University of Cluj-Napoca, Romania
Costin Badica	University of Craiova, Romania
Cristian García Bauza	PLADEMA-UNICEN-CONICET, Argentina
Cristina Caridade	Polytechnic of Coimbra, Portugal
Danish Jamil	Malaysia University of Science and Technology, Malaysia
David Cortés-Polo	University of Extremadura, Spain
David Kelly	University College London, UK
Daria Bylieva	Peter the Great St. Petersburg Polytechnic University, Russia
Dayana Spagnuelo	Vrije Universiteit Amsterdam, Netherlands
Dhouha Jaziri	University of Sousse, Tunisia
Dmitry Frolov	HSE University, Russia
Dulce Mourato	ISTEC - Higher Advanced Technologies Institute Lisbon, Portugal
Edita Butrime	Lithuanian University of Health Sciences, Lithuania
Edna Dias Canedo	University of Brasilia, Brazil
Egils Ginters	Riga Technical University, Latvia
Ekaterina Isaeva	Perm State University, Russia
Eliana Leite	University of Minho, Portugal
Enrique Pelaez	ESPOL University, Ecuador
Eriks Sneiders	Stockholm University, Sweden; Esteban Castellanos ESPE, Ecuador
Fatima Azzahra Amazal	Ibn Zohr University, Morocco
Fernando Bobillo	University of Zaragoza, Spain
Fernando Molina-Granja	National University of Chimborazo, Ecuador
Fernando Moreira	Portucalense University, Portugal
Fernando Ribeiro	Polytechnic Castelo Branco, Portugal
Filipe Caldeira	Polytechnic of Viseu, Portugal
Filippo Neri	University of Naples, Italy
Firat Bestepe	Republic of Turkey Ministry of Development, Turkey
Francesco Bianconi	Università degli Studi di Perugia, Italy
Francisco García-Peñalvo	University of Salamanca, Spain
Francisco Valverde	Universidad Central del Ecuador, Ecuador
Frederico Branco	University of Trás-os-Montes e Alto Douro, Portugal
Galim Vakhitov	Kazan Federal University, Russia

Gayo Diallo | University of Bordeaux, France
Gabriel Pestana | Polytechnic Institute of Setubal, Portugal
Gema Bello-Orgaz | Universidad Politecnica de Madrid, Spain
George Suciu | BEIA Consult International, Romania
Ghani Albaali | Princess Sumaya University for Technology, Jordan
Gian Piero Zarri | University Paris-Sorbonne, France
Giovanni Buonanno | University of Calabria, Italy
Gonçalo Paiva Dias | University of Aveiro, Portugal
Goreti Marreiros | ISEP/GECAD, Portugal
Habiba Drias | University of Science and Technology Houari Boumediene, Algeria
Hafed Zarzour | University of Souk Ahras, Algeria
Haji Gul | City University of Science and Information Technology, Pakistan
Hakima Benali Mellah | Cerist, Algeria
Hamid Alasadi | Basra University, Iraq
Hatem Ben Sta | University of Tunis at El Manar, Tunisia
Hector Fernando Gomez Alvarado | Universidad Tecnica de Ambato, Ecuador
Hector Menendez | King's College London, UK
Hélder Gomes | University of Aveiro, Portugal
Helia Guerra | University of the Azores, Portugal
Henrique da Mota Silveira | University of Campinas (UNICAMP), Brazil
Henrique S. Mamede | University Aberta, Portugal
Henrique Vicente | University of Évora, Portugal
Hicham Gueddah | University Mohammed V in Rabat, Morocco
Hing Kai Chan | University of Nottingham Ningbo China, China
Igor Aguilar Alonso | Universidad Nacional Tecnológica de Lima Sur, Peru
Inês Domingues | University of Coimbra, Portugal
Isabel Lopes | Polytechnic of Bragança, Portugal
Isabel Pedrosa | Coimbra Business School - ISCAC, Portugal
Isaías Martins | University of Leon, Spain
Issam Moghrabi | Gulf University for Science and Technology, Kuwait
Ivan Armuelles Voinov | University of Panama, Panama
Ivan Dunđer | University of Zagreb, Croatia
Ivone Amorim | University of Porto, Portugal
Jaime Diaz | University of La Frontera, Chile
Jan Egger | IKIM, Germany
Jan Kubicek | Technical University of Ostrava, Czech Republic
Jeimi Cano | Universidad de los Andes, Colombia

Jesús Gallardo Casero	University of Zaragoza, Spain
Jezreel Mejia	CIMAT, Unidad Zacatecas, Mexico
Jikai Li	The College of New Jersey, USA
Jinzhi Lu	KTH-Royal Institute of Technology, Sweden
Joao Carlos Silva	IPCA, Portugal
João Manuel R. S. Tavares	University of Porto, FEUP, Portugal
João Paulo Pereira	Polytechnic of Bragança, Portugal
João Reis	University of Aveiro, Portugal
João Reis	University of Lisbon, Portugal
João Rodrigues	University of the Algarve, Portugal
João Vidal de Carvalho	Polytechnic of Porto, Portugal
Joaquin Nicolas Ros	University of Murcia, Spain
John W. Castro	University de Atacama, Chile
Jorge Barbosa	Polytechnic of Coimbra, Portugal
Jorge Buele	Technical University of Ambato, Ecuador; Jorge Gomes University of Lisbon, Portugal
Jorge Oliveira e Sá	University of Minho, Portugal
José Braga de Vasconcelos	Universidade Lusófona, Portugal
Jose M. Parente de Oliveira	Aeronautics Institute of Technology, Brazil
José Machado	University of Minho, Portugal
José Paulo Lousado	Polytechnic of Viseu, Portugal
Jose Quiroga	University of Oviedo, Spain
Jose Silvestre Silva	Academia Military, Portugal
Jose Torres	University Fernando Pessoa, Portugal
Juan M. Santos	University of Vigo, Spain
Juan Manuel Carrillo de Gea	University of Murcia, Spain
Juan Pablo Damato	UNCPBA-CONICET, Argentina
Kalinka Kaloyanova	Sofia University, Bulgaria
Kamran Shaukat	The University of Newcastle, Australia
Katerina Zdravkova	University Ss. Cyril and Methodius, North Macedonia
Khawla Tadist	Morocco
Khalid Benali	LORIA - University of Lorraine, France
Khalid Nafil	Mohammed V University in Rabat, Morocco
Korhan Gunel	Adnan Menderes University, Turkey
Krzysztof Wolk	Polish-Japanese Academy of Information Technology, Poland
Kuan Yew Wong	Universiti Teknologi Malaysia (UTM), Malaysia
Kwanghoon Kim	Kyonggi University, South Korea
Laila Cheikhi	Mohammed V University in Rabat, Morocco
Laura Varela-Candamio	Universidade da Coruña, Spain
Laurentiu Boicescu	E.T.T.I. U.P.B., Romania

Lbtissam Abnane	ENSIAS, Morocco
Lia-Anca Hangan	Technical University of Cluj-Napoca, Romania
Ligia Martinez	CECAR, Colombia
Lila Rao-Graham	University of the West Indies, Jamaica
Liliana Ivone Pereira	Polytechnic of Cávado and Ave, Portugal
Łukasz Tomczyk	Pedagogical University of Cracow, Poland
Luis Alvarez Sabucedo	University of Vigo, Spain
Luís Filipe Barbosa	University of Trás-os-Montes e Alto Douro
Luis Mendes Gomes	University of the Azores, Portugal
Luis Pinto Ferreira	Polytechnic of Porto, Portugal
Luis Roseiro	Polytechnic of Coimbra, Portugal
Luis Silva Rodrigues	Polytencic of Porto, Portugal
Mahdieh Zakizadeh	MOP, Iran
Maksim Goman	JKU, Austria
Manal el Bajta	ENSIAS, Morocco
Manuel Antonio Fernández-Villacañas Marín	Technical University of Madrid, Spain
Manuel Ignacio Ayala Chauvin	University Indoamerica, Ecuador
Manuel Silva	Polytechnic of Porto and INESC TEC, Portugal
Manuel Tupia	Pontifical Catholic University of Peru, Peru
Manuel Au-Yong-Oliveira	University of Aveiro, Portugal
Marcelo Mendonça Teixeira	Universidade de Pernambuco, Brazil
Marciele Bernardes	University of Minho, Brazil
Marco Ronchetti	Universita' di Trento, Italy
Mareca María Pilar	Universidad Politécnica de Madrid, Spain
Marek Kvet	Zilinska Univerzita v Ziline, Slovakia
Maria João Ferreira	Universidade Portucalense, Portugal
Maria José Sousa	University of Coimbra, Portugal
María Teresa García-Álvarez	University of A Coruna, Spain
Maria Sokhn	University of Applied Sciences of Western Switzerland, Switzerland
Marijana Despotovic-Zrakic	Faculty Organizational Science, Serbia
Marilio Cardoso	Polytechnic of Porto, Portugal
Mário Antunes	Polytechnic of Leiria & CRACS INESC TEC, Portugal
Marisa Maximiano	Polytechnic Institute of Leiria, Portugal
Marisol Garcia-Valls	Polytechnic University of Valencia, Spain
Maristela Holanda	University of Brasilia, Brazil
Marius Vochin	E.T.T.I. U.P.B., Romania
Martin Henkel	Stockholm University, Sweden
Martín López Nores	University of Vigo, Spain
Martin Zelm	INTEROP-VLab, Belgium

Mazyar Zand	MOP, Iran
Mawloud Mosbah	University 20 Août 1955 of Skikda, Algeria
Michal Adamczak	Poznan School of Logistics, Poland
Michal Kvet	University of Zilina, Slovakia
Miguel Garcia	University of Oviedo, Spain
Mircea Georgescu	Al. I. Cuza University of Iasi, Romania
Mirna Muñoz	Centro de Investigación en Matemáticas A.C., Mexico
Mohamed Hosni	ENSIAS, Morocco
Monica Leba	University of Petrosani, Romania
Nadesda Abbas	UBO, Chile
Narasimha Rao Vajjhala	University of New York Tirana, Tirana
Narjes Benameur	Laboratory of Biophysics and Medical Technologies of Tunis, Tunisia
Natalia Grafeeva	Saint Petersburg University, Russia
Natalia Miloslavskaya	National Research Nuclear University MEPhI, Russia
Naveed Ahmed	University of Sharjah, United Arab Emirates
Neeraj Gupta	KIET group of institutions Ghaziabad, India
Nelson Rocha	University of Aveiro, Portugal
Nikola S. Nikolov	University of Limerick, Ireland
Nicolas de Araujo Moreira	Federal University of Ceara, Brazil
Nikolai Prokopyev	Kazan Federal University, Russia
Niranjan S. K.	JSS Science and Technology University, India
Noemi Emanuela Cazzaniga	Politecnico di Milano, Italy
Noureddine Kerzazi	Polytechnique Montréal, Canada
Nuno Melão	Polytechnic of Viseu, Portugal
Nuno Octávio Fernandes	Polytechnic of Castelo Branco, Portugal
Nuno Pombo	University of Beira Interior, Portugal
Olga Kurasova	Vilnius University, Lithuania
Olimpiu Stoicuta	University of Petrosani, Romania
Patricia Quesado	Polytechnic of Cávado and Ave, Portugal
Patricia Zachman	Universidad Nacional del Chaco Austral, Argentina
Paula Serdeira Azevedo	University of Algarve, Portugal
Paula Dias	Polytechnic of Guarda, Portugal
Paulo Alejandro Quezada Sarmiento	University of the Basque Country, Spain
Paulo Maio	Polytechnic of Porto, ISEP, Portugal
Paulvanna Nayaki Marimuthu	Kuwait University, Kuwait
Paweł Karczmarek	The John Paul II Catholic University of Lublin, Poland

Pedro Rangel Henriques	University of Minho, Portugal
Pedro Sobral	University Fernando Pessoa, Portugal
Pedro Sousa	University of Minho, Portugal
Philipp Jordan	University of Hawaii at Manoa, USA
Piotr Kulczycki	Systems Research Institute, Polish Academy of Sciences, Poland
Prabhat Mahanti	University of New Brunswick, Canada
Rabia Azzi	Bordeaux University, France
Radu-Emil Precup	Politehnica University of Timisoara, Romania
Rafael Caldeirinha	Polytechnic of Leiria, Portugal
Raghuraman Rangarajan	Sequoia AT, Portugal
Radhakrishna Bhat	Manipal Institute of Technology, India
Raiani Ali	Hamad Bin Khalifa University, Qatar
Ramadan Elaiess	University of Benghazi, Libya
Ramayah T.	Universiti Sains Malaysia, Malaysia
Ramazy Mahmoudi	University of Monastir, Tunisia
Ramiro Gonçalves	University of Trás-os-Montes e Alto Douro & INESC TEC, Portugal
Ramon Alcarria	Universidad Politécnica de Madrid, Spain
Ramon Fabregat Gesa	University of Girona, Spain
Ramy Rahimi	Chungnam National University, South Korea
Reiko Hishiyama	Waseda University, Japan
Renata Maria Maracho	Federal University of Minas Gerais, Brazil
Renato Toasa	Israel Technological University, Ecuador
Reyes Juárez Ramírez	Universidad Autonoma de Baja California, Mexico
Rocío González-Sánchez	Rey Juan Carlos University, Spain
Rodrigo Franklin Frogeri	University Center of Minas Gerais South, Brazil
Ruben Pereira	ISCTE, Portugal
Rui Alexandre Castanho	WSB University, Poland
Rui S. Moreira	UFP & INESC TEC & LIACC, Portugal
Rustam Burnashev	Kazan Federal University, Russia
Saeed Salah	Al-Quds University, Palestine
Said Achchab	Mohammed V University in Rabat, Morocco
Sajid Anwar	Institute of Management Sciences Peshawar, Pakistan
Sami Habib	Kuwait University, Kuwait
Samuel Sepulveda	University of La Frontera, Chile
Sara Luis Dias	Polytechnic of Cávado and Ave, Portugal
Sandra Costanzo	University of Calabria, Italy
Sandra Patricia Cano Mazuera	University of San Buenaventura Cali, Colombia
Sassi Sassi	FSJEGJ, Tunisia

Seppo Sirkemaa	University of Turku, Finland
Sergio Correia	Polytechnic of Portalegre, Portugal
Shahnawaz Talpur	Mehran University of Engineering & Technology Jamshoro, Pakistan
Shakti Kundu	Manipal University Jaipur, Rajasthan, India
Shashi Kant Gupta	Eudoxia Research University, USA
Silviu Vert	Politehnica University of Timisoara, Romania
Simona Mirela Riurean	University of Petrosani, Romania
Slawomir Zolkiewski	Silesian University of Technology, Poland
Solange Rito Lima	University of Minho, Portugal
Sonia Morgado	ISCPSI, Portugal
Sonia Sobral	Portucalense University, Portugal
Sorin Zoican	Polytechnic University of Bucharest, Romania
Souraya Hamida	Batna 2 University, Algeria
Stalin Figueroa	University of Alcala, Spain
Sümeyya Ilkin	Kocaeli University, Turkey
Syed Asim Ali	University of Karachi, Pakistan
Syed Nasirin	Universiti Malaysia Sabah, Malaysia
Tatiana Antipova	Institute of Certified Specialists, Russia
Tatianna Rosal	University of Trás-os-Montes e Alto Douro, Portugal
Tero Kokkonen	JAMK University of Applied Sciences, Finland
The Thanh Van	HCMC University of Food Industry, Vietnam
Thomas Weber	EPFL, Switzerland
Timothy Asiedu	TIM Technology Services Ltd., Ghana
Tom Sander	New College of Humanities, Germany
Tomasz Kisielewicz	Warsaw University of Technology
Tomaž Klobučar	Jozef Stefan Institute, Slovenia
Toshihiko Kato	University of Electro-communications, Japan
Tuomo Sipola	Jamk University of Applied Sciences, Finland
Tzung-Pei Hong	National University of Kaohsiung, Taiwan
Valentim Realinho	Polytechnic of Portalegre, Portugal
Valentina Colla	Scuola Superiore Sant'Anna, Italy
Valerio Stallone	ZHAW, Switzerland
Verónica Vasconcelos	Polytechnic of Coimbra, Portugal
Vicenzo Iannino	Scuola Superiore Sant'Anna, Italy
Vitor Gonçalves	Polytechnic of Bragança, Portugal
Victor Alves	University of Minho, Portugal
Victor Georgiev	Kazan Federal University, Russia
Victor Hugo Medina Garcia	Universidad Distrital Francisco José de Caldas, Colombia
Victor Kaptelinin	Umeå University, Sweden

Viktor Medvedev	Vilnius University, Lithuania
Vincenza Carchiolo	University of Catania, Italy
Waqas Bangyal	University of Gujrat, Pakistan
Wolf Zimmermann	Martin Luther University Halle-Wittenberg, Germany
Yadira Quiñonez	Autonomous University of Sinaloa, Mexico
Yair Wiseman	Bar-Ilan University, Israel
Yassine Drias	University of Algiers, Algeria
Yuhua Li	Cardiff University, UK
Yuwei Lin	University of Roehampton, UK
Zbigniew Suraj	University of Rzeszow, Poland
Zorica Bogdanovic	University of Belgrade, Serbia

Contents

Organizational Models and Information Systems

Implementing Scaled Agile Framework Methodology Principles in the Quality Assurance Process

Raul Ionut Riti, Andreea Cristina Ionica(✉) ⓘ, and Monica Leba ⓘ

University of Petrosani, Universitatii Street, No. 20, Hunedoara County,
332 006 Petrosani, Romania
andreeaionica@upet.ro

Abstract. In the software production realm, implementing the Scaled Agile Framework (SAFe) often presents challenges in harmonizing testing and quality assurance (QA) processes with Agile practices. This paper outlines a structured methodology for implementing SAFe principles in Quality Assurance (QA), aligning with Agile methodologies, and ensuring the delivery of high-quality software products. The methodology progresses through four distinct increment areas: Team, System, Solution, and Release. In the Team Increment, a "Scrum Tea" meeting sets the stage for product clarity and customer-centric QA design thinking, guided by testing principles. The System Increment focuses on updating the test plan/strategy and aligning testing activities with customer needs. The Solution Increment involves component testing, manual and automated testing, and test evidence archiving. Finally, the Release Increment emphasizes early issue identification and continuous integration. A case study illustrates the practical application of this methodology, showcasing how SAFe and Agile principles align with the incremental approach to deliver high-quality software products. This structured methodology provides a comprehensive framework to ensure software quality and customer satisfaction throughout the development lifecycle.

Keywords: Customer-Centricity · Quality Assurance (QA) process · Scaled Agile Framework (SAFe) methodology

1 Introduction

In the realm of software production, working with the SAFe project management methodology often reveals an imbalance in the way testing and quality assurance are carried out. This context underscores the need for a structured and systematic approach to harmonize testing with the rapidly changing software requirements and Agile processes.

The scope of the research revolves around the imperative to map a clear model for testing and QA in software production within the SAFe framework. This entails identifying a structured methodological framework that aligns testing with SAFe principles and practices, ensuring the delivery of high-quality software products.

Á. Rocha et al. (Eds.): WorldCIST 2024, LNNS 989, pp. 3–12, 2024.
https://doi.org/10.1007/978-3-031-60227-6_1

The paper structure will follow a logical and well-defined approach to delve into the subject of implementing SAFe principles in the QA process in software production. It comprises after the overview of the research context, the following sections: (2) Literature Review, that focuses on reviewing existing literature, exploring relevant resources and research related to the implementation of SAFe methodology in the QA process in software production; (3) A Structured Methodology for Implementing SAFe Principles in QA, in which is developed a well-defined methodology for implementing SAFe principles in the QA process, detailing key steps and practices; (4) Implementing SAFe Principles in QA - Case Study, that presents a concrete case study illustrating the practical application of the proposed methodology in a real software production environment; (5) Discussion, in which are analyzed the results of the case study and discussed the implications and outcomes together with the advantages and challenges of implementing SAFe in QA; and (6) Conclusion, the final section summarizing the key conclusions of the research, outlining contributions, and suggesting potential areas for further investigation.

2   Literature Review

The analysis of the existing literature was done having in mind the evolution of SAFe in Software Development, aligning quality assurance objectives with project management principles, challenges, and solutions, and measuring success and impact.

What [1] describes SAFe is how it became popular and is widely used by large-scale software development projects using agile methods. The underlying goal is to enhance alignment, collaboration, and effectiveness in the delivery process. A simplified structure is presented only in the SAFe methodology with integration and processes by [2]. When it comes to how the roles are defined in project management methodologies, [3] partially describes them but not their specific involvement in QA. [4] provides a highly organized, intricate, and elaborated description of the testing/QA process at the generative level enclosing evident samples displaying this integration with project management. [5] defines what working with User Experience entails and the impact that this definition has in a general sense on the human-machine interaction. [6] demonstrates how the impact is improved through team collaboration and pair programming but is not extrapolated to the QA level. [7] presents how at the first level of the Team, multiple members are involved, but it does not delve into a specific area like the initial entry into a clear QA process. On a different note, it is observed that [8] discusses how customer involvement in global agile projects enhances efficiency and quality in various tasks, such as user story creation, feature discussions, and client requirement prioritization. [9] examines how difficulties in the area of prioritization come into play when dealing with a larger team, specific to SAFe.

Therefore, by analyzing all the theoretical information presented above and incorporating personal contributions from production, it can be confidently asserted that a continuous loop in the SAFe methodology on how to best integrate a QA process for the development of exceptionally high-quality software products is identified. To initiate a first step outside of this loop and attempt a narrative closer to best practices, the proposal for this research has been made.

3 Structured Methodology for Implementing SAFe Principles in QA

The methodology will be applied in each Sprint of a Software Development Life Cycle (SDLC) or a Program Increment (PI) from the beginning of the Sprint to its end. The classical elements of the SAFe methodology will be combined but differentiated into the 4 increment areas for the product (see Fig. 1).

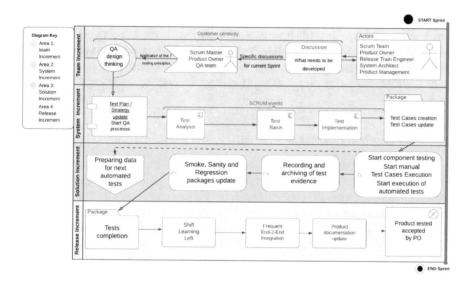

Fig. 1. Block diagram of a structured methodology for implementing SAFe Principles in QA

3.1 Area 1 - Team Increment

The team increment [7] is based on Scrum Tea – a recipe for product clarity and embracing customer-centric QA design thinking [10].

In the dynamic world of software development, ensuring a shared understanding of the product's expectations at the beginning of a sprint is crucial. To achieve this, teams often organize a "Scrum Tea" meeting, where the key ingredients are the Product Owner (PO), the Release Train Engineer (RTE), the System Architect, and members of the Product Management team [2]. This unique blend of roles creates a tea-like infusion of insights, discussions, and shared goals.

Setting the Scene: The team gathers in a collaborative atmosphere, ready to embark on a journey of understanding and alignment. The PO, like a tea connoisseur, is responsible for presenting the product backlog, and the list of items ready for selection, and articulating the vision and priorities. The System Architect discusses the current architecture and any constraints that might influence the sprint. The RTE guides the discussion and facilitates, keeping the conversation on track, just as a well-designed teapot dispenses the tea without spills. The Product Management team brings customer feedback, market analysis, and any changes in product strategy to the table.

Brewing the Scrum Tea - During the Scrum Tea meeting, the team works together to clarify doubts and expectations, to ensure that each backlog item is well-understood, properly prioritized, and aligned with the sprint goal. The outcome is a shared understanding of what the sprint entails, what to prioritize [9], and any technical considerations that might affect the development [8].

Following the initial Scrum Tea meeting and subsequent discussions, the Scrum team proceeds with a specific and collaborative discussion involving the Scrum Master (SM) [3], the QA team, and the PO as they delve into QA design thinking with a strong emphasis on customer-centricity. To guide their approach, they apply the 7 testing principles from the ISTQB Foundation Level Syllabus: Testing shows the presence, not the absence of defects, Exhaustive testing is impossible, Early testing saves time and money, Defects cluster together, Tests wear out, Testing is context-dependent, Absence-of-defects is a fallacy [4].

3.2   Area 2 - System Increment

It's essential to recognize the significance of this phase where the test plan/strategy is updated, signifying the initiation of the QA process. The update process ensures that the test plan remains current, aligned with evolving project requirements, and in harmony with the changing needs of the customer. Here, a series of critical activities come into play, including test analysis, test basis, test design, and test implementation [4], all orchestrated in alignment with Scrum events with the end in test case creation or update.

The QA team conducts thorough test analysis, based on the awareness that the client's perspective is paramount and continues with the basis of testing, referring to the documentation, specifications, and artifacts that form the foundation for creating test cases.

During the test design phase, the QA team meticulously crafts test case templates that mirror real-world user scenarios, and all steps are based on Scrum events serving as the guiding pillars that facilitate coordination, communication, and adaptation.

3.3   Area 3 – Solution Increment

The activities in this phase encompass component testing initiation, manual test case execution, and automated testing. These activities ultimately lead to the recording and archiving of test evidence. Following this, an update can be made to the three test packages: smoke, sanity, and regression, along with the preparation of data for future automation testing.

"Area 3 - Solution Increment" [11] marks the commencement of component testing. This phase focuses on testing individual software components or modules in isolation. Component testing ensures that each part of the software functions correctly before integration.

At the same time, manual test cases are executed meticulously. Testers follow predefined steps and instructions to verify the software's functionality, user interface, and overall performance. Manual testing is an integral part of quality assurance [12] and ensures that the software aligns with customer expectations.

In parallel with manual testing, automated testing is employed. Automated test scripts are executed using testing tools, streamlining the process and allowing for quicker and repeatable testing. Automation helps identify issues early in the development cycle and ensures that the software remains reliable.

As tests are executed, test evidence, including logs, screenshots, and results, is meticulously recorded and archived. This documentation is crucial for traceability, auditing, and addressing any potential issues or defects in the software.

Following the execution of manual and automated tests, the three essential test packages are updated:

Smoke Testing: executed to ensure that the most critical and fundamental functions of the software are working correctly. This helps confirm the software's basic readiness for further testing.

Sanity Testing: performed to verify that specific functionalities or components have been fixed or improved. This testing aids in ensuring that recent changes have not introduced new defects.

Regression Testing: executed to confirm that new changes or features have not adversely affected existing functionalities. It helps maintain the overall stability of the software.

In the end, data and scenarios are prepared for future automation testing. Automation scripts are developed, and test data is organized to be readily available for subsequent testing cycles. This proactive approach ensures that testing remains efficient and repeatable in the future.

3.4   Area 4 – Release Increment

In the final phase, "Area 4 - Release Increment," the focus is on initiating the test completion phase with a well-defined foundation provided by a continuous delivery pipeline. This pipeline comprises three key components:

- Shift Left Learning from SAFe [12]: This concept emphasizes early testing and quality assurance integration into the development process. By identifying and addressing issues as early as possible in the development cycle, the team minimizes risks and ensures a smoother product release.
- Frequent End-to-End Integration: Frequent end-to-end integration involves continually combining and validating various components of the software. This approach ensures that the product functions cohesively as a whole, uncovering integration issues early and allowing for timely resolution.
- Product Documentation Update: Updating product documentation is essential to maintain accuracy and ensure that all product-related information is current and reflects the latest changes and features.

This methodology is designed to conclude at the end of each sprint with a product that has been thoroughly tested and is accepted by the PO. The ultimate goal is to deliver a high-quality product that aligns with customer expectations and is ready for release.

So, the focus shifts from testing execution to test completion and product readiness. It emphasizes the importance of early testing, seamless integration, and accurate documentation to ensure that the product is not only thoroughly tested but also accepted by the PO, marking a successful conclusion to the sprint. This customer-centric approach ensures that the software aligns with user expectations and delivers value.

4   Implementing SAFe Principles in QA - Case Study

The considered project is a project intended for internal use to promote tests in the area of training for employees to obtain certifications, as well as the extraction of reports by company management.

4.1   Area 1 - Team Increment

Setting the Scene: As the dynamics of the project are reduced, the project is one of internal innovation interest, the team is coordinated by the PO who also plays the role of product management and SM. At this stage, the discussions are quite clear as to the factors involved and the articulated vision is also quite clear.

On the RTE position, to a certain extent, are colleagues from the User Interface (UI) / User Experience (UX) department as they play an essential role, the final product being a web-based application, with a major focus in this area, so it is between the general architecture and mockups.

The customer-centricity area is well evidenced by the fact that even the roles described above are directly involved as end users, and the specific discussions for each Sprint have also emphasized the department each role comes from. Thus, the QA team of the project can have a very clear overview of the project and direct involvement of the testing principles.

4.2   Area 2 – System Increment

Having completed the previous discussions, it is more than appropriate for the QA team to do the update in a Confluence space where the testing plan and strategy are structured.

These are directly impacted by all the new information obtained. After this phase, the testing process takes a new shape thanks to the Test Analysis phase with the identification of the test basis.

This test basis is mostly given by the design found with the Figma tool but also by the User Stories - which are stored in a Jira, as a Project Management tool and will lead to a clearer design for the test cases, and in the Implementation phase with a new set of parameters.

It can be concluded in the second phase that test case developments take place based on the new information but also an update of the ones from previous sprints that will be reused.

4.3 Area 3 – Solution Increment

The third stage opens with the team of developers, in this case, two people, in the middle of the Sprint, starting coding for writing Component tests which are then used at this level.

In parallel with them, the QA team works on the manual testing side by executing tests that will be a solid basis for the automated tests in future sprints.

They are also starting to execute automated tests and write new automated tests based on the information from the past sprint.

All the results of the two types of tests are analyzed clearly and specifically so a new basis for the Regression, Sanity, and Smoke test packages can be used successfully - these tests are also grouped in a Confluence space.

4.4 Area 4 – Release Increment

This milestone shows that the official end of the testing process is taking place with a focus on this continuous delivery through the SAFe/Agile prism.

Concretely it is observed that problems found are dealt with early and continuous integration scenarios are realized.

As the information is quite obvious, and approaching the end of the Sprint, it is more than mandatory that the product gets updates for its documentation and the PO can accept and give a final Sign-off for what has been developed in this Sprint in an assumed way.

Each of the four areas has a specificity for quality measures that are objectively achieved through their evaluation (see Fig. 2).

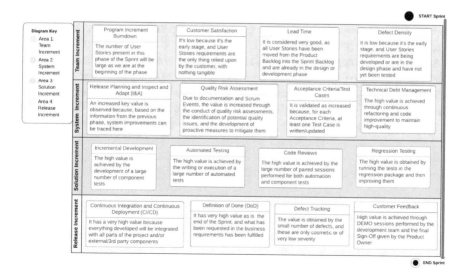

Fig. 2. Quality measures assessed in all four increments

5 Discussion

The examined project is an internal initiative aimed at facilitating employee training and certification while providing insightful reports for company management. In the context of Agile development, this project evolves through distinct increments, each with a unique focus. These increments seamlessly align with Agile and SAFe principles.

Area 1 - Team Increment: In this initial stage, the project dynamics are contained, and the PO wears multiple hats. The UI/UX department plays a pivotal role, emphasizing the web-based nature of the application. The measurable part of this Increment includes the creation of the application design (four wireframes for the login area, user set-up, user page, and reporting area for management), the finalization of discussions between UI/UX and the PO, and the initiation of the writing of test case manuals concerning the design.

Area 2 - System Increment: Building upon the initial discussions, the QA team updates the testing plan and strategy in a Confluence space. This phase leads to a refined test case design. Existing test cases from previous sprints are updated and reused. The increment's focus is on preparing for rigorous testing. The measurable part of this increment includes the adaptation of manual test cases to the final design approved by the PO, the writing of the final steps in the project plan and project strategy, the calibration of the final test base by design, user stories, and the team, and the final realization of the manual test suite.

Area 3 - Solution Increment: In this phase, development and testing occur simultaneously. Component tests are written by developers, while the QA team conducts manual/automation testing. Results are meticulously analyzed, forming the basis for Regression, Sanity, and Smoke test packages, neatly organized in Confluence. The measurable part of this increment includes the execution of all manual tests against the 4 design components, the writing of ~10 component tests by 2 developers – working pair programming, the writing of 5 defect reports, and the creation of regression packages (10 test cases, 5 for smoke, 5 for sanity).

Area 4 - Release Increment: The final stage marks the official conclusion of the testing process within the SAFe framework. Clear and comprehensive information near the end of the sprint necessitates documentation updates. The PO provides final acceptance, signifying the completion of the sprint's objectives. The measurable part of this increment includes the closure and confirmation testing for all 5 defects, a DEMO session for the client, the verification of the correct integration of the design with the specifications provided by the PO, and the acceptance of the final product. Additionally, a new page has been added to the Confluence tool to enhance the understanding of the application's flow.

The project's structured evolution through these increments showcases the practical alignment of SAFe and QA. Each phase plays a crucial role in delivering a high-quality product that meets end-user expectations. This approach not only enhances software quality but also promotes efficient development and testing practices, ultimately benefiting both internal users and management.

6 Conclusion

The proposed structured methodology is a framework that integrates SAFe principles into the QA process, including the following modern approaches:

- Holistic Sprint Integration: This structured methodology aligns with SAFe principles, ensuring that each sprint within a Program Increment (PI) is a cohesive and customer-centric entity.
- Scrum Tea for Clarity: "Area 1 - Team Increment" establishes the importance of early clarity and customer-centricity through the concept of a "Scrum Tea" meeting and is based on robust QA Design Thinking emphasizing a customer-focused approach.
- Structured Test Planning: "Area 2 - System Increment" introduces a well-structured test plan/strategy, marking the commencement of the QA process.
- Efficient Component Testing: "Area 3 - Solution Increment" focuses on component testing and manual and automated testing. Test Package Update and Automation Readiness: The update of test packages, including smoke, sanity, and regression testing, ensures that the software's stability is maintained.
- Shift Left Learning and Continuous Integration: "Area 4 - Release Increment" underlines the importance of "shift left" learning, frequent end-to-end integration, and product documentation updates.

In conclusion, it can be observed that the presented methodology represents a foundation, a pillar-based structure in the field of project management with SAFe, with a focus on high-quality standards. For future enhancement, we envision the incorporation of additional components into our operational framework. This will involve seamlessly integrating external teams, leveraging 3rd-party resources, enhancing management practices, and engaging additional colleagues. This expansion will be facilitated through a holistic approach encompassing continuous evaluation, contextual customization, skill and knowledge expansion, ongoing collaboration, adaptive responses to technological changes, and the systematic sharing of experiences. By embracing these elements, we aim to fortify our capabilities, foster a dynamic and responsive environment, and ensure sustained growth and innovation in our organizational endeavors.

References

1. Knaster R., Leffingwell, D.: SAFe 5.0 Distilled: Achieving Business Agility with the Scaled Agile Framework. Addison-Wesley Professional (2020)
2. Almeida, F., Espinheira, E.: Large-scale agile frameworks: a comparative review. J. Appl. Sci. Manag. Eng. Technol. **2**(1) (2021)
3. Mulcahy, R.: PMP Exam Prep: Review Material, Explanations, Insider Tips, Exercises, Games, and Practice Exams: to Pass PMI's PMP and CAPM Exams, p. 39. RMC Publications, Minnetonka, Minnesota (2020)
4. Albert L., et al.: Certified tester foundation level syllabus v4.0. International Software Testing Qualifications Board, pp. 17–18 (2023)
5. ISO 9241–210: Ergonomics of human system interaction-part 210: Human-centred design for interactive systems. International Standardization Organization (ISO) (2019)

6. Smite, D., Mikalsen, M., Brede Moe, N.M., Stray, V., Klotins, E.: From collaboration to solitude and back: remote pair programming during COVID-19. In: Agile Processes in Software Engineering and Extreme Programming, 22nd International Conference on Agile Software Development, XP 2021 Virtual Event, Proceedings, p. 4 (2021)
7. Beecham, S., Clear, T., Lal, R., Noll, J.: Do scaling agile frameworks address global software development risks? An Empirical Study J. Syst. Softw. **171** (2021)
8. Shameem, M., Khan, A.A., Hasan, Md.H., Akbar, M.A.: Analytic hierarchy process based prioritisation and taxonomy of SuccessFactors for scaling agile methods in global software development. IET Softw. **14**(4), 319–450 (2020)
9. Gustavsson T.: Inter-team Coordination in Large-Scale Agile Software Development Projects, pp. 43–44. Karlstad University Studies (2020)
10. Design Thinking page. https://scaledagileframework.com/design-thinking/. Accessed 29 Oct 2023
11. Framework page. https://scaledagileframework.com/. Accessed 29 Oct 2023
12. Built-In-Quality page. https://scaledagileframework.com/built-In-quality/. Accessed 29 Oct 2023

Using Knowledge Graph and KD-Tree Random Forest for Image Retrieval

Nguyen Thi Dinh[1], Thanh Manh Le[2], and Thanh The Van[3(✉)]

[1] Ho Chi Minh City University of Industry and Trade, HoChiMinh City, Vietnam
dinhnt@huit.edu.vn
[2] University of Sciences, Hue University, Hue, Vietnam
lmthanh@hueuni.edu.vn
[3] HCMC University of Education, HoChiMinh City, Vietnam
thanhvt@hcmue.edu.vn

Abstract. Semantic-based image retrieval has recently become popular and contains many challenges. This work proposes a framework for semantic-based image retrieval by using a Knowledge Graph and a KD-Tree Random Forest. In the preprocessing phase, firstly, we use the R-CNN network to segment the image set into object images, extract features for object images, and classify object images by using the KD-Tree. Secondly, a KD-Tree Random Forest is used to classify the visual relationship of each pair of objects. Thirdly, a Relationship Graph is created based on object images and their visual relationships. Finally, a Knowledge Graph is built and applied to a semantic-based image retrieval system. After segmenting the objects using the R-CNN network for each query image, the KD-Tree is used to classify the objects, and a KD-Tree Random Forest is used to predict their visual relationships. Then, a SPARQL query is generated to retrieve based on a Knowledge Graph and extract similar images. The experiment results with the Oxford Flower-17 and Flickr image sets showed that the precision achieved scores of 0.8586 and 0.8189, respectively. These results are compared with other works to demonstrate the correctness of the proposed method.

Keywords: KD-Tree Random Forest · Relationship graph · Knowledge Graph · Image retrieval

1 Introduction

In recent years, the knowledge graph has attracted the attention of image retrieval and other fields. The applications of the knowledge graph are shown in many problems such as image retrieval, image captioning, information retrieval, visual question answering, etc. [1]. Therefore, a knowledge graph for the image retrieval problem is necessary in the context of digital image data increasing over time. The application of machine learning to computer vision makes the image retrieval problem that is performed by many different methods [1]. At the same time, improving image retrieval accuracy is necessary to adapt to human needs in education, healthcare, transportation, etc.

The main contributions of the article include (1) creating a Relationship Graph to describe image content; (2) building a Knowledge Graph on experimental image sets to describe image semantics; (3) proposing a model for a semantic-based image retrieval system; (4) evaluating image retrieval accuracy of the proposed model on image sets Oxford Flower-17 [15] and Flickr [16].

2   Related Works

The paper performs many stages including segmenting object images, classifying visual relationships between objects, building the scene graph, and building the knowledge graph for the image retrieval problem. Therefore, some related works are surveyed such as:

The Authors Rich Cheng et al. [2] and Madhusri Maity et al. [3] conducted surveys to compare and evaluate the advantages, and disadvantages of several image object segmentation recognition techniques such as the R-CNN and YOLO network. In 2021, Shet Reshma Prakash et al. [4] compared object detection techniques based on the R-CNN and Faster R-CNN networks to identify test objects on PASCAL VOC 2012 and ILSVRC 2016 image sets have good recognition performance for different parameters. Based on these surveyed works and the research group's resources, we used the R-CNN network for image segmentation in this experiment system.

In 2020, Sijin Wang et al. [5] built a scene graph that described the relationship between objects in an image and applied to the text-image retrieval problem. The authors extracted objects, visual relationships between objects, and text to form Visual Scene Graphs (VSG) and Text Scene Graphs (TSG). Finally, the Scene Graph Matching Model (SGM) was designed for the image retrieval problem. Experimental results on the Flickr image set show had high efficiency for image-text retrieval. N. Roopak & Gerard Deepak [6] used knowledge graphs to determine visual relationships between objects on images. In particular, each image is identified, and the relationship between each pair of objects is determined. Dehai Zhang et al. [7] used a knowledge graph to classify images after segmentation on the ImageNet set with higher accuracy than some other works. Then, Wen-tian Zhao et al. [8] built a knowledge graph that was applied to the extracting image captions problem.

Alireza Zareian et al. [9] presented the relationship between a scene graph and a knowledge graph. The authors also presented a unified formulation of these two constructs where a scene graph is seen as an image-conditioned instantiation of a knowledge graph. The work was considered quite detailed in showing the relationship between the scene graph and the knowledge graph. Besides, Wentian Zhao et al. [8] have proposed a novel approach that constructs a multi-modal knowledge graph to associate the visual objects and the relationship between objects simultaneously with the help of external knowledge collected from the web.

Based on the analysis of related works on image segmentation using the R-CNN network, image retrieval using a scene graph and a knowledge graph are feasible. However, the number of image retrieval works that combine these components is limited. Besides, based on inheriting the KD-Tree structure that classifies the visual relationships between objects from the work [10]. In this paper, the proposed model uses the R-CNN network, a KD-Tree Random Forest, and a Knowledge Graph that is applied to semantic-based image retrieval. These theories are presented in Sect. 3, and the image retrieval model is proposed in Sect. 4.

3 Proposed Theoretical

3.1 A KD-Tree Random Forest for Visual Relationship Classification

The original image set is segmented by the R-CNN network into object images and extracted features to build a KD-Tree. The KD-Tree structure for classifying the visual relationship between a pair of objects is inherited from our work [10]. On this basis, a KD-Tree Random Forest has been built and trained from the work [11, 12] and Algorithm 1 presents the process of visual relationship classification. The input data of Algorithm 1 includes the KD-Tree Random Forest and object image; the output data is the visual relationship for each pair of objects. Each pair of objects is classified by multiple KD-Trees that have different results. Figure 1 illustrates the visual relationship classification process using a KD-Tree Random Forest. The process of classifying the visual relationship follows Algorithm 1 with the highest voting result on the KD-Tree Random Forest (Fig. 1).

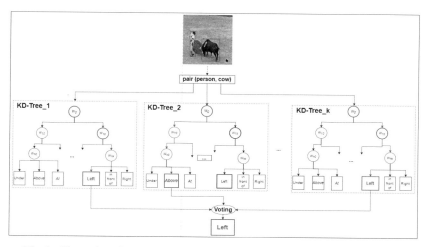

Fig. 1. Illustration visual relationship classification by a KD-Tree Random Forest

Algorithm 1 classifies a visual relationship between object consisting of many KD-Trees. The process of the visual relationship classification between pairs of objects is inherited from the work [10].

Algorithm 1. Classifying visual relationship by using a KD-Tree Random Forest

1.	**Input**: $Vector\ f_{Im}, f_{In}\ of\ object\ image\ I_m,\ I_n$; $k\ KD-Tree$;
2.	**Output**: Visual relationship of $I_m,\ I_n\ (VR_I)$;
3.	**Function CVRKD** $(f_{Im}, f_{In},\ k\ KD-Tree)$
4.	**Begin**
5.	$VR_I = \emptyset$;
6.	**Foreach** $(\ KD-Tree_i\ in\ Random\ Forest)\ do$
7.	Classifying visual relationship by using $KD-Tree_i$;
8.	**EndForeach**;
9.	$FinalVR_I(I_m,\ I_n) = The\ final$ visual relationship $with\ the\ maximum$ $number\ of\ vote\ is\ choosen$;
10.	$Visual\ relationship\ of\ I_n,\ I_m = FinalVR_I(I_m,\ I_n)$;
11.	**Return** $Visual\ relationship\ of\ I_n,\ I_m$;
12.	**End.**

3.2 Process of Building Relationship Graph

In this work, a graph describing the relationship between pairs of objects on an image is named a Relationship Graph. The Relationship Graph is a product that describes the visual relationship between each pair of objects on an image as a triple <objecti>-<predi-cate>-<object>. If the segmented image has many objects, we select up to 5 objects with the largest area. All the objects and the visual relationships between these objects have been used to build a Relationship Graph. In this article, the visual relationship between objects includes On, left, behind, in front of, etc. Figure 2 illustrates a Relationship Graph for the image 1920465.jpg (Flickr).

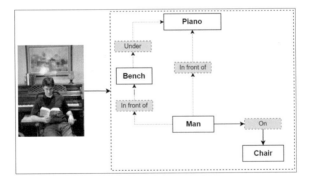

Fig. 2. Illustration a Relationship Graph for the image 1920465.jpg (Flickr)

The output results of Algorithm 1 are the visual relationships of the image that used as input data for Algorithm 2.

Algorithm 2. Creating a Relationship Graph of an image

1. **Input**: $Object\ images\ O_1, ..., O_m$; $visual\ relationships\ VR_1, ..., VR_k$;
2. **Output**: Relationship Graph (RG) of image I;
3. **Function BSG** $(VR_1, ..., VR_k, O_1, ..., O_m)$
4. **Begin**
5. $RG = \emptyset$; $SR_I = \emptyset$;
6. **For** $(int\ i = 0;\ i < m;\ i++)$ **do**
7.   **For** $(int\ j = 0;\ j < k;\ j++)$ **do**
8. $SR_I = O_i + VR_j + O_{i+1}$, ;
9.   **EndFor**;
10. **EndFor**;
11. $RG = RG \cup SR_I$;
12. **Return** RG;
13. **End.**

3.3   Process of Building Knowledge Graph

There are many ways to build a knowledge graph. The knowledge graph has been applied to many problems as information retrieval [13], image captioning [8], visual question answering [14], image retrieval [5], etc. In this paper, a graph storage classes, objects, individuals, visual relationships, and many other information on an image set is named Knowledge Graph. A Knowledge Graph built for semantic-based image retrieval on the Flickr and Oxford Flower-17 image sets. The process of building a Knowledge Graph requires preparing data including (1) classes-superclasses; (2) object images; (3) individuals; (4) visual relationships. Figure 3, illustrates the classes/superclasses and objects images on the Flickr. Figure 4 describes a Knowledge Graph of the Flickr image set.

In this section, a Knowledge Graph construction process is created using a semi-automatic method, in which the Knowledge Graph framework is built automatically and combined with some manual operations such as preparing input data for each experimental image set. Therefore, this Knowledge Graph framework can be extended to many image sets, relationships, and object properties. The new contribution to this research as a Knowledge Graph framework is built by using an automatic method. Based on this, a Knowledge Graph was developed for other image sets and added object attributes to the Knowledge Graph framework.

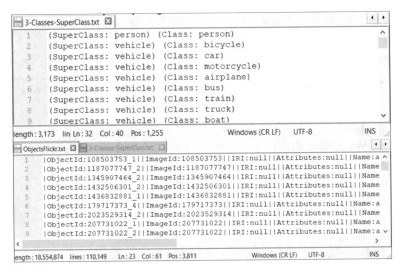

Fig. 3. Illustration classes/superclasses and objects images on the Flickr image set

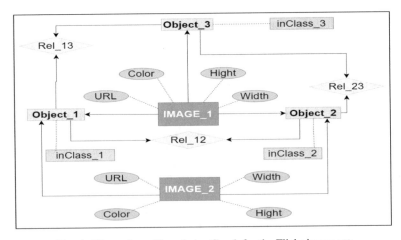

Fig. 4. Illustration a Knowledge Graph for the Flickr image set

Algorithm 3. Building a Knowdledge Graph of image set

1.   **Input**: Set of Relationship Graph on image set $(RG_1, ..., RG_k)$;
2.   **Output**: Knowledge Graph (KG) of image set;
3.   **Function BKG** $(RG_1, ..., RG_k, KG)$
4.   **Begin**
5. $KG = \emptyset$;
6. CI = Create Classes and Individual of classes from image dataset;
7. Ob = Create Objects for Knowledge Graph KG;
8. In = Create Individual of image from relationship graph RG_i;
9. Re = Create Relationships for Knowledge Graph KG;
10. $KG = KG \cup CI \cup Ob \cup In \cup Re$;
11. **Return** KG;
12. **End.**

Algorithm 3 builds a Knowledge Graph for the experimental image set, it is necessary to prepare data: Classes of image sets, individual classes, objects of images, individual images, and relationships of objects. So, lines 6, 7, and 8 in Algorithm 3 do these things.

4   A Model of Proposed Image Retrieval

The proposed model of image retrieval has a preprocessing phase and image retrieval phase (Fig. 5) with the following steps:

(1) Object image is segmented by using the R-CNN network;
(2) Extracting features for the object images;
(3) Using the KD-Tree for object image classification. The KD-Tree inherited from our work [23];
(4) Building a KD-Tree Random Forest to classify visual relationships;
(5) Building a Relationship Graph;
(6) Building a Knowledge Graph for the experimental image set;
(7) A query image is segmented by using the R-CNN network;
(8) Classifying object images by using the KD-Tree. In this step, the KD-Tree inherited from our work [23];
(9) Classifying visual relationships between objects on query image by using the KD-Tree Random Forest;
(10) Creating a SPARQL query by objects and visual relationship;
(11) Retrieving on the Knowledge Graph by using SPARQL query;
(12) Extracting a set of images that are semantically similar to the query image.

This proposed model uses the R-CNN network to segment the object image region. The KD-Tree is used to classify the object image that meets the growing number of classifiers for the experimental image set and is scalable for other image sets.

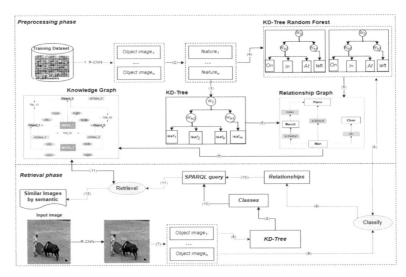

Fig. 5. A proposed model by using Knowledge Graph and KD-Tree Random Forest

Agorithm 4. Retrieving a set of similar images based on the Knowledge Graph

1.	**Input**: Query image I, Knowledge Graph KG;
2.	**Output**: Set of *similar image of I image* (SI);
3.	**Function $SIKG(I, KG)$**
4.	**Begin**
5.	$SI = \emptyset$;
6.	*Extracting objects by using the R-CNN for input image;*
7.	*Extracting features for object images;*
8.	*Classify visual relationships by using the KD-Tree Random Forest;*
9.	*Create a SPARQL query from object images and visual relationship;*
10.	$SI = Retrieve(SPARQL, KG)$;
11.	**Return** SI;
12.	**End.**

5   Experiment and Evaluate Results

The image retrieval system SB-KGF is built on the .NET Framework 4.5 platform the C# programming language. The SB-KGF system is performed on computers with Intel(R) Core i7-5200U processors, CPU 2.70 GHz, RAM 16 GB, and Windows 10 Professional operating systems. Server configuration for training visual relationship classification by using KD-Tree Random Forest: Xeon(R) Gold 6258R CPU 2.70 GHz CPU, 1024 GB SSD, 16 GB RAM, Server Datacenter 2019 operating system. The experimental data are described in Table 1.

Figure 6 describes the 189 component features that are extracted from the image sets.

Table 1. Describe experimental image sets

Data sets	Number of images	Training images	Testing images	Validation images
Oxford Flower-17 [15]	1,360	1,000	360	360
Flickr [16]	31,783	29,000	1,783	1,000

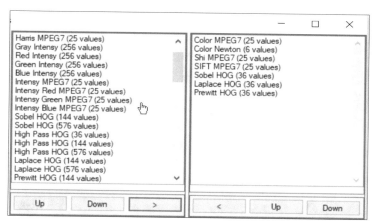

Fig. 6. Illustration of feature extraction for a Flickr image set

Figure 7 describes the process of building a Knowledge Graph (*Knowledge Graph processing*) and includes the steps: Create classes-superclasses and individuals; create object images and object properties; create visual relationships; create data and a Knowledge Graph. Then, the Knowledge Graph is loaded (*Load Knowledge Graph*). A query image is segmented by using the R-CNN (*R-CNN*); an object image is classified by the KD-Tree; visual relationships are classified by using a KD-Tree Random Forest (*Visual Relationship Classification*). A SPARQL query is created (*Create SPARQL*) and retrieved on the Knowledge Graph (*Image Retrieval*).

Figure 8 describes a set of similar images to the query image 109982467.jpg (Flickr). The experimental results of image retrieval by using the Knowledge Graph and KD-Tree Random Forest on Oxford Flower-17 and Flickr image sets with validation images and precision, recall, F-measure, and average time query that are presented in Table 2.

The results of image retrieval by using the Knowledge Graph and KD-Tree Random Forest have some limitations: (1) classifying visual relationships between objects by using the KD-Tree random Forest does not reach 100% performance; (2) the process of building a Relationship Graph and Knowledge Graph depending on visual relationship classification results; (3) the process of retrieving for similar image sets on Knowledge Graph has errors from classifying relationships between objects on the input image.

Therefore, to improve image retrieval performance, we need to do: (1) increase KD-Tree Random Forest training for better visual relationship classification results; (2) add

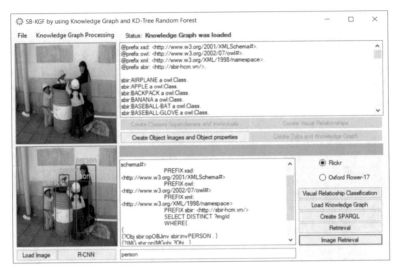

Fig. 7. Illustration of the SB-KGF image retrieval system

Fig. 8. The results of similar images for the image 109982467.jpg (Flickr)

Table 2. Experimental results of the SB-KGF image retrieval system

Image sets	Validation images	Precision	Recall	F-measure	Average time query (ms)
Oxford Flower-17 [15]	360	0.8586	0.7349	0.7919	45.02
Flickr [16]	1,000	0.8189	0.6677	0.7356	79.18

object image attributes and visual relationships to the Knowledge Graph to enrich image semantics for retrieving similar image sets more accurately.

The ROC graphs on the Flickr and Oxford Flower-17 image sets were performed by Matlab 2015 as shown in Fig. 9. Each curve on the map description of retrieval results with precision and coverage of an image subject in image sets.

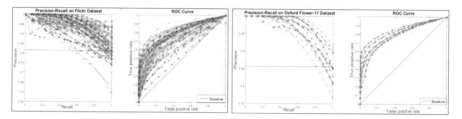

Fig. 9. Precision, Recall and ROC curve on Flickr and Oxford Flower-17 image sets

Table 3 compares test image retrieval precision with other methods.

Table 3. Comparison of Mean Average Precision (MAP) on experimental image sets

Method (authors)	Image sets	MAP
IRSGS-GCN, TopK = 5, (Yoon, S., et al., 2020) [17]	Flickr	0.5670
CAMP, TopK = 5, (Wang Z., et al., 2019) [18]	Flickr	0.7710
BGAN-48 bit (Song et al., 2018) [19]	Flickr	0.7030
SB-KGF	**Flickr**	**0.8189**
CBIR (S. Gao, et al., 2014) [20]	Oxford Flower-17	0.7343
CBIR (Ahmed, et al., 2019) [21]	Oxford Flower-17	0.7710
CBIR-BoVW, (Samy Bakheet, et al., 2023) [22]	Oxford Flower-17	0.8494
SB-KGF	**Oxford Flower-17**	**0.8586**

The experimental results in the SB-KGF system are higher than some of the works chosen for comparison because (1) The KD-Tree Random Forest is used to classify visual relationships between objects and the KD-Tree is used to classify object images; (2) Combining the Knowledge Graph with the KD-Tree Random Forest contributes to improving the accuracy of semantically similar image retrieval. The selected works for comparison in Table 3 are mostly performed by using machine learning techniques. Therefore, building a Knowledge Graph for semantic-based image retrieval is a development step and improving the ontology from our previous works.

In the author's work [10], we experimented on the same Flickr image set with an average retrieval precision of 0.9163; the result of using a relationship KD-Tree to determine the relationship between objects on the image. From there, a set of similar images based on triples are retrieved on the ontology. This work is quite complicated, and at the same time improves the limitations of ontology in works [10]. However, the accuracy results of [11] work are higher on the Flickr image set, but the cost of training the KD-Tree Random Forest is quite high. At the same time, this work has improved the overall query time compared to the work [11]. However, in the process of experimenting with the Flickr image set, we have seen some limitations in the process of building a Knowledge Graph that needs to be improved in further development.

6   Conclusion and Development Orientation

In this paper, we built an image retrieval system by using a Knowledge Graph and a KD-Tree Random Forest. The newer system's precision scores of the Flickr and Oxford Flower-17 are 0.8189 and 0.8586, respectively. The proposed image retrieval method can be applied to multi-object image sets and the objects in the image are related to each other. Some nature and landscape image sets can be used for this experiment as MS-COCO and Visual Genome based on this Knowledge Graph framework. The Knowledge Graph for semantic-based image retrieval is a major contribution to this work. The set of similar images is an initial result of using this Knowledge Graph. There is still a problem with the improvement and development of a Knowledge Graph to improve semantic-based image retrieval accuracy and image description. A Visual Question-answering problem that our research group is also aiming to perform.

Acknowledgment. The authors would like to thank the Faculty of Information Technology, University of Sciences, Hue University for their professional advice for this study. We would also like to thank HCMC University of Industry and Trade, HCMC University of Education which are sponsors of this research. We also thank anonymous reviewers for their helpful comments on this paper.

References

1. Zou, X.: A survey on application of knowledge graph. J. Phys. Conf. Ser. **1487**(1), 012016 (2020)
2. Cheng, R.: A survey: comparison between convolutional neural network and YOLO in image identification. J. Phys. Conf. Ser. **1453**(1), 012139 (2020)
3. Maity, M., Banerjee, S., Chaudhuri, S.S.: Faster R-CNN and yolo based vehicle detection: a survey. In: 2021 5th International Conference on Computing Methodologies and Communication (ICCMC), pp. 1442–1447. IEEE (2021)
4. Prakash, S.R., Singh, P.N.: Object detection through region proposal based techniques. Mater. Today Proc. **46**, 3997–4002 (2021)
5. Wang, S., Wang, R., Yao, Z., Shan, S., Chen, X.: Cross-modal scene graph matching for relationship-aware image-text retrieval. In: Proceedings of the IEEE/CVF Winter Conference on Applications of Computer Vision, pp. 1508–1517 (2020)
6. Roopak, N., Deepak, G.: OntoKnowNHS: ontology driven knowledge centric novel hybridised semantic scheme for image recommendation using knowledge graph. In: Villazón-Terrazas, B., Ortiz-Rodríguez, F., Tiwari, S., Goyal, A., Jabbar, M.A. (eds.) Knowledge Graphs and Semantic Web. CCIS, vol. 1459, pp. 138–152. Springer, Cham (2021). https://doi.org/10.1007/978-3-030-91305-2_11

7. Zhang, D., et al.: Knowledge graph-based image classification refinement. IEEE Access **7**, 57678–57690 (2019)
8. Zhao, W., Wu, X.: Boosting entity-aware image captioning with multi-modal knowledge graph. IEEE Trans. Multimedia (2023)
9. Zareian, A., Karaman, S., Chang, S.-F.: Bridging knowledge graphs to generate scene graphs. In: Vedaldi, A., Bischof, H., Brox, T., Frahm, J.-M. (eds.) Computer Vision – ECCV 2020. LNCS, vol. 12368, pp. 606–623. Springer, Cham (2020). https://doi.org/10.1007/978-3-030-58592-1_36
10. Dinh, N.T., Van, T.T., Le, T.M.: Semantic relationship-based image retrieval using KD-tree structure. In: Vedaldi, A., Bischof, H., Brox, T., Frahm, J.M. (eds.) ECCV 2020, vol. 12368, pp. 455–468. Springer, Cham (2022). https://doi.org/10.1007/978-3-031-21743-2_36
11. Dinh, N.T., Nhi, N.T.U., Le, T.M., Van, T.T.: A model of image retrieval based on KD-tree random forest. Data Technol. Appl. (2023)
12. Le, T.M., Dinh, N.T., Van, T.T.: Developing a model semantic-based image retrieval by combining KD-tree structure with ontology. Exp. Syst., e13396 (2023)
13. Liu, Z., Xiong, C., Sun, M., Liu, Z.: Entity-duet neural ranking: understanding the role of knowledge graph semantics in neural information retrieval. arXiv preprint arXiv:1805.07591 (2018)
14. Sawant, U., Garg, S., Chakrabarti, S., Ramakrishnan, G.: Neural architecture for question answering using a knowledge graph and web corpus. Inf. Retrieval J. **22**, 324–349 (2019)
15. Oxford Flower-17 Image set. https://www.robots.ox.ac.uk/~vgg/data/flowers/17/. Accessed 20 Aug 2023
16. Flickr 2017 Image set. https://www.kaggle.com/datasets/hsankesara/flickr-image-dataset. Accessed 20 Aug 2023
17. Yoon, S., et al.: Image-to-image retrieval by learning similarity between scene graphs. arXiv preprint arXiv 4322, 2012.14700 (2020)
18. Wang, Z., et al.: Camp cross-modal adaptive message passing for text-image retrieval. In: Proceedings of the IEEE/CVF International Conference on Computer Vision, pp. 5764–5773 (2019)
19. Song, J., He, T., Gao, L., Xu, X., Hanjalic, A., Shen, H.T.: Binary generative adversarial networks for image retrieval. In: Proceedings of the AAAI Conference on Artificial Intelligence, vol. 32, no. 1 (2018)
20. Gao, S., Tsang, I.W.-H., Ma, Y.: Learning category-specific dictionary and shared dictionary for fine-grained image categorization. IEEE Trans. Image Process. **23**(2), 623–634 (2013)
21. Ahmed, K.T., Ummesafi, S., Iqbal, A.: Content based image retrieval using image features information fusion. Inf. Fusion **51**, 76–99 (2019)
22. Bakheet, S., Al-Hamadi, A., Soliman, E., Heshmat, M.: Hybrid bag-of-visual-words and FeatureWiz selection for content-based visual information retrieval. Sensors **23**(3), 1653 (2023)
23. Dinh, N.T., Le, T.M., Van, T.T.: An improvement method of KD-tree using k-means and K-NN for semantic-based image retrieval system. In: Rocha, A., Adeli, H., Dzemyda, G., Moreira, F. (eds.) WorldCIST 2022. LNNS, vol. 469, pp. 177–187. Springer, Cham (2022). https://doi.org/10.1007/978-3-031-04819-7_19

Blockchain in the Portuguese Public Electronic Service

José A. M. Victor[1]([✉]), Teresa Guarda[2,3], and Ana Dopico-Parada[3]

[1] N2i, Polytechnic Institute of Maia, Maia, Portugal
javemor@ismai.pt
[2] Universidad Estatal Península de Santa Elena, La Libertad, Ecuador
[3] Universidade de Vigo, Vigo, Spain
adopico@uvigo.es

Abstract. The use of blockchain technology in public electronic services is necessary to guarantee a continuous recording service of data sets, for the common good of citizens. In Portugal, the non-incorporation of blockchain as part of Digital Transformation (DT) makes it difficult to obtain documentary evidence in a timely manner, in strategic areas of nature: spatial, sociodemographic, health, water, energy, mobility and climate and meteorology. For this reason, the aim is to find out how blockchain technology is integrated into the Portuguese government's electronic service. To verify the facts and obtain documentary evidence, we resorted to a bibliographical review, by consulting the Scocus and ScienceDiretct databases. For each selected article, the number of repetitions of the word blockchain is counted. We conclude that blockchain technology fits into funded scientific research projects.

Keywords: Blockchain · government · implementation · electronic service

1 Introduction

Blockchain is an innovative and disruptive technology [1], secure and transparent [2–7]. It can function as a means of shared public record, for on-chain and secure transactions, being an immutable accumulated storage system, which reduces the risk of tampering [8]. Blockchain technology creates trust between untrustworthy parties [9]. Therefore, it is a support to the problem of Government responsibility, in which the parties cooperate for a result [10], and, where technology makes the Public Administration more reliable, more effective in meeting time, more sustainable, fulfilling duties towards citizens. And the high-value data sets generated encourage responsibility in taking measures to protect the environment and human beings.

The electronic public service when integrating blockchain technology as part of Digital Transformation (DT) has mandatory challenges. These are: the lack of regulation of the use of blockchain, compliance with the General Data Protection Regulation (GDPR), the associated benefits and risks, interoperability between different blockchains to be effective in TD, and scalability, due to the large data transaction volume efficiently. In context with the Portuguese case, the electronic public service is still developing and adapting to current demands. Therefore, when adopting measures for TD, the big data concept is a real phenomenon, where the amount of information to be stored reaches massive levels. This increase can present challenges and opportunities, from issues related to privacy and data security to the ability to efficiently process and analyse these vast amounts of information. Including security measures to manage, store and analyse these gigabytes of information become essential for anyone dealing with the growing complexity of the digital environment.

The challenging problem for Portugal, of not incorporating blockchain as part of DT, will make it difficult, in the future, to obtain documentary evidence in a timely manner, in essential areas of governance, at the level: spatial, sociodemographic, health, water, energy, mobility and climate and meteorology. To understand the current state of implementation of blockchain technology by the Portuguese government, a bibliographical review is carried out on the projects that include the major strategic areas for the proper functioning of the electronic service. This review is carried out in the Scocus and ScienceDiretct databases, through keyword research, relating blockchain to the selected strategic areas, identifying articles with Portuguese projects. We show that the Portuguese government has funded scientific research projects to implement blockchain technology.

2 Blockchain in Portuguese Electronic Service

Blockchain technology is an innovative and disruptive application that is growing [1], is immutable, and reduces the risk of tampering [8]. Blockchain is a technology that, when implemented, creates trust between parties [9]. The blockchain concept stands out for its security [2–7], a non-negotiable characteristic, although it requires efforts to manage the network and detect attacks or anomalies in the system. Blockchain is a tool we must guarantee continuous recording of data sets. This technology meets the application of European Union (EU) Directive 2019/1024, transposed into Portuguese legislation in April 2021, and which rewrites the rules on open data and the reuse of public sector information [11], Public Administration open data portal (dados.gov), and Publications Office of the European Union (2019). It is an important step in the Portuguese electronic service, sharing data produced by Portuguese public entities. As an example, we highlight high-value data sets: geospatial, earth and environmental observation, meteorological, statistics, companies and company properties and mobility. EU Directive 2019/10324 complements the 2007 Infrastructure for Spatial Information in Europe (INSPIRE) Directive [12], which regulates the recording and maintenance of historical data, over time and in real time. The transposition of the INSPIRE Directive into Portuguese legislation occurred in 2009, with Decree-Law No. 180/2009 [13].

High-value datasets are necessary to ensure investigations, with the aim of protecting the environment and humans. In Portugal, the incorporation of blockchain technology as part of Digital Transformation (DT) presents several challenges. Some of the main challenges include: the lack of clear regulation of the use of blockchain technology, representing an obstacle to its implementation, compliance with the General Data Protection Regulation (GDPR); adoption, obtaining training on the benefits and risks associated with it, interoperability between different blockchains to be effective in DT capable of working with different systems and technologies, and scalability due to the large volume of data transactions.

The Portuguese electronic public service is in a TD process, with several actions for the digital transition in the most diverse government sectors. According to some researchers [14, 15], which inspired others [16], the digital progression of public services is visible in the sharing of data outside the physical space where they are generated, such as data from Health Information Systems (SIS), especially since the coronavirus disease (Covid-19).

To gather information on the major areas of governance as strategic for the proper functioning of Portuguese electronic government, with the integration of blockchain, we gathered data on how the State has intervened so that this technology adds benefits to TD. The integration of blockchain in projects in which the government participates with public investment is a strategy to communicate unique and innovative value propositions to satisfy citizens' needs, increasing trust in services, in the short, medium and long term. Digital records for major strategic areas are fundamental.

3 Method and Strategic Areas

Data sets from a country, such as Portugal, must necessarily be shared. Certainly, the increase in the volume of digital information in gigabytes is a natural consequence of the increasing use of digital technologies. The results obtained from these data help to take more sustainable measures, including the following dimensions: environmental, economic and social. Therefore, we selected some of the major strategic areas for the proper functioning of a nation: space, sociodemographic, health, water, energy, mobility, and climate and meteorology. The information is obtained from projects developed for the Portuguese case, described in scientific articles, available in the Scocus and ScienceDiretct databases, for the period between 2019–2024. Projects that received public funding from the *Fundação da Ciência e da Tecnologia* (FCT) were highlighted. For each article analyzed, the number of repetitions of the word blockchain is quantified.

3.1 Spatial

The spatial area allows learning through data, interpreting it, but to do this it is necessary to have a digital asset, preferably automated, accessible and secure. Therefore, the digital electronic service involves the implementation of public projects. One of them is the *Portal Digital Único* (PDU), which has already been tested and implemented in the European Union [17], launched in December 2020. This is a public electronic service web (eportugal.gov, 2020), where it brings together all the information necessary for citizens of the European Union (EU) space, about living, studying, working, or opening a company. Another project is the platform Balcão Único do Prédio (BUPi), 2017, for digital registration of identification of rustic and mixed properties (public, community and private areas), currently taking place in 137 municipalities out of a total of 308, for Georeferenced Graphic Representation (RGG). This tool allows for the centralized management of information about properties and buildings, including data related to land records, licenses, and other relevant documents on property rights. The BUPi (portugal.gov.pt, 2023) is a project financed by the Recovery and Resilience Plan (PRR). However, BUPi data is not open access for researchers or ordinary citizens. We found that the two large-scale projects mentioned never mentioned the integration of blockchain technology.

The projects that relate to a specific geographic area are immense. Some of them, although small, and which directly or indirectly share data with government bodies, already integrate blockchain technology, showing some of them (Table 1): control of the medicine distribution chain [19], control of citizen data [20], digital assets [21], manufacturing [22]. All these projects include automating the value of intellectual work, opening new areas of added value for the accumulated capital of data sets. The electronic public service requires governments to respond to the invasion of automation in the domain of intellectual work, as this radical change presents new and unique challenges to the security and value of human work.

Geographic distribution data, which we call spatial, forms a network of connections that allows us to reduce uncertainty. Uncertainty is a quantitative measure that affects human confidence, which can be modelled probabilistically, unlike precision, which is a distance or proximity to the true value [21]. For this reason, the development, customization and implementation of systems such as digital manufacturing have become increasingly complex as they require an integrated vision, which includes both digital and physical space [22].

Table 1. Portuguese projects that relate the space area with blockchain technology, obtained in systematic review studies.

Source and pages	Project	Blockchain (included in titles, keywords and body text) and pages with number of repetitions	Public financing and observations
[19, pp. 1, 7, 10–12]	Control of the medicine distribution chain during the Covid-19 pandemic	The term repeats on pages 1 (3), 7 (1), 10 (6), 11 (7), 12 (2) and in the bibliography (4)	-, refers to 6 Portuguese documents, but it is a non-national study
[20, pp. 2–7, 11, 18]	Data control, person-centric and intelligent	The term repeats on pages 2 (1), 3 (14), 4 (1), 5 (1), 6 (1), 7 (6), 11 (1), 18 (1), and in bibliography 38	Yes
[21, pp. 2, 4, 8]	Digital assets, patterns in data, artificial intelligence	The term repeats on pages 2 (1), 4 (1), 8 (2)	-

(*continued*)

Table 1. (*continued*)

Source and pages	Project	Blockchain (included in titles, keywords and body text) and pages with number of repetitions	Public financing and observations
[22, pp. 5, 10]	Digital manufacturing	The term is repeated on pages 5 (2), 10 (1), and in bibliography 3	-

Source: Own elaboration based on the cited bibliography.

We have a set of information organized by units and geographic areas and the structuring of data sets is in the way of TD and, at the same time, we are faced with gigantic amounts of data that increase every day. Although it is not completely true that the more data, the better the accuracy, there is always random uncertainty, but the irreducible uncertainty in the data will not be affected. The value of data is proportional to its richness, and at the same time, for territorial sustainability it is intrinsically valuable.

3.2 Demography

Demography is another major strategic area selected for this study. In this segment, we have several projects identified (Table 1), with or without reference to the blockchain concept. From the research carried out, we identified an investigation [23] on the mobility of population fluctuations in the City of Porto. But it never mentions the blockchain concept.

In Portugal, the resident population is 20.6% elderly [24]. Therefore, it makes sense for the Government to invest in support projects in this area (Table 1), financing a municipal assistance platform for the elderly [25].

In Portugal there are residents of Portuguese nationality, migrants and refugees. Therefore, it makes sense to refer to the research project, "Blockchain technology and universal health coverage: health data space in global migration" [26], which deals with national digital identity, including migrants and refugees. The project incorporates blockchain technology to comply with and enforce different human rights regulations, with each citizen owning their own data.

Anthropometric databases on the dimensions of the human body [27], updated, for the Portuguese working population, allow furniture to be adjusted to working conditions.

It is confirmed that the government invested, by financing the various research projects, gentrification [23], assistance to the elderly [25], and anthropometric databases [27], with health being a sector of concern for national policies.

Table 2. Portuguese projects that connect the population with blockchain technology.

Source and pages	Project	Blockchain (included in titles, keywords and body text) and pages with number of repetitions	Public financing
[23]	Gentrification in Porto, Portugal	-	Yes
[25, pp. 314, 319]	Community-dwelling elderly - service platform	The term repeats on pages 314(1), 319 (2) and in the bibliography 1 time	Yes
[26, pp. 1–5]	Digital identity data of migrants and refugees	The term repeats on pages 1 (4), 2 (21), 3 (9), 4 (12, 4 in the body of text and the rest in the bibliography), 5 (1, in the bibliography)	-
[27]	Anthropometric database	-	Yes

Source: Own elaboration based on the cited bibliography.

3.3 Health

The health projects identified are described and summarized in the Table 2. The reality is that in Portugal, health policy makers recognize the role of the Health Information System (SIS), creating the National Strategy for the Health Information Ecosystem [28] 20–22, referred to in the investigation of Teixeira et al. [16]. Data obtained from national health units needs to be unified. Then, Cunha, Duarte, Guimarães, & Santos [30] recommend blockchain technology and the Electronic Health Record (openEHR) interoperability standard, which is an international standard for electronic health records (RES openEHR). The technology, openEHR, is an internationally known database [30–32]. With openEHR technology, unification is guaranteed, and access is restricted by users. Combining openEHR and blockchain we obtain integrity, security, immutability, verifiability and privacy of confidential medical data [30]. We have the *Centro Hospital Universitário do Porto* (CHUP), which adopted the openEHR model as its structure, considered to be interoperable, organized, fast and easy to use [32]. Health data projects have research and implementation teams to improve the management of this information [33], an intelligent system, where there is great interaction between health units.

There are already technologies that make it possible to overcome the difficulties of supporting users, in outermost regions, such as the Autonomous Region of the Azores, using telesurgery [34].

Portugal has already used blockchain technology (Table 2) in some health unit database systems.

3.4 Water

Water is a basic necessity and there are many sets of data involved and they are an engine for the good functioning of a nation, therefore, water-related projects are described

Table 3. Portuguese projects linked to the health sector and which addressed blockchain technology.

Source and pages	Project	Blockchain (included in titles, keywords and body text) and pages with number of repetitions	Public financing
[16]	Health	The term is only mentioned in a bibliographical consultation	Yes
[30, pp. 242, 243, 244, 245, 246]	Health	The term is referred to 34 times: 242 (4); 243 (3), 244 (18), 245 (5) and 246 (4). Implementation of blockchain technology in healthcare units	Yes
[31, pp. 708, 709, 710, 711, 712]	Health	The term is referred to 64 times: 708 (7), 709 (14), 7010 (12), 711 (21), 712 (10) Implementation of blockchain technology in the healthcare sector	Yes
[32]	Health (openEHR)	-	Yes
[33]	Health (openEHR)	-	-
[34]	Telesurgery	-	-

Source: Own elaboration based on the cited bibliography.

(Table 3). Furthermore, the human factor is the key to new interconnected ecosystems [19], without forgetting the obligation to comply with the EU GDPR. It is a user-cantered data acquisition process.

TD, in agricultural production, includes Industry 5.0, as a human-cantered solution [35]. Process automation via artificial intelligence (AI), machine learning, sensors and robotics informed by IoT-Cloud are a virtual reality, verified in the study [35].

A project [36], indirectly related to the quality of water for consumption, in which Portugal did not participate, but was mentioned, and a model of daily reports and food safety complaints was replicated. This model recommends a system using blockchain, to allow traceability, and to combat food fraud throughout the chain. The situation is confirmed [37], in the study of a food labelling system for sustainable production and healthy and responsible consumption, when a certificate is issued by the traceability system.

The recycling of water (and energy) by the cosmetics industry (example Lush and Weleda), is concerned with educating the consumer [38], by using cosmetics sustainably.

Table 4. Portuguese projects linked to the water sector and which addressed blockchain technology.

Source and pages	Project	Blockchain (included in titles, keywords and body text, excluding bibliography) and pages with number of repetitions	Public financing
[35, pp. 2, 3, 4, 5, 8, 15, 16]	Crop TD, including environmental protection	The term is only mentioned in a bibliographical consultation. The term is referred to 15 times: 2 (1), 3 (1), 4 (5), 5 (1), 8 (1), 15 (4), 16 (1)	Yes
[36, pp. 25, 27, 29]	Big data in food safety	The term is referred to 12 times: 25 (1), 27 (9) and 29 (2)	Yes, but without Portuguese support
[38, p. 22]	Sustainable cosmetics	The term is mentioned once, on page. 22	Yes

The following point refers to the energy sector, a market that is a fundamental and strategic pillar of governance, generating value for domestic consumers and companies or organizations.

3.5 Energy

The energy sector has several investigations at Portuguese universities (Table 4). The energy area has evolved. Industry 4.0 pointed to a human-friendly approach to digitalization; while Industry 5.0 has a future-oriented and intersectoral approach [39]. According to the study [39], on bibliometric analysis of a set of Scopus articles, we are in the era of sustainable optimization of energy consumption, taking advantage of TD. The energy transition is being a local social innovation [40], with new forms of governance, social configurations, policies and regulations, with new business models. These difficulties were confirmed [41]. Energy communities, as key players in the energy transition, were also analysed in a book chapter [42]. The authors [42], wrote that the "Clean Energy for All Europeans" legislative package focuses European energy policy on citizens and communities, to promote local energy production, consumption and trade. Socialization and the formation of energetic community groups combines the common good with profit [43]. The investigation [43] reinforces the previous ones, and was financed, by the EU's Horizon 2020 program. Another investigation [42], confirms that collective energy systems cantered on citizens and communities benefit local electricity trade.

Among renewable energies, wind energy generates uncertainty in demand, as it depends on the time of year with wind, compared to traditional supply systems [44].

In eight studies (Table 5), for energy quantification and the continuous collection of data sets, TD is incorporated and, in the examples, blockchain technology is a technology to be incorporated into TD.

Table 5. Portuguese projects linked to the energy sector and which addressed blockchain technology.

Source and pages	Project	Blockchain (included in titles, keywords and body text, bibliography) and pages with number of repetitions	Public financing
[41, p. 20]	Use of energy supply to buildings in shared communities	The term is referred to 1 time: 16 (1)	Yes
[40, p. 7]	Energy communities	The term is referred to 1 time: 7 (1)	Yes
[42]	Energy communities		
[42]	Energy communities	-	Yes
[43, pp. 4, 5]	Energy communities	The term is referred to 2 times: 4 (1), 5 (1)	Yes
[45, p. 6]	Renewable energy	The term is referred to 3 times: 6 (1); 21 (2): bibliography	Yes
[46]	Energy use	-	Yes
[39, pp. 779, 778, 783, 784, 786, 786, 877]	Energy sector more focused on humans	The term is referred to 11 times: 779 (2), 781 (1), 783 (1), 784 (1), 786 (2), 787 (1); 787 (2), 788 (1): bibliography	Yes

We confirm (Table 5), that blockchain technology in the energy sector in Portugal is financed through public funds.

3.6 Mobility

Mobility is an area where the Government invests in public policies to promote sustainability in travel (Table 5). Turning to drones for mobility purposes has become emerging. The project [47] and the articles co-cited by him, certify that drones are a technology in an advanced stage of development, and it is planned to use these means of transport for commercial deliveries [47, 48]. Disruptive technology can meet global strategic requirements, such as combating climate change and traffic management. The project [47], relates drones to satellite systems, quantum blockchain and data record keeping. This record keeping requires transparency, immutability, security and processing and availability of information [47–49]. What is intended with optimized mobility, with drones, is in the near future, for supplier services [50], with more efficient use of resources [51]. And, satellites could be a way for people to stay connected [52].

For an outermost region, such as the Autonomous Region of the Azores, for the purposes of surgery, and without transport conditions, telesurgery allows the necessary security [34]. As a result, drone transportation can improve the conditions of telemedicine healthcare services. Associated with mobility, there is the Internet of Things (IoT) to connect with intelligent transport [53]. Blockchain, IoT, machine learning (ML) and smart services protection systems are seen as requirements for a security approach.

In the maritime transport sector (Table 5), enthusiasm is evident for blockchain technology for electronic data processing. Blockchain can improve any transaction related to mobility contracts, as a form of global trade TD [54].

Table 6. Portuguese projects and studies linked to the mobility sector and which addressed blockchain technology.

Source and pages	Project	Blockchain (included in titles, keywords and body text, excluding bibliography) by pages and number of repetitions	Public financing
[47, pp. 7, 8, 30, 33]	Drones	The term is only mentioned in bibliographical consultations. The term is referred to 6 times: 7 (3), 8 (1), 30 (1) and 33 (1)	Yes
[54, pp. 807–813]	Sea transport	The term is mentioned 69 times: every page	-

There are more projects that we could describe, even if they are not Portuguese. To this end, and because there is the possibility of being replicated [55], the Exonum platform is an example that integrates blockchain technology to increase cybersecurity by creating a secure and reliable system for sending parameters of each vehicle's current state using signals from neighbouring vehicles.

3.7 Climate and Meteorology

For the broad area that includes climate and meteorology, we present some investigations that may be useful for this work (Table 6). According author [56], we can classify extreme climatic and meteorological events by the Universal Climate Index (UTCI) into different contexts, from extremely cold (< -40 °C), very cold]$-40, -27$], cold]$-27, -13$], moderately cold]$-13, 0$], slightly cold]$0, 9$], no heat stress]$0, 26$], moderately hot [$26, 32$[, warm [$32, 38$[, very hot [$38, 46$[, to extremely hot (≥ 46 °C) (Table 6). As examples of events, we have extreme cold with or without precipitation and with or without wind, heat waves, and these, with or without association with earthquakes. Events associated with natural disasters have grown in both frequency and intensity, says a co-citation [57]. It is [58], in the last 15, the increase was 2% per year. Water-related risks in urbanized areas were the cause of around 90% of disastrous events in the last century, and losses have increased fivefold since 1980 [59]. Blockchain technology is an intelligent management tool for climate and meteorological data sets.

Climate change affects ecosystems. Oyster production in the Tagus Estuary [60]. It is affected by the climate, as the maintenance of the biomass of the phytoplankton community is at stake. As water temperature increases, dissolved oxygen concentrations decrease [60] and consequent decrease in nutrients. The authors [60] obtained the in situ datasets, biological and environmental data, Aquaculture in the Sado estuary (AQUASADO) project.

We know that there is a relationship between climate change and carbon dioxide (CO_2) emissions. And, we are struggling to limit global warming to 1.5 °C [61]. The authors [61], carried out the project in Ireland (Table 6), which, among other organizations, had the participation of the *Centro de Ciências do Mar* (CCMAR) of the University of Algarve, which took advantage of TD for the eco-innovation of peatlands - fossil fuels.

The sustainable agricultural food industry is concerned with immobilizing CO_2 in the soil or forest [62]. This was a research project at the University of Évora, which investigated forms of food sustainability, using distributed Leger (decentralized) technologies, to track the authenticity of product certificates through blockchain technology.

In addition to the previous studies, we selected another on changes in sea surface temperature. We found that, due to the lack of long in situ data time series, to expand the temporal and spatial coverage of the study, the authors [63], used satellite remote sensing products. They also used in situ atmospheric time series of air temperature, wind and total precipitation, provided by the *Instituto Português do Mar e da Atmosfera* (IPMA) and considered two meteorological stations on the Galician coast [64], with historical data provided by MeteoGalicia, in nine stations distributed along the Portuguese coast (Table 6).

Next, a project that involved the study of tropical cyclone precipitation [65], and researchers from the University of Lisbon participated (Table 6). On average, precipitation, between 1980 and 2018, related to tropical cyclones represented~4.2% of the average precipitation between August and November, although September, with~7.1%, has presented the largest contribution [66]. Similarly, the highest rainfall related to tropical cyclones was found in the west and north of the Euro-region Galicia-North of Portugal. In these last two articles, we did not find any reference to blockchain.

For research through Instituto Dom Luiz (IDL), Faculty of Sciences, University of Lisbon [67] about a flood event on the Rivillas River, between November 5th and 6th, 1997 (Badajoz, Spain), the main data set was provided by the State Meteorology Agency (AEMET). These data were collected at ten-minute intervals and were used to define the continuous precipitation data chart. Poor territorial management was identified, similar to this day, as the event's peaks are similar today. The three main factors that increased the negative effects of the event were: (1) the amount of precipitation recorded from 5 pm on November 5th to 1 am on November 6th; (2), high levels of soil moisture content; (3), the poor state of conservation of the bridges, partially blocked during the event.

Table 7. Portuguese climate-related projects and studies that addressed blockchain technology.

Source and pages	Project description	Blockchain (included in titles, keywords and body text, bibliography) by pages and number of repetitions	Portuguese public financing
[56]	Heat waves	-	Yes
[63]	Changes in sea surface temperature on the Iberian West Coast. With data collection at 8 oceanographic stations, from State Ports (Spain) and from Hydrographic Institute (Portugal): Estaca de Bares, Vilão Sirsagas, Cabo Silleiro, Leixões, Nazaré, Sines, Faro, Cádiz	-	Yes
[60]	Ensure the sustainability of the ecosystem, in this case in aquaculture in the Tagus Estuary	-	Yes
[65]	Precipitation from North Atlantic tropical cyclones linked to their tracks: sources of moisture. The data is from the Atlantic Hurricane Database (HURDAT2) between 1980 and 2018	-	Yes, partially
[66]	Discovery of whether precipitation in the Euro region Galicia - North of Portugal is due to cyclones of tropical origin. The amount of precipitation that contributes to tropical cyclones	-	Yes, partially, same project as the previous line
[61, pp. 2, 3, 4, 5, 8, 16, 16, 17]	TD of peatland eco-innovations ('Paludiculture'): Enabling a paradigm shift towards real-time sustainable production of 'eco-friendly' products and services. Project that had the participation of the Center for Marine Sciences (CCMAR), University of Algarve. Approaches blockchain for Cybersecurity	Referred 15 times: 2(1), 3(1), 4 (4), 5(1), 8(1), 15 (4), 16(1), 17 (2)	Yes, partially, because it is European funding

(*continued*)

Table 7. (*continued*)

Source and pages	Project description	Blockchain (included in titles, keywords and body text, bibliography) by pages and number of repetitions	Portuguese public financing
[62, pp. 1, 3, 6, 8]	Indexing Distributed Ledger Technologies for Food Sustainability. The dataset used was emissions in open fields, provided by the Centro de Demostracion y Transferencia Agraria el Mirador S.COOP (CDTA El Mirador)	Mentioned 41 times: 1(2), 3(4), 6(3), 8(1); the remaining 31 times belonging to the bibliography	Yes, for the EU
[67]	The Rivillas flood of November 5th to 6th, 1997 (Badajoz, Spain) revisited: an approach based on Iber + modeling	-	Yes
[68]	Database of coastal floods from 1980 to 2018 for the Portuguese continental coastal zone	-	
[68]	The weather, the damage and the value of the insurance policy	-	Yes

Moving on to coastal flooding in mainland Portugal [68] data were obtained from newspaper sources. Nowhere in the article was blockchain technology mentioned (Table 6). "Hemerographic analysis allowed the identification of 650 occurrences of coastal flooding and overcoming between 1980 and 2018, with high temporal and spatial variability" [68, p. 5].

Indirectly, the values of climate or weather conditions, when they are intense or of high magnitude, cause consequences. We know that in Portugal [68], during the period from 1980 to 2019, the consequences due to floods and other meteorological events were not covered by insurance companies in 91%. The study in no paragraph referred to blockchain technology. This project had financial support from national institutions: the Portuguese Association of Insurers (APS).

A non-Portuguese project, with the title "climate change management: a resilience strategy for flood risk using Blockchain tools [69], shows concern about the insurance market. The investigation states that the Public Administration is responsible for the cost of repairing services damaged by meteorological phenomena, whether they are public goods or private property. This study is fundamental due to the possibility of being replicated in the Portuguese case, to quantify the flood risk mitigation measure, via blockchain, with the possibility of implementing an event management platform, either through data collection or certification of them at the various stages of the process.

In the presence of intense meteorological events, with human consequences, blockchain technology can act in terms of certifying that data comes from reliable sources. The problem that arises is the random distribution of damages, causing difficulty for insurers in quantifying or determining the consequent adjustment of the premium level for policy coverage [68]. This is what happens in Portugal.

We check in investigations [60, 63, 65, 66, 67, 70], that data sets can be obtained through public bodies, which register them and store long series of data; or through specific projects. We identified that, in the broad area of climate and meteorology, blockchain is still in its infancy or the technology is not yet used.

We identify research that contributes to combating climate change [61, 62], through the sustainability of the production industry and, resorted to blockchain, as an innovation strategy, which puts human beings at the centre of attention. One of the discoveries was that, although studies on the application of blockchain tools for managing databases on extreme meteorological events, such as floods, are not yet known in Portugal, this technology already exists in other countries [69].

4 Discussion

Table 8. Summary matrix of the priority given to blockchain in the Portuguese government's electronic service, based on the bibliography cited.

Seven major strategic areas	Criteria for the degree of priority given to blockchain: implementation	fonts	repetitions 4 – Very high 3 – high, 2 - medium, 1 - low			Sources (no.)	No reference to Blockchain
Spatial	1	1	1	two*; 4	two; 0
Sociodemographic	1	1	1	4	two
Health	two	two	3	5	3
Water	two	two	3	4	0
Energy	3	two	3	8	1
Mobility	two	1	two	two	0
Climate and meteorology	1	1	two	10	7
Total	12	10	15	39	15

* Two projects: PDU and BUPi.

In general, in the investigations presented, we certify that the Portuguese government has contributed financial support to implement blockchain technology. Comparing the seven major selected areas (Table 7), the implementation of blockchain in the Portuguese e-government service, in general terms, represents: a low level (9; 42%), medium (8; 38%), high (4; 19%) and very high (0; 0%). The synthesis matrix (Table 7) represents the 39 sources of citations, of which 15 (38.46%) made no reference to blockchain. It appears that the projects, for the most part, had support financed by the Portuguese Government, with the exception of three of them in the space area [19, 21, 22], one referring to demography [26], and two to health [33, 34] and one referring to mobility [54]. It is confirmed that in the seven major strategic areas, blockchain technology, directly or indirectly, is under development (Table 8).

Some international projects are included in this study, as blockchain technology has been implemented and these can be replicated in Portugal, such as the example of research [71], by including in the process of data collection and certification of datasets. The Portuguese governance system can replicate international projects for governance and sustainable development, financing scientific research work that adopts blockchain technologies. The integration of blockchain in major strategic areas in the electronic public service is necessary and public investment in scientific research projects validates the proper functioning of services, including satisfying citizens' needs, increasing trust in them, in the short, medium and long term. Deadlines. Furthermore, it was possible to prove that electronic public services are sources of data collection, as some authors mentioned [60, 63, 65–67].

5   Conclusion

The main stages of this investigation to study blockchain technology in Portuguese electronic services required bibliographical research to identify and select the major strategic areas where the blockchain concept is incorporated in the various sectors and processes. Therefore, the answer to the main objective, to find out how blockchain technology has been integrated into the Portuguese government's electronic service, has been achieved. The main discovery was that blockchain technology has been integrated into the electronic service and is financed by the Portuguese Government.

The identified problem of the consequences of not incorporating blockchain, in the seven major areas of governance, at the level: spatial, sociodemographic, health, water, energy, mobility, and climate and meteorology, is being addressed with public funding through scientific research. The incorporation of the blockchain concept has the potential to significantly transform several sectors, offering greater security, transparency and efficiency in different processes.

The difficulties experienced during the process of this investigation are essentially due to the embryonic state in some of the major areas mentioned, and there is therefore little or no information about the incorporation of blockchain technology. As future work, it is intended, in the short and medium term, to present evidence that blockchain technology is essential for the development of the Portuguese electronic service, through a case study linked to the area of climate and meteorology, contributing to national sustainable development policies, including reducing the risk of catastrophes.

References

1. Abuhashim, A.A., Tan, C.C.: Improving smart contract search by semantic and structural clustering for source codes. Blockchain Res. Appl., 100117 (2022). https://doi.org/10.1016/j.bcra.2022.100117
2. Albany, M., Alsahafi, E., Alruwili, I., Elkhediri, S.: A review: secure internet of thing system for smart houses. Procedia Comput. Sci. **201**, 437–444 (2022). https://doi.org/10.1016/j.procs.2022.03.057
3. Alves, I.M., Carvalho, L.M., Peças Lopes, J.A.: Modeling demand flexibility impact on the long-term adequacy of generation systems. Int. J. Electr. Power Energy Syst. **151**, 109169 (2023). https://doi.org/10.1016/j.ijepes.2023.109169
4. Alves, O., Garcia, B., Rijo, B., et al.: Market opportunities in Portugal for the water-and-waste sector using sludge gasification. Energies (19961073) **15**, 6600–6600–6615 (2022). https://doi.org/10.3390/en15186600
5. Alzahrani, B., Oubbati, O.S., Barnawi, A., et al.: UAV assistance paradigm: state-of-the-art in applications and challenges. J. Netw. Comput. Appl. **166**, 102706 (2020). https://doi.org/10.1016/j.jnca.2020.102706
6. Anacleto Filho, P.C., da Silva, L., Mattos, D., et al.: Establishing an anthropometric database: a case for the Portuguese working population. Int. J. Ind. Ergon. **97**, 103473 (2023). https://doi.org/10.1016/j.ergon.2023.103473
7. Balcão Único do Prédio - BUPiI (2017) Plataforma
8. Barata, J., Kayser, I.: Industry 5.0 – past, present, and near future. Procedia Comput. Sci. **219**, 778–788 (2023). https://doi.org/10.1016/j.procs.2023.01.351
9. Biguino, B., Antunes, C., Lamas, L., et al.: 40 years of changes in sea surface temperature along the Western Iberian Coast. Sci. Total. Environ. **888**, 164193 (2023). https://doi.org/10.1016/j.scitotenv.2023.164193
10. Brito, A.C., Pereira, H., Picado, A., et al.: Increased oyster aquaculture in the Sado Estuary (Portugal): how to ensure ecosystem sustainability? Sci. Total. Environ. **855**, 158898 (2023). https://doi.org/10.1016/j.scitotenv.2022.158898
11. Capocasale, V., Danilo, G., Perboli, G.: Comparative analysis of permissioned blockchain frameworks for industrial applications. Blockchain Res. Appl., 100113 (2022). https://doi.org/10.1016/j.bcra.2022.100113
12. Cardoso, R.M., Lima, D.C.A., Soares, P.M.M.: How persistent and hazardous will extreme temperature events become in a warming Portugal? Weather Climate Extremes **41**, 100600 (2023). https://doi.org/10.1016/j.wace.2023.100600
13. Corte-Real, A., Nunes, T., Santos, C., Rupino da Cunha, P.: Blockchain technology and universal health coverage: health data space in global migration. J. Forensic Leg. Med. **89**, 102370 (2022). https://doi.org/10.1016/j.jflm.2022.102370
14. Cunha, J., Duarte, R., Guimarães, T., et al.: Blockchain analytics in healthcare: an overview. Procedia Comput. Sci. **201**, 708–713 (2022). https://doi.org/10.1016/j.procs.2022.03.095
15. Cunha, J., Duarte, R., Guimarães, T., Santos, M.F.: Permissioned blockchain approach using open data in healthcare. Procedia Comput. Sci. **210**, 242–247 (2022). https://doi.org/10.1016/j.procs.2022.10.144
16. Cunha, J., Duarte, R., Guimarães, T., Santos, M.F.: OpenEHR and business intelligence in healthcare: an overview. Procedia Comput. Sci. **220**, 874–879 (2023). https://doi.org/10.1016/j.procs.2023.03.118
17. Dall-Orsoletta, A., Cunha, J., Araújo, M., Ferreira, P.: A systematic review of social innovation and community energy transitions. Energy Res. Soc. Sci. **88**, 102625 (2022). https://doi.org/10.1016/j.erss.2022.102625

18. Di Francesco Maesa, D., Ricci, L., Sastry, N.: Blockchain: protocols, applications, and transactions analysis. Blockchain Res. Appl. **3**, 100071 (2022). https://doi.org/10.1016/j.bcra.2022.100071

19. Diallo, E., Dib, O., Al Agha, K.: A scalable blockchain-based scheme for traffic-related data sharing in VANETs. Blockchain Res. Appl. **3**, 100087 (2022). https://doi.org/10.1016/j.bcra.2022.100087

20. Diário da República: Decreto-Lei n.o 180/2009, de 7 de agosto (2009)

21. Ducrée, J.: Research – a blockchain of knowledge? Blockchain Res. Appl. **1**, 100005 (2020). https://doi.org/10.1016/j.bcra.2020.100005

22. Ecclesia, M.V., Santos, J., Brockway, P.E., Domingos, T.: A comprehensive societal energy return on investment study of Portugal reveals a low but stable value. Energies (19961073) **15** (2022). https://doi.org/10.3390/en15103549

23. EUR-Lex: A infraestrutura da União Europeia de informação geográfica (Inspire) (2007)

24. Fernández, A.J., Sicard, M., Costa, M.J., et al.: Extreme, wintertime Saharan dust intrusion in the Iberian Peninsula: Lidar monitoring and evaluation of dust forecast models during the February 2017 event. Atmos. Res. **228**, 223–241 (2019). https://doi.org/10.1016/j.atmosres.2019.06.007

25. Ferreira, C.S.S., Moruzzi, R., Isidoro, J.M.G.P., et al.: Impacts of distinct spatial arrangements of impervious surfaces on runoff and sediment fluxes from laboratory experiments. Anthropocene **28**, 100219 (2019). https://doi.org/10.1016/j.ancene.2019.100219

26. Foti, M., Vavalis, M.: What blockchain can do for power grids? Blockchain Res. Appl. **2**, 100008 (2021). https://doi.org/10.1016/j.bcra.2021.100008

27. Frimpong, S.A., Han, M., Boahen, E.K., et al.: RecGuard: an efficient privacy preservation blockchain-based system for online social network users. Blockchain Res. Appl., 100111 (2022). https://doi.org/10.1016/j.bcra.2022.100111

28. Gkogkos, G., Lourenço, P., Pechlivani, E.M., et al.: Distributed ledger technologies for food sustainability indexing. Smart Agric. Technol. **5**, 100312 (2023). https://doi.org/10.1016/j.atech.2023.100312

29. de Godoy, J., Otrel-Cass, K., Toft, K.H.: Transformations of trust in society: a systematic review of how access to big data in energy systems challenges Scandinavian culture. Energy AI **5**, 100079 (2021). https://doi.org/10.1016/j.egyai.2021.100079

30. Gökalp, E., Kayabay, K., Gökalp, M.O.: Leveraging digital transformation technologies to tackle COVID-19: proposing a privacy-first holistic framework. In: Emerging Technologies During the Era of COVID-19 Pandemic, pp. 149–166 (2021). https://doi.org/10.1007/978-3-030-67716-9_10

31. González-Cao, J., Fernández-Nóvoa, D., García-Feal, O., et al.: The Rivillas flood of 5–6 November 1997 (Badajoz, Spain) revisited: an approach based on Iber+ modelling. J. Hydrol. **610**, 127883 (2022). https://doi.org/10.1016/j.jhydrol.2022.127883

32. Graydon, M., Parks, L.: Connecting the unconnected': a critical assessment of US satellite internet services. Media Cult. Soc. **42**, 260–276 (2020). https://doi.org/10.1177/0163443719861835

33. Güneralp, B., Güneralp, İ, Liu, Y.: Changing global patterns of urban exposure to flood and drought hazards. Glob. Environ. Chang. **31**, 217–225 (2015). https://doi.org/10.1016/j.gloenvcha.2015.01.002

34. Gyongyosi, L., Imre, S.: Opportunistic entanglement distribution for the quantum internet. Sci. Rep. **9**, 2219 (2019). https://doi.org/10.1038/s41598-019-38495-w

35. Hak, F., Oliveira, D., Abreu, N., et al.: An openEHR adoption in a Portuguese healthcare facility. Procedia Comput. Sci. **170**, 1047–1052 (2020). https://doi.org/10.1016/j.procs.2020.03.075

36. Jin, C., Bouzembrak, Y., Zhou, J., et al.: Big data in food safety-a review. Curr. Opin. Food Sci. **36**, 24–32 (2020). https://doi.org/10.1016/j.cofs.2020.11.006

37. Kaiser, J., McFarlane, D., Hawkridge, G., et al.: A review of reference architectures for digital manufacturing: classification, applicability and open issues. Comput. Ind. **149**, 103923 (2023). https://doi.org/10.1016/j.compind.2023.103923

38. Kalajdjieski, J., Raikwar, M., Arsov, N., et al.: Databases fit for blockchain technology: a complete overview. Blockchain Res. Appl., 100116 (2022). https://doi.org/10.1016/j.bcra.2022.100116

39. Kalogirou, V., Stasis, A., Charalabidis, Y.: Assessing and improving the national interoperability frameworks of European union member states: the case of Greece. Gov. Inf. Q. **39**, 101716 (2022). https://doi.org/10.1016/j.giq.2022.101716

40. Karandikar, N., Abhishek, R., Saurabh, N., et al.: Blockchain-based prosumer incentivization for peak mitigation through temporal aggregation and contextual clustering. Blockchain Res. Appl. **2**, 100016 (2021). https://doi.org/10.1016/j.bcra.2021.100016

41. Kazim, E., Fenoglio, E., Hilliard, A., et al.: On the sui generis value capture of new digital technologies: the case of AI. Patterns **3**, 100526 (2022). https://doi.org/10.1016/j.patter.2022.100526

42. Kumar, A., de Jesus, A., Pacheco, D., Kaushik, K., Rodrigues, J.J.P.C.: Futuristic view of the internet of quantum drones: review, challenges and research agenda. Veh. Commun. **36**, 100487 (2022). https://doi.org/10.1016/j.vehcom.2022.100487

43. Kwan, C., Kish, L., Saez, Y., Cao, X.: Low cost and unconditionally secure communications for complex UAS networks, pp. 5895–5900. IEEE (2018)

44. Leal, M., Hudson, P., Mobini, S., et al.: Natural hazard insurance outcomes at national, regional and local scales: a comparison between Sweden and Portugal. J. Environ. Manage. **322**, 116079 (2022). https://doi.org/10.1016/j.jenvman.2022.116079

45. Lisa Clodoveo, M., Tarsitano, E., Crupi, P., et al.: Towards a new food labelling system for sustainable food production and healthy responsible consumption: the med index checklist. J. Funct. Foods **98**, 105277 (2022). https://doi.org/10.1016/j.jff.2022.105277

46. Loeza-Mejía, C.-I., Sánchez-DelaCruz, E., Pozos-Parra, P., Landero-Hernández, L.-A.: The potential and challenges of health 4.0 to face COVID-19 pandemic: a rapid review. Heal. Technol. **11**, 1321–1330 (2021). https://doi.org/10.1007/s12553-021-00598-8

47. Lorenz-Meyer, F., Santos, V.: Blockchain in the shipping industry: a proposal for the use of blockchain for SMEs in the maritime industry. Procedia Comput. Sci. **219**, 807–814 (2023). https://doi.org/10.1016/j.procs.2023.01.354

48. Martins, A.M., Marto, J.M.: A sustainable life cycle for cosmetics: from design and development to post-use phase. Sustain. Chem. Pharm. **35**, 101178 (2023). https://doi.org/10.1016/j.scp.2023.101178

49. Narbayeva, S., Bakibayev, T., Abeshev, K., et al.: Blockchain technology on the way of autonomous vehicles development. Transp. Res. Procedia **44**, 168–175 (2020). https://doi.org/10.1016/j.trpro.2020.02.024

50. Pérez-Alarcón, A., Coll-Hidalgo, P., Fernández-Alvarez, J.C., et al.: Climatological variations of moisture sources for precipitation of North Atlantic tropical cyclones linked to their tracks. Atmos. Res. **290**, 106778 (2023). https://doi.org/10.1016/j.atmosres.2023.106778

51. Pérez-Alarcón, A., Fernández-Alvarez, J.C., Sorí, R., et al.: How much of precipitation over the Euroregion Galicia – Northern Portugal is due to tropical-origin cyclones? A Lagrangian approach. Atmos. Res. **285**, 106640 (2023). https://doi.org/10.1016/j.atmosres.2023.106640

52. Pinto, F., Ferreira da Silva, C., Moro, S.: People-centered distributed ledger technology-IoT architectures: a systematic literature review. Telematics Inform. **70**, 101812 (2022). https://doi.org/10.1016/j.tele.2022.101812

53. Pires Klein, L., Krivoglazova, A., Matos, L., et al.: A novel peer-to-peer energy sharing business model for the Portuguese energy market. Energies (19961073) **13**, 125–125–125 (2020). https://doi.org/10.3390/en13010125

54. PORDATA: População residente com 65 e mais anos, estimativas a 31 de dezembro: total e por grupo etário (2022)
55. Portal de dados abertos da Administração Pública (dados.gov) (2021) Diretiva (UE) 2019/1024
56. Rahman, M.S., Khalil, I., Atiquzzaman, M.: Blockchain-powered policy enforcement for ensuring flight compliance in drone-based service systems. IEEE Network **35**, 116–123 (2021). https://doi.org/10.1109/MNET.011.2000219
57. Reis, I., Gonçalves, I., Lopes, M., Henggeler Antunes, C.: Business models for energy communities: a review of key issues and trends. Renew. Sustain. Energy Rev. **144**, 111013 (2021). https://doi.org/10.1016/j.rser.2021.111013
58. Rivadeneira, J.E., Sá Silva, J., Colomo-Palacios, R., et al.: User-centric privacy preserving models for a new era of the Internet of Things. J. Netw. Comput. Appl. **217**, 103695 (2023). https://doi.org/10.1016/j.jnca.2023.103695
59. Rosa, M., Faria, C., Barbosa, A.M., et al.: A platform of services to support community-dwelling older adults integrating FHIR and complex security mechanisms. Procedia Comput. Sci. **160**, 314–321 (2019). https://doi.org/10.1016/j.procs.2019.11.085
60. Rowan, N.J., Murray, N., Qiao, Y., et al.: Digital transformation of peatland eco-innovations ('Paludiculture'): enabling a paradigm shift towards the real-time sustainable production of 'green-friendly' products and services. Sci. Total. Environ. **838**, 156328 (2022). https://doi.org/10.1016/j.scitotenv.2022.156328
61. Serre, D., Heinzlef, C.: Assessing and mapping urban resilience to floods with respect to cascading effects through critical infrastructure networks. Int. J. Disaster Risk Reduction **30**, 235–243 (2018). https://doi.org/10.1016/j.ijdrr.2018.02.018
62. Silva, J.P., Santos, C.J., Torres, E., et al.: A double-edged sword: residents' views on the health consequences of gentrification in Porto, Portugal. Soc. Sci. Med. **336**, 116259 (2023). https://doi.org/10.1016/j.socscimed.2023.116259
63. de Medeiros Sousa, G., Santos, A.M.P.: The viability of Telesurgery service in the autonomous region of the Azores, supported by the 5G Network. Procedia Comput. Sci. **219**, 422–430 (2023). https://doi.org/10.1016/j.procs.2023.01.308
64. SPMS: ENESIS 20-22 (2017)
65. Tavares, A.O., Barros, J.L., Freire, P., et al.: A coastal flooding database from 1980 to 2018 for the continental Portuguese coastal zone. Appl. Geogr. **135**, 102534 (2021). https://doi.org/10.1016/j.apgeog.2021.102534
66. Teixeira, L., Cardoso, I., Oliveira e Sá, J., Madeira, F.: Are health information systems ready for the digital transformation in Portugal? Challenges future perspectives. Healthcare (2227–9032) **11**, 712–712–731 (2023). https://doi.org/10.3390/healthcare11050712
67. Teixeira, L., Ferreira, C., Santos, B.S.: Characterization of Portuguese Haemophilia patients based on the national registry data. Procedia Comput. Sci. **181**, 995–1001 (2021). https://doi.org/10.1016/j.procs.2021.01.273
68. Vannucci, E., Pagano, A.J., Romagnoli, F.: Climate change management: a resilience strategy for flood risk using Blockchain tools. Decisions Econ. Finan. **44**, 177–190 (2021). https://doi.org/10.1007/s10203-020-00315-6
69. Vannucci, J., Capozzi, R., Vinci, D., et al.: Concomitant intubation with minimal cuffed tube and rigid bronchoscopy for severe Tracheo-Carinal obstruction. J. Clin. Med. **12** (2023). https://doi.org/10.3390/jcm12165258
70. Wittmayer, J.M., Avelino, F., Pel, B., Campos, I.: Contributing to sustainable and just energy systems? The mainstreaming of renewable energy prosumerism within and across institutional logics. Energy Policy **149**, 112053 (2021). https://doi.org/10.1016/j.enpol.2020.112053
71. Zhou, X., Nehme, A., Jesus, V., et al.: AudiWFlow: confidential, collusion-resistant auditing of distributed workflows. Blockchain Res. Appl. **3**, 100073 (2022). https://doi.org/10.1016/j.bcra.2022.100073

Social Media Use: How Does Critical Consumption of News Impact Voting Persuasion in Angola?

Ivânia Silva and Mijail Naranjo-Zolotov[(⊠)] [iD]

NOVA Information Management School (NOVA IMS), Universidade NOVA de Lisboa,
Campus de Campolide, 1070-312 Lisbon, Portugal
`mijail.naranjo@novaims.unl.pt`

Abstract. The study provides important findings regarding the impact of external variables on Angolan's political participation. More specifically, it is analysed how critical news consumption on social media influences voting persuasion behaviour. The model was evaluated with a sample of 200 adult Angolan citizens to achieve significant conclusions. Through the provided answers, it was concluded that Angolans, further than being critical of the information shared on social media, try to influence others to vote online or offline. The collected sample also allowed us to conclude that age is correlated with voting persuasion. The older people are, the higher the probability they will persuade others to vote. Through this study, policymakers have the possibility to understand citizens better and target political campaigns on social media platforms, always employing respectful and trustful behaviour. This study also reinforces the need for proper education so citizens critically interact with news.

Keywords: Angola · Incidental News Exposure · Intentional News Exposure · Political participation · social media · Voting persuasion

1 Introduction

"The people are full of fear" [1]. This is how a Roman Catholic bishop described the emotional state of Angolan citizens before the first attempt to democratise Angola. This has marked the beginning of Angola's first steps to build a fair republic, with fair elections and giving the citizens the right to vote in the party they find suitable for the country's needs.

Previous studies suggest that group communication influences the degree to which people agree on a topic [2], affecting the decisions of citizens on political elections [3]. Recently, social media has had a great impact on politics, polarising international opinions [4]. However, that impact may be interpreted as not beneficial for society if citizens are constantly exposed to online disinformation and propaganda [5].

Considering that people interact with their relatives and friends on social media to get and share information, the question "What is the impact of critical consumption in

Angolan's intention to persuade others to vote?" appears as a field to be studied, recurring to information management tools and techniques. To the best of the authors' knowledge, this study is the first contribution to research analysing critical consumption of news on social media and the effect on voting persuasion in Angola.

Through a multi-group analysis, we estimate the differences between Angolan men and women, demonstrating key existing contrasts when comparing both groups. The results of this study also allow Angolan society in general to understand the importance of carefully reading information on social media as it will have an impact on the arguments shared with others and, possibly, on election results.

2   Political Participation

The definition of Political Participation can be simplified as the citizens' activities that affect politics [6]. Through political participation, it is possible to assess whether a country has a healthy democracy or not. For Verba & Nie (1987), democracies in which citizens present more representativity in decision-making are seen as the ones with more quality. However, due to inequalities in society, the different layers of the socioeconomic pyramid do not have the same level of democracy. Income inequality negatively impacts political engagement [8], which demonstrates that countries with unbalanced powers face lower levels of voting by low-income citizens.

In the specific case of Angola, there are existent disparities not just in the socioeconomic sphere but also in healthcare and life opportunities [9]. Those inequalities are geographically reinforced [10]. Still, in Angola, even though it is a democratic country, there is a lack of political participation options for citizens. The top-down distribution of power in this country translates into a direct involvement of the president in the daily affairs of the state [11].

3   Hypotheses Development

The model presented in Fig. 1 represents voting persuasion behaviour. The aim is to analyse the drivers that most impact the citizens' will to try to convince others to vote.

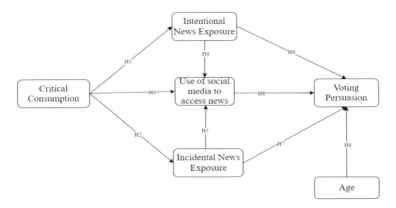

Fig. 1. Research model

3.1   Critical Consumption and News Exposure

Most literature refers to critical consumption as an alias for news literacy and scepticism [12] and the solution for misinformation avoidance [13]. Yet, Ku et al. (2019), combining various understandings, define critical consumption as "the first line of defence when information cannot always be trusted". Hence, critical consumption of social media implies the ability to identify patterns of low-quality arguments [13] and the subsequent fact-checking through trustful sources of information (Ku et al., 2019), mitigating the likelihood of creating misperceptions.

H1: Critical consumption is positively associated with Intentional News Exposure.

While using social media, users may find messages they did not expect to interact with. Researchers gave the name "Incidental News Exposure" to this "phenomenon", describing it as "coming across news when online for other reasons" [15]. It is proved that people who most frequently interact with news per accident, due to a lack of interest in particular information, are less critical when consuming SNS content [16].

H2: Critical consumption is positively associated with Incidental News Exposure.

Encountering news on social media can be a predictor for spending more time getting information through it [17], as critical consumers of information are characterised by the continuous process of information seeking [13]. As previously mentioned, critical consumption leads people to search for information actively and proactively. Then, it is expected that this behaviour translates into more time spent reading news on social media, utilising SNS as a source of information.

H3: Critical Consumption is positively associated with the Use of social media to access news.

Also, news exposure, either intentional or incidental, creates an environment for the increasing hours of news consumption, as a person's decision could be to use the time on social media to read the encountered news [18]. It is estimated that around 62% of adults get news on SNS, and the percentage is growing [16], reflecting the importance social media embodies in the spread of news.

H4: Intentional News Exposure is positively associated with the Use of social media to access news.

H5: Incidental News Exposure is positively associated with the Use of social media to access news.

3.2   News Exposure and Voting Persuasion

Voting persuasion is defined as the act an information sender starts to influence other people to adopt certain attitudes or behaviours regarding a political event in an environment of free will [19]. The proposed model addresses the impact that the time one spends reading or watching news on social media has on voting persuasion. In university students, it is already verified that the time spent on social media and other factors, such as political affiliation, are significant to online political persuasion [19].

H6: Intentional News Exposure is positively associated with Voting Persuasion.

H7: Incidental News Exposure is positively associated with Voting Persuasion.

H8: Use of social media to access news is positively associated with Voting Persuasion.

3.3 Age and Voting Persuasion

Previous studies detected that the perception generations have about political participation and political duties has changed over the years [20, 21], emphasising that emerging media platforms are responsible for shaping political engagement [22]. Harrison (2020) argues that there is a tendency for older people to tolerate more vote persuasion of opposite voters, while young voters are considered not to easily accept this type of inducement. It was also identified that, for older generations, voting is understood as a duty, as they not only were taught that it is a way of exercising their entitlement but also lived the time when people fought for that right.

H9: Age is positively associated with Voting Persuasion.

4 Methodology

Data was collected using an online survey from a convenience sample. The survey was composed of questions regarding the Use of social media to access news [24], Incidental News Exposure [25], Intentional News Exposure [26], Voting Persuasion [27] and Critical consumption [28]. The target of this study was Angolan people, and the aim of this study was to collect data from citizens with similar experiences and realities in political matters. Consequently, the questionnaire was delivered in Portuguese (as it is Angola's mother language) and answered by Angolans with age higher than 18 years old. The 200 completed answers were analysed through SmartPLS 3.0. Sample characteristics are presented in Table 1.

Table 1. Demographic data of respondents (N = 200)

Demographic Variable	Types	Frequency
Gender	Male	81
	Female	115
	Other	4
Age	<22	16
	22–30	91
	31–39	43
	40–48	23
	49–57	11
	58–66	13
	>66	3
Education	Primary Education	7
	High School	30
	Technical-Professional Course	20

(continued)

Table 1. (*continued*)

Demographic Variable	Types	Frequency
	Bachelor's degree	102
	Post-graduate degree	15
	Master's degree	24
	Doctorate degree	2

5 Results

In statistics, two variables that measure different constructs are discriminately valid when the level of correlation between them demonstrates that they are measuring distinct constructs [29]. The discriminant validity for this study was based on the Heterotrait-Monorait Ratio (HTMT). All calculated values for HTMT are less than 0.85, which is the minimum threshold referred to in the research of Hamid et al. (2017). Considering this, it is valid to assume that there is no evidence of a lack of discriminant validity.

Fornell-Larcker criterion compares the square root of AVE of each construct with the correlation of another existent construct in the model; this one should be lower than the first, and the threshold varies between 0.85 and 0.9 [31]. Considering this limit and comparing it to values in Table 2, it is possible to accept that the model has good discriminant validity, and the construct validity is confirmed (see Table 3).

Table 2. Fornell-Larcker discriminant validity

	Age	Critical Consumption	Incidental Exposure	Intentional Exposure	Social Media Use	Voting Persuasion
Age	1.000					
Critical Consumption	0.026	0.809				
Incidental Exposure	-0.095	0.412	0.888			
Intentional Exposure	-0.074	0.445	0.545	0.871		
Social Media Use	0.040	0.311	0.241	0.333	1.000	
Voting Persuasion	0.164	0.174	0.251	0.295	0.258	0.952

Multicollinearity is assessed through the Variance Inflation Factor (VIF). In the literature, the authors mention different thresholds for VIF. As 3.3 is the lowest limit

Table 3. Construct validity

	Cronbach's alpha	Composite reliability (rho_a)	Composite reliability (rho_c)	Average variance extracted (AVE)
Critical Consumption	0.824	0.840	0.883	0.654
Incidental Exposure	0.733	0.740	0.882	0.789
Intentional Exposure	0.841	0.843	0.904	0.759
Voting Persuasion	0.899	0.922	0.951	0.907

[31], this is the value to target when comparing the VIF values calculated by SmartPLS. Our model shows no multicollinearity among the constructs, as all values rely below the established maximum.

R-square measures the importance an independent variable has on the dependent variable's variance. The strength of this relationship is calculated on a scale from 0% to 100%. An independent variable is considered to explain the dependent variable when its R-square is higher than 50% [32]. However, a low percentage of R-square is not necessarily a sign of a non-fitted model. In social science research, if some or most of the independent variables statistically explain the dependent variable, the R-square is acceptable [33]. For this area of research, r-square values between 10% and 50% are considered valid due to the volatility of human behaviour, making it difficult to accurately foresee changes. Thus, all R-square values in our model are valid. Therefore, all the predictors are statistically significant and explain the dependent variable (Voting persuasion). Figure 2 shows the structural model evaluation.

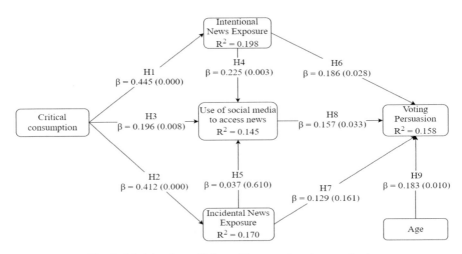

Fig. 2. Model path coefficients. Note: p-values in parenthesis.

5.1 Multigroup Analysis

In statistics, multigroup analysis is used to test existing differences between groups of participants [32]. There are differences between the calculated values for men and women, which are the groups that split the collected sample. Both the first (males) and second (females) groups validate the paths created for hypotheses H1, H2 and H4 ($p <$ 0.1); please see Table 4.

Table 4. Multigroup analysis

	Males		Females	
	β	p-Value	β	p-Value
Age → Voting Persuasion	0.252	0.028	0.087	0.371
Critical Consumption → Incidental Exposure	0.419	0.000	0.419	0.000
Critical Consumption → Intentional Exposure	0.492	0.000	0.416	0.000
Critical Consumption → Social Media Use	0.055	0.610	0.278	0.002
Incidental Exposure → Social Media Use	0.178	0.111	−0.071	0.496
Incidental Exposure → Voting Persuasion	0.203	0.081	0.051	0.756
Intentional Exposure → Social Media Use	0.258	0.052	0.209	0.042
Intentional Exposure → Voting Persuasion	0.13	0.290	0.210	0.071
Social Media Use → Voting Persuasion	0.146	0.287	0.189	0.054

6 Discussion

The proposed model explains that the more people select the information that is shared, the higher the probability of searching for updates on social media and, subsequently, persuading other people to vote. People who seek news develop strong priors regarding politics. So, it is more likely that they will exchange their opinion regarding voting choices. In contrast, people who unintentionally interact with news on social media are less willing to influence people's voting behaviour. The same is not valid for men who, based on news they unintentionally view, try to influence other people to vote.

Another finding enabled by the collected data is that critical thinking against information on social media leads only users who intentionally seek information to spend more time on social media reading and watching news. Citizens with competencies to evaluate the reliability of news may, at first, proceed to a critical analysis of the shared content and find the need to collect more information to validate it. After this process, social media users create robust arguments to persuade others to vote.

Table 5 shows that intentional news exposure is the variable that most impact voting persuasion ($\beta = 0.222$), followed by critical consumption ($\beta = 0.187$). Angolans not only seek updated information but are also critical of it. The ones who search for information are also the ones who try to influence the will of other people to vote.

Table 5. Total Effects

	Age	Critical Consumption	Incidental Exposure	Intentional Exposure	Social Media Use	Voting Persuasion
Age						0.183
Critical Consumption			0.412	0.445	0.311	0.185
Incidental Exposure					0.037	0.135
Intentional Exposure					0.225	0.222
Social Media Use						0.157
Voting Persuasion						

In the current study, it was not analysed how people influence others to vote. In Nigeria, people also use social media to share their opinions regarding voting choices [34]. In the study conducted by Uwalaka and Amadi (2020), it was also discovered that voting persuasion is done by appealing to social media followers' emotions. They verified that the church is used to reach people and influence their voting decisions. Voting is seen as an exaltation of God.

The collected data highlights that the older people are, the more they tend to influence people to vote. This is possibly explained by the war experience and the recent country's socio-economic evolution. Also, when persuading others to vote, older men present more tendency to do it ($\beta = 0.2520$; p $= 0.28$) when compared to women ($\beta = 0.0870$; p $= 0.3710$). The analysis of the model shows a negative relation between INCNE and SMU for women. Thus, in general, there is a tendency for women to spend less time reading news on social media when they accidentally encounter information regarding political issues.

The collected data proves that, in general, Incidental News Exposure has no impact on the time people spend reading news online. This shows that, usually, people do not spend time on news when they accidentally find it on social media. Most Angolans may consider some news unreliable when shared on specific platforms and in certain formats, so they decide not to spend time reading it.

7   Conclusions

This study was conducted to understand Angolan citizens' behaviour in a specific type of political participation, considering external variables. People are critical of the information they consume and have an interest in interacting with news, as they intentionally and unintentionally are exposed to it. The present study allows us to understand that people consume news when they use social media and that the more they deliberately consume it, the more they are going to encourage people to vote. If policymakers intend to be closer to citizens and make their campaigns more effective, they can start making use of social media to achieve their goals, always implementing respectful and trustful behaviour.

References

1. Paris, R.: At War's End: Building Peace After Civil Conflict. Cambridge University Press (2004)
2. Zhan, M., Liang, H., Kou, G., Dong, Y., Yu, S.: Impact of social network structures on uncertain opinion formation. IEEE Trans. Comput. Soc. Syst. **6**, 670–679 (2019). https://doi.org/10.1109/TCSS.2019.2916918
3. Yin, X., Wang, H., Yin, P., Zhu, H.: Agent-based opinion formation modelling in social network: a perspective of social psychology. Phys. Stat. Mech. Its Appl. **532**, 121786 (2019). https://doi.org/10.1016/j.physa.2019.121786
4. Hong, S., Kim, S.H.: Political polarisation on twitter: implications for the use of social media in digital governments. Gov. Inf. Q. **33**, 777–782 (2016). https://doi.org/10.1016/j.giq.2016.04.007
5. Tucker, J.A., et al.: Social media, political polarization, and political disinformation: a review of the scientific literature (2018). https://papers.ssrn.com/abstract=3144139. https://doi.org/10.2139/ssrn.3144139
6. van Deth, J.W.: What is political participation? In: Oxford Research Encyclopedia of Politics (2016). https://doi.org/10.1093/acrefore/9780190228637.013.68
7. Verba, S., Nie, N.H.: Participation in America: Political Democracy and Social Equality. University of Chicago Press (1987)
8. Schäfer, A., Schwander, H.: 'Don't play if you can't win': does economic inequality undermine political equality? Eur. Polit. Sci. Rev. **11**, 395–413 (2019). https://doi.org/10.1017/S1755773919000201
9. Rosário, E.V.N., Severo, M., Francisco, D., Brito, M., Costa, D.: Examining the relation between the subjective and objective social status with health reported needs and health-seeking behaviour in Dande. Angola. BMC Public Health. **21**, 979 (2021). https://doi.org/10.1186/s12889-021-11003-4
10. Fernandes, E.C.B., de Castro, T.G., Sartorelli, D.S.: Associated factors of malnutrition among African children under five years old, *Bom Jesus*. Angola. Rev. Nutr. **30**, 33–44 (2017). https://doi.org/10.1590/1678-98652017000100004
11. Amundsen, I.: Always top-down: constitutional reforms in Angola. Chr. Michelsen Institute (2021)

12. Vraga, E.K., Tully, M.: News literacy, social media behaviors, and skepticism toward information on social media. Inf. Commun. Soc. **24**, 150–166 (2021). https://doi.org/10.1080/1369118X.2019.1637445
13. Tully, M., Maksl, A., Ashley, S., Vraga, E.K., Craft, S.: Defining and conceptualizing news literacy. Journalism **23**, 1589–1606 (2022). https://doi.org/10.1177/14648849211005888
14. Ku, K.Y.L., Kong, Q., Song, Y., Deng, L., Kang, Y., Hu, A.: What predicts adolescents' critical thinking about real-life news? The roles of social media news consumption and news media literacy. Think. Ski. Creat. **33**, 100570 (2019). https://doi.org/10.1016/j.tsc.2019.05.004
15. Schäfer, S.: Incidental news exposure in a digital media environment: a scoping review of recent research. Ann. Int. Commun. Assoc. **0**, 1–19 (2023). https://doi.org/10.1080/23808985.2023.2169953
16. Kim, A., Dennis, A.R.: Says Who? The effects of presentation format and source rating on fake news in social media (2018). https://papers.ssrn.com/abstract=2987866. https://doi.org/10.2139/ssrn.2987866
17. Ahmadi, M., Wohn, D.Y.: The antecedents of incidental news exposure on social media. Soc. Media Soc. **4**, 2056305118772827 (2018). https://doi.org/10.1177/2056305118772827
18. Erdelez, S., Makri, S.: Information encountering re-encountered: a conceptual re-examination of serendipity in the context of information acquisition. J. Doc. **76**, 731–751 (2020). https://doi.org/10.1108/JD-08-2019-0151
19. Ahmad, S.: Political behavior in virtual environment: role of social media intensity, internet connectivity, and political affiliation in online political persuasion among university students. J. Hum. Behav. Soc. Environ. **30**, 457–473 (2020). https://doi.org/10.1080/10911359.2019.1698485
20. Grasso, M.T., Farrall, S., Gray, E., Hay, C., Jennings, W.: Socialization and generational political trajectories: an age, period and cohort analysis of political participation in Britain. J. Elections Public Opin. Parties. **29**, 199–221 (2019). https://doi.org/10.1080/17457289.2018.1476359
21. Diehl, T., Barnidge, M., Gil de Zúñiga, H.: Multi-platform news use and political participation across age groups: toward a valid metric of platform diversity and its effects. J. Mass Commun. Q. **96**, 428–451 (2019). https://doi.org/10.1177/1077699018783960
22. Fisher, P.: Generational cycles in American politics, 1952–2016. Society **57**, 22–29 (2020). https://doi.org/10.1007/s12115-019-00437-7
23. Harrison, S.: A vote of frustration? Young voters in the UK general election 2019. Parliam. Aff. **73**, 259–271 (2020). https://doi.org/10.1093/pa/gsaa032
24. Valenzuela, S., Halpern, D., Katz, J.E., Miranda, J.P.: The paradox of participation versus misinformation: social media, political engagement, and the spread of misinformation. Digit. J. **7**, 802–823 (2019). https://doi.org/10.1080/21670811.2019.1623701
25. Tewksbury, D., Weaver, A.J., Maddex, B.D.: Accidentally informed: incidental news exposure on the world wide web. J. Mass Commun. Q. **78**, 533–554 (2001). https://doi.org/10.1177/107769900107800309
26. Heiss, R., Knoll, J., Matthes, J.: Pathways to political (dis-)engagement: motivations behind social media use and the role of incidental and intentional exposure modes in adolescents' political engagement. Communications **45**, 671–693 (2020). https://doi.org/10.1515/commun-2019-2054
27. Yamamoto, M., Morey, A.C.: Incidental news exposure on social media: a campaign communication mediation approach. Soc. Media Soc. **5**, 2056305119843619 (2019). https://doi.org/10.1177/2056305119843619
28. Koc, M., Barut, E.: Development and validation of New Media Literacy Scale (NMLS) for university students. Comput. Hum. Behav. **63**, 834–843 (2016). https://doi.org/10.1016/j.chb.2016.06.035

29. Rönkkö, M., Cho, E.: An updated guideline for assessing discriminant validity. Organ. Res. Methods **25**, 6–14 (2022). https://doi.org/10.1177/1094428120968614
30. Hamid, M.R.A., Sami, W., Sidek, M.H.M.: Discriminant validity assessment: use of fornell & Larcker criterion versus HTMT criterion. J. Phys. Conf. Ser. **890**, 012163 (2017). https://doi.org/10.1088/1742-6596/890/1/012163
31. Rasoolimanesh, S.M.: Discriminant validity assessment in PLS-SEM: a comprehensive composite-based approach, pp. 1–8 (2022)
32. Hair, J.F.H., Jr., Hult, G.T.M., Ringle, C.M., Sarstedt, M.: A Primer on Partial Least Squares Structural Equation Modeling (PLS-SEM). Sage Publications (2017)
33. Ozili, P.K.: The acceptable R-square in empirical modelling for social science research (2022). https://papers.ssrn.com/abstract=4128165. https://doi.org/10.2139/ssrn.4128165
34. Uwalaka, T., Amadi, B.N.C.: Hashtag activism: exploring the church must vote campaign in Nigeria. Covenant J. Commun. (2020)

Exploring the Performance of Large Language Models for Data Analysis Tasks Through the CRISP-DM Framework

Nurlan Musazade, József Mezei[(✉)], and Xiaolu Wang

Faculty of Social Sciences, Business and Economics, and Law, Åbo Akademi University, Turku, Finland
Jozsef.Mezei@abo.fi

Abstract. This paper investigates the impact of Large Language Models (LLMs), specifically GPT, on data analysis tasks within the framework of CRISP-DM (Cross-Industry Standard Process for Data Mining). In order to assess the efficiency of text-to-code language models in data-related tasks, we systematically examine the performance of LLMs in the stages of the data mining process. GPT models are tested against a series of Python programming and SQL tasks derived from a Master's program's curriculum. The tasks focus on data exploration, visualization, preprocessing, and advanced analytical tasks like association rule mining and classification. The findings show that GPT models exhibit proficiency in Python programming across various CRISP-DM stages, particularly in Data Understanding, Preparation, and Modeling. They adeptly utilize Python libraries for data manipulation and visualization, demonstrating potential as effective tools in data science. However, the study also uncovers areas where the GPT Text-to-code model shows partial correctness, highlighting the need for human oversight in complex data analysis scenarios. This research contributes to understanding how AI can augment traditional data analysis methods, particularly under the CRISP-DM framework. It reveals the potential of LLMs in automating stages of data analysis, suggesting an acceleration in analytical processes and decision-making. The study provides valuable insights for organizations integrating AI into data analysis, balancing AI strengths with human expertise.

Keywords: Large Language Models · CRISP-DM · GPT · Decision Support

1 Introduction

The integration of Artificial Intelligence (AI) has witnessed a large number of applications across diverse industries (Loureiro et al., 2021). This adoption of AI technologies has the potential to significantly influence both the operational paradigms of businesses and the decision-making processes. These developments can have far-reaching implications for the society and labour market. For instance, AI and Machine Learning technologies have demonstrated remarkable efficiency gains in productivity across various sectors, which raises concerns about the potential displacement of human labor (West, 2018).

One of the important effects of technological changes is on employment dynamics. The current trends in the digital economy show the declining demand for the labor force by the businesses, e.g. as observed in the US (West, 2018). Yet, the same technological changes cause the creation of new jobs and professions (West, 2018, World Economic Forum, 2023). Accordingly, there is an ever-rising demand for expertise in data-related professions, such as data scientist, data engineer (World Economic Forum, 2023). Skill requirements experience changes continuously, especially in the emerging data-related positions (Musazade, 2023).

The introduction of new language models (e.g. GPT) with the text-to-code capabilities generates the need for studying the role of these technologies and related capabilities in automating data-related tasks. There are already some studies indicating that these tools can significantly increase the efficiency of knowledge workers (Dell'Acqua et al., 2023). At the same time, the studies focusing on data related skills (e.g. Musazade, 2023, Smaldone et al., 2022) have been conducted without incorporating possible introduction or application of these innovative technologies with the text-to-code features. The primary objective of this study is to delineate the shifts in skills demanded following the introduction of text-to-code language models and the consequent automation of data-related tasks. Derived from this objective, the following research question can be formulated:

To what extent and how efficiently do LLMs, such as GPT, perform data-related tasks, and how does this impact the expected skill set of data analysts?

The rest of the paper is structured as follows. In the next section, we discuss related literature on modeling data-related processes in organizations, and LLMs. Then we present the research methodology and describe the experiments used to test the capabilities of LLMs. The results are discussed, and future recommendations are provided with regards to the future utilization of related tools in the industry.

2 Literature Review

In order to understand how LLMs can aid organizations in data science tasks, we will utilize a commonly used framework, the CRoss-Industry Standard Process for Data Mining or CRISP-DM (Chapman et al., 2000). Most of the tasks performed on a regular basis by a data analyst can be classified into the stages of the CRISP framework. In this section, we will also discuss the basic ideas of LLMs.

2.1 CRISP-DM

CRISP-DM has been studied in different fields or industries, and is a technology-independent model and selected by considering it as more organized, simple, and valid, in comparison with other models (Schröer et al., 2021). Based on the CRISP-DM model, data-related activities can be structured into six phases (Chapman et al., 2000).

The process starts with the *Business Understanding*: defining and comprehending business objectives, and project-specific assessments and requirements. In this phase, engagement of all affected organizational stakeholders and domain knowledge are crucial for constructing and developing business objectives and defining criteria for future evaluations (Nisbet et al., 2009, Abbott, 2014).

There are four generic tasks in the *Data Understanding* phase. The phase begins with the collection of data, and possible data loading. Secondly, data description, a generic level data understanding, including the specification of variables, format, etc. Thirdly, an exploratory task requires more profound analyzes, including data visualization, statistical analysis and correlation between features. The final task is the data quality verification (Chapman et al., 2000). In addition, the analysis includes detecting anomalies, grouping numeric variables, and the creation of new variables. The analysis requires appropriate visualizations regarding the above listed operations and relationships (Larose, 2015).

The main tasks in *Data preparation* are concerned with the selecting and cleaning variables, and possibly generating new features (Abbott, 2014). In case of predictive modeling, this phase is considered as the most laborious, e.g. 60–90% of the process, requiring the utilization of domain knowledge (Abbott, 2014). Data preparation phase has five generic tasks: (i) selecting data (e.g., selection criteria, data sampling, subsets), (ii) cleaning data (e.g., handling missing data, noise reduction), (iii) constructing data, (iv) integrating data, and (v) converting data to suitable format for modeling. The order of phases and tasks are alterable (Larose, 2015).

The *Modeling* phase starts with the selection of the appropriate techniques (Chapman et al., 2000). The next task is formulating test design as an assurance of the existence of an approach for examining model's effectiveness, capacity and quality. Then a model is created, by specifying model parameters and technical documentations (Nisbet et al., 2009). A good practice is building several different algorithms for obtaining various viewpoints and more accurate predictions (Nisbet et al., 2009). Finally, technical performance evaluation of the models is conducted based on predefined criteria, e.g. accuracy (Chapman et al., 2000).

The *Evaluation* phase refers to assessing and comparing the outcome with a predefined business objective. In this phase, the optimal realization of processes and opportunities for improvements are identified (Chapman et al., 2000). The final phase is *Deployment*, which starts with the deployment plan, also related to maintenance and monitoring tasks. A final report on the project should be created, including documentation of the experiences for possible benefits in the next projects (Chapman et al., 2000).

2.2 Large Language Models

Natural Language Processing (NLP) can be applied for various objectives, including translations, summarizing documents, sentiment analysis, and question answering. The technological and computational advancements, as well as progress and introductions in NLP techniques and architectures, have enabled advancements of more powerful language models (Sabharwal & Agrawal, 2021; Jain, 2022).

The existence of architecture and models capable of processing and predicting textual data by considering the whole context can be considered a deterministic success factor in the field. With the development and application of neural networks within the last decade, in particular Recurrent Neural Networks (RNN), the performance of models radically increased. Yet, RNNs have deficiencies, such as extraction of the longer dependencies, which have been overcome via architectural advancements (Jain, 2022). The development of the Transformer architecture (Vaswani et al., 2017) and models trained with this architecture have also resulted in substantial advancements in the field. The new architecture has improved efficiency in computational processing and identification of long-range relationships between words. High performance of the Transformer model followed by the development of the other successful pretrained models, including BERT and GPT.

The BERT model utilizes bi-directional self-attention that enables incorporating both side contexts. Pretrained models can be fine-tuned for the particular NLP tasks by training on the labeled task-specific data and adjusting model's parameters (Devlin et al., 2018). Another pretrained Transformer based model GPT (Radford et al., 2018) has a unidirectional architecture and predicts from left context to right (Devlin et al., 2018). New GPT models have been introduced recently with larger capacities, pre-trained data size and parameters. GPT-3 model with few-shot "learning" has achieved closer results to the fine-tuned models in a number of NLP tasks, such as some translation and question answering (Brown et al., 2020).

The latest GPT-4 model has achieved improved performance on the tasks compared to GPT-3, and higher performance than most of the individuals in many exams (OpenAI, 2023). A possible utilization of LLMs is to generate programming code. The models, including GPT Codex (Chen et al., 2021) and the latest GPT-4, have been studied in the context of coding, in particular in Python programming language. GPT Codex has been fine-tuned based on the open-source codes, whereas its deployment discontinued after availability of chat models (e.g. GPT-3.5).

In 2023, ChatGPT introduced ChatGPT Code Interpreter, which is termed as the Advanced Data Analysis later, that enables users to input datasets to the system. The function enables reading the dataset and presenting the results of analysis, including interpretations and graphs. Currently, there is a limited number of studies that systematically analyse LLMs' capabilities from the perspective of data analysis. In a review of ChatGPT's impacts, Aggarwal (2023) briefly posits how ChatGPT can be used in development and improvement of machine learning models, as well as in data pre-processing. Borger et al. (2023) review the role of ChatGPT in the coding and data analytics from the perspective of bioinformatics studies conducted by Piccolo et al., (2023). Based on the review, Borger et al., (2023) suggest that mathematical knowledge, creativity and logical thinking may be more crucial, since bioinformaticians expected to be exempted from routine tasks. However, within the mid-term horizon, testing and enhanced insight will be more prioritized requirements than correct coding skills (Borger et al., 2023).

There has been some research already on the general code generating capabilities of LLMs already before the wide availability of ChatGPT (Vaithilingam et al., 2022). However, research contributions increased in particular within the last year, as a variety of AI-assisted coding tools have emerged, each with its distinct focus and capabilities, both commercial and open-source. For example, Phind utilizes the GPT-3.5 model for general software development tasks, while Codex, the engine behind GitHub Copilot, assists in generating code snippets and entire functions. For data analysis, existing tools offer integrated environments that combine the power of AI with data science workflows. While one can find some results in articles about the use of LLMs for quantitative and qualitative (Xiao et al., 2023; Maddigan & Susnjak, 2023) data analysis, there is still a lack of systematic research that considers the potential use in all stages of the process of utilizing data in organizations.

3 Methodology

The current study adopts a computational methodology to assess the performance of LLMs, specifically GPT, when used for tasks typically performed by a data analyst. In order to do so, we make use of tasks that closely align with the curriculum of a Master's program specializing in the impact of digitalization on organizations and aimed at providing the main skills currently required for many data-related positions in the job market. Top frequent skills that have been found on the job advertisements (Musazade, 2023) are the basis for the analysis of criticality of skills and tools in the future. Furthermore, results of the research are compared, studied and generalized from the CRISP-DM framework perspective as discussed in the previous section.

The primary criteria governing the evaluation of ChatGPT's performance center on the accuracy of the responses it generates. While the primary emphasis is on the correctness of the solutions produced, the evaluation process also takes into consideration whether ChatGPT adheres to the requisite problem-solving steps. These steps include considerations such as data scaling, missing value treatment, use of training and test sets when applicable. To facilitate a thorough evaluation of ChatGPT's problem-solving capabilities within the tasks, the study employed two versions of ChatGPT: ChatGPT 3.5 and ChatGPT 4. Additionally, ChatGPT was supplemented with the advanced data analysis plugin, enhancing its capacity to engage with and respond to the assignment tasks effectively. In line with the findings of previous research identifying the top tools required from professionals in data related positions, the tasks used in this research test the LLMs' ability to solve problems using Python programming language and SQL. The assignment tasks used for this comprehensive evaluation were sourced from four distinct courses within the curriculum of the Master's program:

- Course 1: a general introduction to programming concepts using Python. The assignment tasks encompass the essential principles of programming.
- Course 2: programming tools for data analysis in Python. The assignment tasks are related to data visualization, data manipulation, data preprocessing, as well as the basics of building and evaluating machine learning models.
- Course 3: advanced machine learning concepts, NLP and applications with Python. The assignment tasks within this course challenge students with more complex and specialized problem-solving scenarios.
- Course 4: data management and SQL. Students learn about entity-relationship (ER) diagrams, and querying databases with SQL. The assignment tasks in this course emphasize practical querying skills.

The assignment tasks were presented to ChatGPT in a manner consistent with their formulations utilized in the assignments. To maintain the authenticity of the course context, each assignment, comprising multiple tasks, was consolidated within a single comment. In instances where ChatGPT sought clarification or additional information, responses were provided mirroring the types of queries typically posed by students. To ensure the reliability of the results, the answers generated by ChatGPT underwent a validation process performed by the instructors who teach the courses.

4 Results

In the following, we discuss some specific observations from the results of the tests. As in the experiments our focus is on assessing the capabilities of LLMs with regard to more technical tasks, some phases of CRISP-DM are represented more than others[1]. In particular, most of the tasks can be classified as belonging to *Data Understanding*, *Data Preparation* and *Modelling*. Additionally, the other tasks, *Business Understanding*, *Evaluation*, and *Deployment*, in practice typically require the involvement of multiple stakeholders, not only the work of a single data analyst.

Regarding *Data Understanding*, the results show that GPT models excelled in tasks like data exploration and analysis. Considering the specific tasks related to this phase we observed the following:

- data description: the tool effectively utilized Python libraries like Pandas, and Numpy, which are crucial for data manipulation and visualization. Specifically, the GPT models have no problem in understanding what summary statistics to generate for different variable types.
- exploratory data analysis tasks: the models were adept at creating visual representations of data, using libraries like Matplotlib and Seaborn. This also addresses tasks related to understanding the data distribution and identifying patterns or outliers visually
- data quality verification: the GPT models can assess the completeness and correctness of the data, specifically identifying missing values

We note here that the Code interpreter was correct in all tasks, while the GPT Text-to-code model was only partially correct in some instances of descriptive analysis. For instance, it did not generate the requested number of visualizations, or incorrectly used the correlation measure.

In the *Data Preparation* phase, both GPT versions showed proficiency, especially in handling datasets using Pandas. For instance, they could perform operations like data cleaning and transformation. In text data preparation, the models were adept at using NLTK for tasks like tokenization and stop-word removal. The tools aptly handled tasks related to selecting and cleaning data with the aim of increasing its value, integrating data, and converting the data to suitable format for modeling.

The *Modeling* phase is a critical stage in the CRISP-DM process where various statistical or machine learning models are built. This stage often involves selecting appropriate modeling techniques, building models, and assessing their effectiveness. The GPT-models demonstrated strong capabilities in complex tasks like association rule mining and classification, utilizing for example the Sklearn library. They could build and interpret models effectively. Furthermore, for the machine learning-based tasks, such as regression analysis, the solutions' performance was accurate, suggesting a good understanding of statistical modeling. Although generally successful, there were instances where the GPT Text-to-code model provided only partially correct solutions, particularly in advanced modeling tasks. This indicates a limitation in handling certain complex nuances of statistical modeling.

---

[1] A complete list of the tasks can be found at: https://doi.org/10.6084/m9.figshare.24624501.v1.

Complexity in this situation can be approached from different perspectives. For instance, from the data format perspective, GPT Text-to-code's incorrect split of the ID format (expecting to return "-" in the list while using function .split("-")) in a question generated incorrect solutions in data preparation and understanding, which may had an impact also on the modeling phase.

Secondly, complexity can refer to the steps and descriptions of the tasks. In our experiments, some of the steps and guidance were missed while performing specific tasks, in particular by the text-to-code model. For instance, when asked to exclude stopwords specific to a given context, GPT Advanced Data Analysis model still included words that were less meaningful. These and other examples illustrate that the instructions to the LLMs should be as clear as possible.

Moreover, some modules are unavailable in the ChatGPT advanced data analysis model. For example, the model lacked access to the tokenization of the nltk library, and used instead the split function, which may be insufficient in some cases. The models were also unable to initialise nltk stopwords, and the model defined them manually. In some cases, human intervention is required related to necessary domain knowledge. For example, when analysing news articles about data science and machine learning, although the model removed functional words (e.g., "would", "one", "also"), the terms "data", "learning" are retained incorrectly. As another example, the GPT assumes that NaNs imply non-existence of (text) data and replaces them with 0. In other words, domain knowledge, critical thinking and decision-making may be required in some cases.

Finally, some modelling tasks require parameter optimizations. Although the models used Grid Search, human engagement may contribute in defining parameters and metrics that are most suitable for the task. The experiments in this study may be insufficient to generalise the mentioned observations. Yet, the mentioned points indicate nuances that are beyond the simple coding tasks and should be considered in practice and studied further.

In summary, while the GPT models showed impressive capabilities in Python programming tasks across various stages of CRISP-DM, particularly in Data Understanding, Preparation, and Modeling, there were areas, especially in advanced statistical modeling, where the GPT Text-to-code model showed partial correctness. This suggests room for improvement in handling complex tasks and nuances in data analysis and modeling.

4.1 Discussion

The detailed results discussed above provide a systematic preliminary evaluation of GPT-based models in executing a range of Python programming and SQL-based tasks aligned with the stages of the CRISP-DM process. In the crucial phases of *Data Understanding and Preparation*, the GPT models demonstrated proficiency in tasks like data exploration, visualization, and preprocessing using essential Python libraries such as Pandas, Matplotlib, and NLTK. The *Modeling* phase highlighted their capability in handling complex analytical tasks, including association rule mining, classification, and regression modeling. However, the study also identified areas of partial correctness, particularly in the GPT Text-to-code model during advanced modeling tasks and certain aspects of descriptive analysis. For example, the results show that while GPT 3.5 succeeded in accomplishing SQL tasks involving working with only a single table, it was more prone to commit errors in assignments requiring joining the tables. Moreover, the assignments involve detailed description of the methods and stages as a guide for performing the tasks. These suggest a nuanced understanding of data analysis processes but also underscores the need for human oversight, especially in complex or critical analytical scenarios.

Introducing ChatGPT advanced data analysis, which enables importing dataset, increased the performance in accomplishing tasks correctly by enabling direct engagement. In particular, in the outcome of GPT text-to-code experiments most of the failures (e.g., incorrect answers, errors, incompleteness) have been addressed by additional questions or clarifications. However, ChatGPT advanced data analysis enables running the LLM generated code over the datasets, and automatically reinitiating the script when errors are encountered. Secondly, ChatGPT advanced data analysis enabled interpretations of the results, as well as performing and finalizing the interdependent tasks. In other words, a shift to less human involvement can be observed, whereas it is still crucial, considering how the tiny errors or misinterpretations may cost the businesses. Therefore, aligned with the Borger et al.'s (2023) forecasts, enhanced understanding and assessment may be more crucial than accurate programming skills. Considering that the analyzed phases may constitute the most time spent by the data professionals, the experiment indicates that LLMs can be anticipated to streamline data mining tasks. As mentioned earlier, the data preparation phase may constitute more than half of the time spent for the predictive modeling, and may reach 90% of time spent. Our observations are consistent with the expectations of the Borger et al. (2023), that GPT may effect by exempting from regular routine tasks and professionals with logical thinking and creativity may be preferred.

4.2 Main Contributions

Academically, this study adds significant value to the field of AI, data mining and data science by empirically testing and documenting the capabilities and limitations of LLMs in data analysis. It contributes to the understanding of how AI can augment traditional data analysis methodologies, particularly in the context of the CRISP-DM framework. Practically, the findings have substantial implications for businesses and data analysts. They reveal the potential of AI models, like GPT, in automating or aiding various stages of data analysis, thereby accelerating the analytical process and potentially enhancing decision-making. However, the study also highlights some limitations with different versions that practitioners need to be aware of. The results indicate that there still should be a role played by human expertise in guiding, interpreting, and validating AI-generated analysis, especially in complex or nuanced data scenarios. This balanced understanding is crucial for organizations aiming to integrate AI into their data analysis practices effectively.

5   Conclusion

The emergence of LLMs, has sparked significant interest in their potential applications across various domains, including data analysis. The motivation for this study stems from the increasing need to understand how these advanced AI tools may be integrated in organizational settings, and increase the level of automation. This research aimed to empirically evaluate the capabilities of GPT-based models in performing data analysis tasks within the CRISP-DM framework, a standard in the industry. The results provide valuable insights into how LLMs can augment human data analysts and potentially reshape data-driven decision-making in organizations.

The study revealed that GPT models are proficient in understanding and executing a range of Python programming tasks, notably in Data Understanding, Preparation, and Modeling stages. They demonstrated adeptness in utilizing diverse Python libraries for data manipulation, visualization, and machine learning, indicating their utility as powerful tools in data science. In conclusion, this study contributes to the academic discourse on the role of AI in data analysis by providing an empirical assessment of LLMs in this field. Practically, it serves as an invaluable reference for organizations looking to integrate AI into their data analysis practices, balancing the strengths of AI with the critical need for human oversight. As AI technologies continue to advance, their role in data analysis is poised to grow. Still, this research underscores the importance of a synergistic relationship between AI capabilities and human expertise to achieve the most accurate and contextually relevant insights.

While this research provides valuable insights into the capabilities of GPT-based models in data analysis tasks, it is important to acknowledge its limitations. First, the study did not present results for all the stages of the CRISP-DM process. This incomplete coverage limits the understanding of how GPT models might perform in the initial conceptual stages of a data project and the final implementation of the model outputs. Furthermore, the techniques in data understanding and preparation, and type of modeling tasks covered in this study can be considered as incomplete. Second, the experiments predominantly dealt with structured or semi-structured data, offering limited insights into the performance of GPT models with unstructured data. This is a notable gap, as unstructured data forms a significant portion of organizational data and poses distinct challenges. Finally, some of the assignments or datasets are publicly available (e.g., in Kaggle). Although the main variables (e.g., column names, datasets) have been partially changed or masked when possible and mostly in GPT 3.5, if the relevant tasks, solutions and datasets have been included in the training of the model, it can result in the better performance of the GPT.

References

Loureiro, S.M.C., Guerreiro, J., Tussyadiah, I.: Artificial intelligence in business: state of the art and future research agenda. J. Bus. Res. **129**, 911–926 (2021)

West, D.M.: Future of Work: Robots, AI, and automation. Brookings Inst (2018)

World Economic Forum: The Future of Jobs Report 2023 (2023). https://www3.weforum.org/docs/WEF_Future_of_Jobs_2023.pdf

Musazade, N.: Tools and technologies utilized in data-related positions: an empirical study of job advertisements. In: 36th Bled eConference, vol. 155 (2023)

Dell'Acqua, F., et al.: Navigating the jagged technological frontier: field experimental evidence of the effects of AI on knowledge worker productivity and quality. In: Harvard Business School Technology & Operations Management Unit Working Paper, pp. 24–013 (2023)

Smaldone, F., Ippolito, A., Lagger, J., Pellicano, M.: Employability skills: profiling data scientists in the digital labour market. Eur. Manag. J. **40**(5), 671–684 (2022)

Chapman, P., et al.: CRISP-DM 1.0: step-by-step data mining guide. SPSS (2000)

Schröer, C., Kruse, F., Gómez, J.M.: A systematic literature review on applying CRISP-DM process model. Procedia Comput. Sci. **181**, 526–534 (2021)

Nisbet, R., Elder, J.I., Miner, G.D., Elder, J., Elder, J.I., Miner, G.D.: Handbook of Statistical Analysis and Data Mining Applications. Elsevier Science & Technology (2009)

Abbott, D.: Applied Predictive Analytics: Principles and Techniques for the Professional Data Analyst. Wiley (2014)

Larose, D.T.: Data Mining and Predictive Analytics. Wiley, Incorporated (2015)

Sabharwal, N., Agrawal, A.: Hands-on Question Answering Systems with BERT: Applications in Neural Networks and Natural Language Processing. Apress (2021)

Jain, S.M.: Introduction to Transformers for NLP, 1st edn. Apress (2022)

Vaswani, A., et al. Attention is all you need (2017). arXiv.org. https://arxiv.org/abs/1706.03762

Devlin, J., Chang, M.-W., Lee, K., Toutanova, K.: BERT: pre-training of deep bidirectional transformers for language understanding (2018). https://doi.org/10.48550/arXiv.1810.04805

Radford, A., Narasimhan, K., Salimans, T., Sutskever, I.: Improving Language Understanding by Generative Pre-Training (2018)

Brown, T.B., et al.: Language Models are Few-Shot Learners (2020)

OpenAI: GPT-4 Technical Report (2023). arXiv:2303.08774v3

Chen, M., et al.: Evaluating large language models trained on code (2021). arXiv preprint arXiv: 2107.03374

Aggarwal, S.: A review of ChatGPT and its impact in different domains. Int. J. Appl. Eng. Res. **18**(2), 119–123 (2023). https://doi.org/10.37622/ijaer/18.2.2023.119-123

Borger, J.G., et al.: Artificial intelligence takes center stage: exploring the capabilities and implications of ChatGPT and other AI-assisted technologies in scientific research and education. Immunol. Cell Biol. (2023)

Piccolo, S.R., Denny, P., Luxton-Reilly, A., Payne, S., Ridge, P.G.: Many bioinformatics programming tasks can be automated with ChatGPT. ArXiv.org (2023)

Vaithilingam, P., Zhang, T., Glassman, E.L.: Expectation vs. experience: evaluating the usability of code generation tools powered by large language models. In: CHI Conference on Human Factors in Computing Systems Extended Abstracts, pp. 1–7 (2022)

Xiao, Z., Yuan, X., Liao, Q.V., Abdelghani, R., Oudeyer, P.Y.: Supporting qualitative analysis with large language models: combining codebook with GPT-3 for deductive coding. In: Companion Proceedings of the 28th International Conference on Intelligent User Interfaces, pp. 75–78) (2023)

Maddigan, P., Susnjak, T.: Chat2VIS: generating data visualisations via natural language using ChatGPT, codex and GPT-3 large language models. IEEE Access **11**, 45181–45193 (2023)

An Exploratory Big Data Approach to Understanding Commitment in Projects

Narasimha Rao Vajjhala[1]([⊠]) [iD] and Kenneth David Strang[2,3] [iD]

[1] University of New York Tirana, Tirana, Albania
narasimharao@unyt.edu.al
[2] W3 Research , New York, USA
[3] University of the Cumberlands, Williamsburg, KY, USA

Abstract. This study addresses the twin challenges of talent retention and high project failure rates (40–70%) by harnessing machine learning (ML) techniques to analyze retrospective big data. The study's objective was to ascertain whether project performance indicators can be a reliable gauge of project manager (PM) organizational commitment. This approach sidesteps the inherent bias and small effect sizes associated with survey self-report responses. Our innovative methodology leverages secondary big data, transforming the values into structured features that predict PM organizational commitment. This study proposes a novel conceptual framework, focusing on actual behavioral evidence rather than traditional, self-reported attitudes to assess the fuzzy predictors of organizational commitment. Among the three developed ML models, one demonstrated a significant 24% effect size, uncovering key features correlating PM tenure and organizational commitment with success. The insights gained from this research have broad implications for global stakeholders in projects and programs, offering a more objective and big data-driven understanding of PM commitment.

Keywords: Project Management · Organizational Commitment · Machine Learning · Big Data Analysis · Talent Retention · Project Failure Rates · Data-Driven Management · Predictive Modeling · Stakeholder Implications · Program Management · Performance Indicators

1 Introduction

Organizational commitment refers to an employee's psychological attachment and loyalty to their organization [1] and has been a focal point in organizational behavior research for decades [2–4]. Indicators of organizational commitment are the length of tenure, experience, collaboration, and continued participation with the organizational teams. In today's project-centric business world, grasping the nuances of organizational commitment within project management is increasingly vital. Research in this area has linked organizational commitment to a host of positive organizational outcomes, including lower staff turnover, enhanced job performance, and reduced absenteeism [5–7]. These findings suggest a clear benefit to organizations when employees are deeply committed: a boost in overall organizational health and performance.

© The Author(s), under exclusive license to Springer Nature Switzerland AG 2024
Á. Rocha et al. (Eds.): WorldCIST 2024, LNNS 989, pp. 66–75, 2024.
https://doi.org/10.1007/978-3-031-60227-6_6

In project management, the indicators of success typically include delivering projects on time, within budget, in alignment with objectives, and to the satisfaction of stakeholders [8]. Projects marked by higher levels of organizational commitment may have a greater likelihood of success [9, 10]. However, researchers must thoroughly examine the direct relationship between organizational commitment and project success. Well-established factors in successful project execution, such as effective communication, strong leadership, team cohesion, and stakeholder satisfaction, have been extensively explored in project management studies [11, 12]. There are indications of these factors influencing organizational commitment. For instance, effective leadership has been associated with heightened employee commitment [13, 14], and clear communication is recognized as a critical contributor to commitment in team settings [15]. Considering the complex nature of what makes a project successful, exploring the factors that foster organizational commitment can offer valuable insights for project managers seeking to improve project results and employee dedication.

Artificial Intelligence (AI) is the ability of a machine to mimic intelligent human behavior, encompassing diverse subfields from logical reasoning to specific tasks like language translation or autonomous driving [16]. Its essence lies in creating machines capable of tasks that typically require human intelligence. A critical subset of AI, ML, empowers computer systems to perform tasks through algorithms and statistical models without explicit instructions, relying on pattern recognition and inference [17]. Despite the potential of AI and ML, the high failure rates of projects, hovering between 40–70%, persist across various industries [18, 19]. While ML has seen widespread organizational adoption for analyzing project performance, its incorporation into project management (PM) has stagnated since 2017 [20]. This stagnation occurs despite significant advancements in AI and ML, leading to questions about their underutilization in PM for performance enhancement. Intriguingly, project failure rates remain unchanged even with AI and ML applications in PM, like ChatGPT's use in automating project communications [21]. This study uses extensive big data to explore the predictive relationship between project characteristics, including indicators of organizational commitment such as experience, collaboration, and continued participation with the organizational team. Our study focuses on whether PMs' organizational commitment in successful projects correlates with other project attributes (RQ1) and identifying attributes in successful projects that predict PMs' commitment tenure (RQ2).

2   Review of Literature

Meyer and Allen's [1] tri-dimensional model of organizational commitment describes three types: affective commitment (an emotional attachment to the organization), continuance commitment (awareness of the costs associated with leaving), and normative commitment (feeling of obligation to stay). This model provides insight into why employees choose to stay with an organization and how their commitment types influence their behavior, such as going above and beyond job requirements or doing the bare minimum due to perceived high costs of leaving [1]. However, the model has faced criticism for the small effect sizes in empirical studies, questioning its practical significance [22]. For example, research has shown modest relationships between commitment types and outcomes like turnover and job performance, suggesting limited real-world implications, particularly for continuance commitment [7, 23, 24]. Moreover, questions about the distinctiveness of the commitment components have been raised, with studies indicating overlap between affective and normative commitment and potential bi-dimensionality in the continuance commitment scale [25, 26].

Further, the model's applicability varies across cultural contexts and organizational changes because the model's dimensions may manifest differently in non-Western cultures due to cultural nuances [27]. Also, with technological advancements and changing workforce dynamics, the evolving nature of workplaces calls for a re-examination and potential adaptation of the model. This includes considering the impact of external variables like job satisfaction, leadership behavior, and work-life balance on commitment and expanding the model to include occupational commitment [28, 29].

ML has evolved significantly in PM, transitioning from early optimization-focused efforts to recent advancements in predictive analysis. Initially, ML applications in PM were not centered on predictive analysis, with early instances dating back to Hosley [30] focusing primarily on optimization. More recently, the emphasis has shifted to predictive ML, as demonstrated by Wauters and Vanhoucke [31], who used Support Vector Machines (SVM) for forecasting critical paths in projects, marking a significant step in applying ML to evaluate and enhance PM performance. These advancements reflect a broader integration of ML in PM, particularly in decision-making, outcome prediction, task automation, and overall efficiency enhancement. Risk prediction and management in PM have been revolutionized by ML, as seen in the works of Mahdi et al. [32] with ensemble learning approaches, Asadi, Alsubaey and Makatsoris [33] in predicting construction delays, and Shoar, Chileshe and Edwards [34] using Random Forest algorithms for risk factor identification. This trend is further evidenced by efforts to integrate Deep Learning (DL) and Digital Twins (DT) in construction [35], the development of smart PM frameworks [36], and the classification of risks in construction projects [37].

3   Research Methodology

In this study, the researchers have adopted a pragmatic and interpretative approach, blending elements of post-positivism with a more flexible methodology. While post-positivist principles guide the pursuit of factual, quantitative data, the researchers employ a tailored approach in applying analytical ML techniques. This flexibility allows for adapting the sequence of ML methods in response to emerging results, focusing on interpreting the data to derive meaningful insights beyond mere statistical reporting. The research framework aligns with pragmatic ideology, typically guided by research questions or propositions rather than pre-formulated hypotheses, as seen in positivist paradigms. This approach is particularly relevant when employing ML techniques, where hypotheses may not always be explicitly stated. For instance, in techniques like K-Means cluster analysis or Principal Component Analysis (PCA), the aim is often exploratory, such as uncovering predominant factor groups or determining the number of factors with significant loadings without predefined hypotheses.

In line with the exploratory nature of the study, the specific ML techniques to be utilized are not entirely predetermined but are influenced by the data types and research questions, as suggested by Liu and Yu [38]. These factors will guide the decision between classification or prediction/learning methods within the realm of ML. The selection of ML techniques will be further refined during the study, particularly in the procedures section, post sample data identification. Efforts will be made to collect and utilize quantitative data, employing data cleaning and transformation as part of ML pre-processing when necessary. Given the study's pragmatic and exploratory design, a confidence level of 90% will be targeted for any statistical tests, whether nonparametric or parametric.

In this research project, ethical clearance was obtained, and the study exclusively used secondary declassified government public data, specifically from a large U.S.-based big data repository containing historical metrics of thousands of IT projects, primarily from U.S. military divisions. Despite the ambiguity in discerning whether all IT projects involved military personnel or subcontractors, the data was well-documented, including descriptions of names, types, and lengths of various fields. The study focused on a sample of successful projects, defined by meeting criteria in scope, schedule, cost, and quality as per sponsors' evaluations. The intent was to retrospectively analyze successful projects to identify key features contributing to project management (PM) effectiveness and organizational commitment, avoiding confounding factors like resistance impacting outcomes. The research encompassed 1230 features from over 500 IT projects (2015–2023), covering aspects such as PM characteristics, team metrics, and project-specific details like budget and quality. PM tenure was a critical measure for organizational commitment. After rigorous data cleaning, which included removing invalid records, the final sample size was reduced to 439 projects.

The exploratory data analysis directed the research towards two main analytical paths: classification and regression analysis. The choice of specific ML techniques, whether for classification or regression, was influenced by the nature of the target variable data types, with the transformation of alpha-ordinal target variables into numeric ordinals to allow a broader application of ML techniques. Supervised ML techniques were utilized due to the labeled nature of the fields, offering a more reliable foundation for meaningful predictive modeling. The regression analysis involved the application of various algorithms like LASSO, Ridge, or elastic net, focusing on input features regressed on a single response variable, thereby generating coefficients and effect sizes for predictive hypotheses. In contrast, the classification analysis involved techniques like random forest, nearest neighbor, and neural network algorithms, each offering unique approaches to predicting the dependent variable based on feature values.

The study also employed advanced ML techniques, such as neural networks with multi-layer perception algorithms and SVM, to develop predictive models. These techniques involved building a constant baseline model, followed by full regression models, and using decision trees in the case of random forests. The evaluation of these models was carried out using traditional regression model estimates like MSE, RMSE, MAE, and the coefficient of determination (r^2 effect size), in addition to cross-validation methods for accuracy assessment. The final comparison of scoring across all ML techniques included considerations of MSE, RMSE, MAE, and r^2 effect size, providing a comprehensive view of the predictive accuracy and reliability of the models in relation to the large sample size of the study.

4 Findings

In this study, data cleaning involved removing records with corrupted values, missing information, or confidential fields, setting the stage for an organizational-level analysis. At the organizational level of analysis, each project was completed by a different subcontractor and team. The projects were approximately equally spread across ocean/water, ground/military bases, and air space types of work. The average project team size was found to be 24, with a standard deviation of 11.4, and the mean final budget for projects was approximately $29.9 million, with a significant level of dispersion. At the individual PM level, the average age was 40.2 years, and most PMs (54.3%) were male with substantial relevant experience (an average of 32.1 years). A majority had at least a college degree, and nearly a third were PMI certified, highlighting a well-educated and experienced demographic profile. The average Project Manager (PM) commitment tenure was 11.08 years, ranging from 1 to 25 years.

The exploratory analysis began with identifying relevant features from over 1,000 attributes, with 11 features ultimately being selected based on their relation to PM commitment tenure. These included factors like PM experience, professional certification, education level, gender, project budget and contract values, mandatory timesheets, and previous project success counts. The analysis aimed to understand the impact of these variables on PM commitment tenure, with a focus on identifying patterns and relationships. The radial analysis revealed interesting patterns, such as certain features being associated with higher levels of PM commitment tenure and others correlating with project breaches or final budget figures. These insights laid the groundwork for further statistical analysis using various ML techniques.

The study then progressed to Multi-Dimensional Scaling (MDS), corroborating findings from the radial analysis and revealing distinct clusters of features in relation to PM commitment tenure. This clustering suggested conditional relationships between the features. However, the silhouette scores indicated weak clustering, prompting a shift towards ML regression analysis. This regression analysis compared various ML techniques, assessing their accuracy through metrics like r2 and Mean Absolute Error (MAE). The results, while varied, provided a robust measure of the models' accuracy, offering valuable insights into the relationships between PM characteristics, project attributes, and commitment tenure.

The analysis of various Machine Learning (ML) techniques revealed that the Linear Regression (LR) model was the most effective in predicting Project Manager (PM) commitment tenure. With an effect size of 23.8% and a Mean Absolute Error (MAE) of only 5.3%, the LR model demonstrated moderate success according to ML benchmarks, such as those suggested by Hohman et al. (2020). In comparison, the Random Forest (RF) model showed a lower effect size of 10.8% and a slightly higher MAE of 5.97%, while the Support Vector Machine (SVM) model lagged with the smallest effect size of 4.7% and the highest MAE of 6.27%. This performance assessment indicates that both LR and RF models were reasonably accurate in their predictions, with LR emerging as the most effective model. Table 1 lists the important accuracy scores of the quantitative ML regression techniques.

The next phase of the study focused on evaluating the scoring rank coefficients of the highest-performing ML regression models, specifically to assess the strength of

Table 1. ML regression feature prediction of project commitment scores (N = 439)

Model	MSE	RMSE	MAE	r^2
Baseline (Constant)	53.801	7.3349	6.4034	−0.001
Linear Regression	40.957	6.3998	5.3051	0.238
Random Forest	47.939	6.9238	5.9719	0.108
SVM	51.263	7.1599	6.2692	0.047
Baseline (Constant)	53.801	7.3349	6.4034	−0.001

association between each feature and the target variable, PM commitment tenure. This evaluation, using the relative relief coefficient rank due to the extensive continuous data (notably in budget and experience fields), enabled an in-depth analysis of the relationship between features and the dependent variable. The LR model, with its superior performance metrics, was chosen for this feature score ranking analysis. The coefficients of the LR model, as listed in descending order of effectiveness, provided a clear indication of which features most strongly influenced PM commitment tenure. Table 2 lists the LR scoring coefficient estimates of the 10 features regressed on the target variable PM commitment tenure, sorted in descending order (better coefficients are at the top). The quantitative fields in this analysis were marked as 'NA' in frequency counts, differentiating them from alpha fields where frequencies of categories were applicable.

In Table 2, the type of project (line of business) was the most influential field. Here we can see that ocean-based projects were more successful, as compared to air space and land-based. The PM experience was the next important feature, followed by use of internet tools for network collaboration, and cross-industry team membership. Contracted PMs were more successful as compared to in-house (on staff) PMs. The remaining features were much less impactful, but worthy of mentioning, namely: having an online project management office (PMO), having the sponsor co-located with the PM on the site, using timesheets, male gender, providing formal in-house training to teams, using PM software, and requiring the teams to collaborate regularly online. However, remote participation by the PM was also correlated with success. From a statistical perspective, the remaining features with less than 0.15 relative relief function coefficient estimates would have negligible association or impact on the project outcome, thus, these were not discussed here. Overall, it can be observed that experience, network collaboration, cross industry collaboration, sponsor co-location (assuming increased collaboration), all support the proposition of increased PM commitment being associated with successful projects.

Table 2. Linear Regression scoring coefficients to predict PM commitment tenure

#	Feature	Counts	Relative Relief Function Coefficient
1	LineOfBusiness	4.0	0.3598683802115484*
2	Experience	NA	0.2933840007056992*
3	NetworkCollab	3.0	0.2766551615534942*
4	CrossIndustry	3.0	0.27047151678144743*
5	Contract	3.0	0.22191487599278825*
6	OnlinePMO	3.0	0.17692792605462082*
7	SponsorClose	2.0	0.17661907724188222*
8	Timesheet	2.0	0.17372536014740814*
9	Gender	2.0	0.17121451660147974*
10	Training	3.0	0.17061866990368293*
11	Other	2.0	0.15666343444053557*
12	PMsw	3.0	0.15387165339160203*
13	TeamOnline	3.0	0.14181805677185277*
14	Remote	3.0	0.1403387363857974*
15	PMP	2.0	0.13880176897460472*
16	ContractK	NA	0.1378930932379765*
17	BudgetK	NA	0.13038407847261024*
18	Education	2.0	0.062454816391712305
19	Breach	2.0	<0

Table notes: $* = p < .05$

5   Future Research Directions

Future research in this field could expand beyond the retrospective analysis of big data to include real-time data from various project management tools, offering dynamic insights into organizational commitment patterns. Combining structured survey data with unstructured big data could yield a more comprehensive understanding of these dynamics, bridging traditional and modern analytical approaches. Given the international scope of many projects, focusing on how cultural differences affect organizational commitment would be enlightening. Using ML models to analyze commitment levels across different cultural and regional contexts could reveal distinct patterns and influencing factors. A more detailed exploration of the less tangible predictors of organizational commitment, including specific behaviors, patterns, and triggers, is recommended. Investigating how commitment evolves throughout the life cycle of projects could identify critical periods where commitment is most at risk. Advancements in ML algorithms, including deep learning, transfer learning, and ensemble methods, could further refine the predictive accuracy of these studies. As the reliance on big data and predictive analytics grows, research must also address the ethical and privacy implications of data usage. External factors like macroeconomic trends and global events, such as pandemics, are likely to impact project manager commitment significantly, and future research should strive to quantify and incorporate these elements into predictive models.

This study needs to be replicated using different samples, and with triangulation of methods (additional methods besides ML). Additionally, establishing feedback mechanisms where insights from ML models continuously inform organizational strategies could be an intriguing study area. Research could also explore targeted interventions based on predictive insights to enhance commitment levels proactively. Extending the research scope beyond project managers to include other project roles could offer a more rounded view of commitment dynamics within project management. Investigating commitment patterns among team members, stakeholders, or senior leadership could reveal broader organizational trends and contribute to a more holistic understanding of commitment within project environments.

6   Conclusion

In today's dynamic organizational management landscape, the value of human capital is paramount, with a critical focus on retaining experienced talent and reducing high project failure rates. This study has leveraged advanced ML techniques on a vast array of retrospective big data to forge a connection between project performance and organizational commitment, moving beyond the traditional, speculative approaches often based on survey data. The large effect size of 24%, and the relative relief coefficients, in the ML LR model indicated that several features in the secondary big data correlated with successful projects. For example, the type of project (line of business) was the most influential field. Here we can see that water-based projects were more successful, as compared to air space and land-based.

The PM experience was the next important feature, followed by use of internet tools for network collaboration, and cross-industry team membership. Contracted PMs were more successful as compared to in-house (on staff) PMs. The remaining features were much less impactful, but worthy of mentioning, namely: having an online project management office (PMO), having the sponsor co-located with the PM on the site, using timesheets, male gender, providing formal in-house training to teams, using PM software, and requiring the teams to collaborate regularly online. However, remote participation by the PM was also correlated with success. Overall, it can be observed that experience, network collaboration, cross industry collaboration, sponsor co-location (assuming increased collaboration), all support the proposition of increased PM commitment being associated with successful projects.

References

1. Meyer, J.P., Allen, N.J.: A three-component conceptualization of organizational commitment. Hum. Resour. Manag. Rev. **1**(1), 61–89 (1991)
2. Allen, N.J., Meyer, J.P.: Construct validation in organizational behavior research: the case of organizational commitment. In: Goffin, R.D., Helmes, E. (eds.) Problems and Solutions in Human Assessment: Honoring Douglas N. Jackson at seventy, pp. 285–314. Springer, Boston (2000). https://doi.org/10.1007/978-1-4615-4397-8_13

3. Purwanto, A., et al.: Effect of transformational leadership, job satisfaction, and organizational commitments on organizational citizenship behavior. Inovbiz: Jurnal Inovasi Bisnis **9**, 61–69 (2021)
4. Grego-Planer, D.: The relationship between organizational commitment and organizational citizenship behaviors in the public and private sectors. Sustainability **11**(22), 6395 (2019)
5. Norawati, S., et al.: The effect of supervision, work motivation, and interpersonal communication on employee performance and organizational commitment as variables intervening. IJEBD (Int. J. Entrep. Bus. Dev.) **5**(1), 92–104 (2022)
6. Irabor, I.E., Okolie, U.C.: A review of employees' job satisfaction and its affect on their retention. Ann. Spiru Haret Univ. Econ. Ser. **19**(2), 93–114 (2019)
7. Guzeller, C.O., Celiker, N.: Examining the relationship between organizational commitment and turnover intention via a meta-analysis. Int. J. Cult. Tourism Hosp. Res. **14**(1), 102–120 (2020)
8. Herath, S., Chong, S.: Key components and critical success factors for project management success: a literature review. Oper. Supply Chain Manag. Int. J. **14**(4), 431–443 (2021)
9. Covin, J.G., et al.: Individual and team entrepreneurial orientation: scale development and configurations for success. J. Bus. Res. **112**, 1–12 (2020)
10. Wu, G., et al.: Role stress, job burnout, and job performance in construction project managers: the moderating role of career calling. Int. J. Environ. Res. Public Health **16**(13), 2394 (2019)
11. Shaukat, M.B., et al.: Revisiting the relationship between sustainable project management and project success: the moderating role of stakeholder engagement and team building. Sustain. Dev. **30**(1), 58–75 (2022)
12. Imam, H., Zaheer, M.K.: Shared leadership and project success: the roles of knowledge sharing, cohesion and trust in the team. Int. J. Project Manage. **39**(5), 463–473 (2021)
13. Saleem, M.A., et al.: Enhancing performance and commitment through leadership and empowerment: an emerging economy perspective. Int. J. Bank Mark. **37**(1), 303–322 (2019)
14. Hadian Nasab, A., Afshari, L.: Authentic leadership and employee performance: mediating role of organizational commitment. Leadersh. Organ. Dev. J. **40**(5), 548–560 (2019)
15. Giudici, M., Filimonau, V.: Exploring the linkages between managerial leadership, communication and teamwork in successful event delivery. Tour. Manag. Perspect. **32**, 100558 (2019)
16. Chahar, V.: The impact of artificial intelligence on innovation. Best J. Innov. Sci. Res. Dev. **2**(7), 199–235 (2023)
17. Sarker, I.H.: Ai-based modeling: techniques, applications and research issues towards automation, intelligent and smart systems. SN Comput. Sci. **3**(2), 158 (2022)
18. Abioye, S.O., et al.: Artificial intelligence in the construction industry: a review of present status, opportunities and future challenges. J. Build. Eng. **44**, 103299 (2021)
19. Lee, I., Shin, Y.J.: Machine learning for enterprises: applications, algorithm selection, and challenges. Bus. Horiz. **63**(2), 157–170 (2020)
20. Chui, M., et al.: The state of AI in 2022—and a half decade in review (2022). https://www.mckinsey.com/capabilities/quantumblack/our-insights/the-state-of-ai-in-2022-and-a-half-decade-in-review
21. Wachnik, B.: Analysis of the use of artificial intelligence in the management of Industry 4.0 projects. The perspective of Polish industry. Prod. Eng. Arch. **28**(1), 56–63 (2022)
22. Benkarim, A., Imbeau, D.: Organizational commitment and lean sustainability: literature review and directions for future research. Sustainability **13**(6), 3357 (2021)
23. Shoss, M.K., et al.: The joint importance of secure and satisfying work: Insights from three studies. J. Bus. Psychol. **35**, 297–316 (2020)
24. Rawashdeh, A.M., Tamimi, S.A.: The impact of employee perceptions of training on organizational commitment and turnover intention: an empirical study of nurses in Jordanian hospitals. Eur. J. Train. Dev. **44**(2/3), 191–207 (2020)

25. Hameed, K., et al.: Exploring the employee's commitment through interpretative phenomeno-logical analysis (IPA) approach: evidences from private sector organizations of Pakistan. iRASD J. Manag. **3**(2), 156–170 (2021)
26. McCormick, L., Donohue, R.: Antecedents of affective and normative commitment of organisational volunteers. Int. J. Hum. Resour. Manag. **30**(18), 2581–2604 (2019)
27. Koch, J., et al.: The affective, behavioural and cognitive outcomes of agile project management: a preliminary meta-analysis. J. Occup. Organ. Psychol. (2023)
28. Young, D.K., et al.: Examining the influence of occupational characteristics, gender and work-life balance on IT professionals' occupational satisfaction and occupational commitment. Inf. Technol. People **36**(3), 1270–1297 (2023)
29. Oyewobi, L.O., et al.: Influence of organizational commitment on work–life balance and organizational performance of female construction professionals. Eng. Constr. Archit. Manag. **26**(10), 2243–2263 (2019)
30. Hosley, W.N.: The application of artificial intelligence software to project management. Proj. Manag. J. **18**, 73–75 (1987)
31. Wauters, M., Vanhoucke, M.: Support vector machine regression for project control forecasting. Autom. Constr. **47**, 92–106 (2014)
32. Mahdi, M.N., et al.: Software project management using machine learning technique—a review. Appl. Sci. **11**(11), 5183 (2021)
33. Asadi, A., et al.: A machine learning approach for predicting delays in construction logistics. Int. J. Adv. Logist. **4**(2), 115–130 (2015)
34. Shoar, S., et al.: Machine learning-aided engineering services' cost overruns prediction in high-rise residential building projects: application of random forest regression. J. Build. Eng. **50**, 104102 (2022)
35. Kor, M., et al.: An investigation for integration of deep learning and digital twins towards construction 4.0. Smart Sustain. Built Environ. **12**(3), 461–487 (2023)
36. Pan, Y., Zhang, L.: A BIM-data mining integrated digital twin framework for advanced project management. Autom. Constr. **124**, 103564 (2021)
37. Anysz, H., et al.: Quantitative risk assessment in construction disputes based on machine learning tools. Symmetry **13**(5), 744 (2021)
38. Liu, H., Yu, L.: Toward integrating feature selection algorithms for classification and clustering. IEEE Trans. Knowl. Data Eng. **17**(4), 491–502 (2005)

Low-Cost System for Monitoring Water Quality Parameters in Lentic and Lotic Ecosystems

Soila Benguela, Filipe Caetano[✉], and Clara Silveira

Polytechnic Institute of Guarda, Guarda, Portugal
{caetano,mclara}@ipg.pt

Abstract. Water consumption is vital for human survival and plays a crucial role in the healthy functioning of the human body. However, the use of drinking water for human consumption has raised growing concerns due to water scarcity and droughts. Therefore, it is necessary to implement projects aimed at the effective use of water, including waste reduction and the reuse of alternative sources. This paper aims to describe the development of a low-cost water quality parameter monitoring system named APPT (Drinking Water for All), with a focus on the chlorine dosing system. Monitoring will be conducted in two reservoirs, one for storing rainwater and river water and the other for treating the stored water. The system provides automatic control of residual chlorine dosage and visualization of pH, water level, and temperature parameters through a website, using technologies such as Apache, Lolin D1 Mini Pro, MySQL, PHP, JavaScript, HTML, CSS, Node.js, and Twilio. Additionally, APPT includes a real-time notification system via SMS and the website to alert users about water shortages in the water treatment reservoir and when pH and temperature values are outside defined limits. The manuscript also presents preliminary results of the system: in the conducted experiments, correct monitoring of pH, chlorine, water level, and temperature parameters was observed. Notifications via SMS and the web were also successfully tested.

Keywords: Water parameter monitoring · Arduino · Chlorine dosing · Web Application

1 Introduction

Water is a vital resource and plays a fundamental role in many human activities. However, the availability of freshwater is becoming increasingly scarce in many regions of the world due to factors such as population growth, urbanization, and climate change [1].

Considering this, the implementation of sustainable practices for the proper use and control of water quality is essential. In this context, this project aims to develop a water quality parameter monitoring system, named Drinking Water for All - APPT, in households that use rainwater and river water as an alternative to conventional water supply in an economical manner.

This approach is particularly relevant in regions near Lentic ecosystems (e.g., lakes and ponds) and Lotic ecosystems (e.g., rivers and streams). One of the major challenges

Á. Rocha et al. (Eds.): WorldCIST 2024, LNNS 989, pp. 76–85, 2024.
https://doi.org/10.1007/978-3-031-60227-6_7

in harnessing water from environments such as rivers and rainwater is ensuring that it is free from contamination and safe for human consumption. In this regard, the control of residual chlorine plays a crucial role in the water treatment process. Residual chlorine is an important indicator of the effectiveness of water treatment and the maintenance of water quality during storage and transportation [2].

Based on this need, an automatic chlorine dosing system has been developed for the APPT system. The system will provide real-time data on the concentrations of this parameter in the water. This paper is divided into six chapters: after the "Introduction" comes the "State of the Art" section that presents related works with a summary table; in Sect. 3 "Requirements Analysis" the use case diagram and SWOT analysis are presented; the "Implementation" section includes a description of the architecture and technologies; in Sect. 5 "Simulation and tests", the main tests carried out are described. Finally, the conclusion and future work are presented.

2 State of the Art

Currently, there are some water quality parameters monitoring systems. Some of these systems will be described below, as they share similarities with the system to be developed. These articles address the use of technologies such as Arduino and the Message Queuing Telemetry Transport (MQTT) protocol in monitoring water quality parameters. They provide practical and efficient solutions for real-time and remote monitoring, as well as data transmission for analysis and decision-making.

According to the work in [3], water quality measurement for monitoring is conducted in various areas using sensors such as temperature, turbidity, and pH sensors. Arduino (nano) is employed to collect these parameters and transmit them to the Raspberry Pi3 through the Wi-Fi module. Utilizing the MQTT protocol, information is transferred between the Raspberry Pi. Sensors can be installed in commercial areas, wastewater sewage, and urban pipelines with human intervention. The collected data is analysed using the Naive Bayes algorithm to obtain information about water quality. Another study [4], presents a wireless sensor network system based on LoRa and MQTT for monitoring water quality in maritime environments. The system was designed by combining two forms of networks, with the data transfer from sensor nodes to the gateway via LoRa and then to the server's visualization platform using MQTT. The backend management server continuously updated the monitoring website.

Other studies utilize artificial neural networks to explore and predict residual chlorine concentrations in drinking water distribution systems. The model developed in [5] showed promising results in accurately estimating residual chlorine concentrations. The case study was conducted in the water distribution system of Hope Valley, located in Adelaide, Australia. The collected data were obtained from specific sampling points, and various network parameters were considered, including water temperature, chlorine concentrations, dissolved organic carbon quantity, and absorption of rays. Data division was carried out based on the method proposed by Bowden [6] using the Self-Organizing Map (SOM) technique by Kohonen.

In [7] the authors address the use of wireless sensors and machine learning technique to measure the quality of drinking water in building plumbing systems, emphasizing the importance of real-time monitoring of residual chlorine levels. Finally, there is a study focused on the development and application of an IoT system for continuous water monitoring in a lentic ecosystem in southern Brazil, contributing to the understanding of environmental parameters in this specific context [8]. These studies emphasize the importance of real-time monitoring of water quality parameters, including residual chlorine, in drinking water distribution networks [5, 9]. The use of wireless sensor networks allows continuous data acquisition and real-time monitoring of residual chlorine levels in water.

Table 1 provides a concise comparison of the systems mentioned with the system to be developed.

Table 1. Comparison Table of All Mentioned Systems

Key Contributions	System						
	APPT	[3]	[4]	[5]	[6]	[7]	[8]
Development of a water quality monitoring system based on IoT.	✓	✓	✓	✓	✓	✓	✓
Use of artificial neural networks to model chlorine residuals in water distribution systems.	✗	✗	✗	✓	✗	✗	✗
SMS notifications.	✓	✗	✗	✗	✗	✗	✗
Chlorine dosing (actuator to activate the dosing).	✓	✗	✗	✗	✗	✗	✗
Mechanism to open and close the reservoir lid.	✓	✗	✗	✗	✗	✗	✗

After analysing the functionalities of each of the systems, the APPT System aims to monitor water quality parameters in a cost-effective manner, considering the following objectives:

- The system will include two water reservoirs, one for water storage and the other for treatment with residual chlorine.
- Continuously collect data from sensors: temperature, pH, water level, and rain detection sensors.
- Send SMS notifications to inform the user about parameters, such as the water level.
- Create a mechanism to control a chlorine dosing system.
- Create a mechanism to open and close the reservoir responsible for storing rainwater and river water.

3   Requirements Analysis

The requirements' analysis is presented in the form of use cases. The use cases are aligned with the objectives of the actors involved in the APPT system. The Use Case Diagram (Fig. 1) follows, summarizing who interacts with the system and the provided functionalities.

In this way, the actors that compose the system are presented, clearly outlining their goals and distinct functions. The use cases in Fig. 1 are described below.

- **Generate a PDF with Sensor Information**: The user selects the option to generate the PDF on the APPT system interface. The system processes sensor data and generates a PDF file with relevant information.

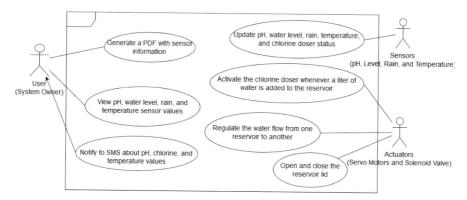

Fig. 1. Use Case Diagram of the APPT System

- **View pH, Reservoir Level, and Temperature Values**: The user accesses the APPT system interface and can view the displayed pH, reservoir level, and temperature values.
- **Notify to SMS about pH, chlorine, and temperature values:** The APPT system continuously monitors water levels and pH values. If the water level is below the defined limit or pH values are outside the established limits, the system sends alert notifications via SMS and displays information on the website.
- **Update pH, water level, rain, temperature, and chlorine doser status**: The sensors capture readings of water quality parameters.
- **Activate the chlorine doser whenever a liter of water is added to the reservoir**: The chlorine dosing system is activated whenever a liter of water is added to the treatment reservoir. The actuators move the chlorine dosing system as needed.
- **Regulate the water flow from one reservoir to another**: The solenoid valve actuator is activated whenever a liter of water decreases in the treatment reservoir, regulating the water flow to maintain the appropriate level.
- **Open and close the reservoir lid**: Based on reservoir conditions, the actuators open the lid when the reservoir is empty or raining and close the lid when the reservoir is full.

In this context, it is relevant to develop a SWOT analysis that assesses the internal and external aspects of the APPT system, providing an overview of its strengths, weaknesses, opportunities, and threats.

Strengths: The cost-effective approach of APPT makes it accessible, promoting its implementation in deprived communities and contributing to sustainable water resource management; Automation in residual chlorine dosing highlights its effectiveness in maintaining water quality standards and ensuring consumer safety; The use of technologies like Arduino, and protocols like Twilio offers a comprehensive and integrated solution for monitoring and notification	**Weaknesses:** Dependency on multiple technologies may create a learning curve for less familiar users, increasing the need for training; Regular maintenance and calibration of sensors can pose challenges, requiring technical understanding and dedicated time
Opportunities: The financial accessibility of APPT creates opportunities for expansion in communities facing challenges in accessing clean water; Alignment with governmental initiatives can open doors to partnerships and financial support, strengthening system implementation	**Threats:** Environments subjected to extreme weather conditions may impact the system's operation, requiring additional protective measures

4 Implementation

This chapter describes the technologies used, the architecture, and the development of the APPT system.

The architecture of the APPT system can be visualized in the diagram in Fig. 2. The hardware components of the system are described below:

- **Arduino UNO:** Microcontroller board that plays the central role in the system, coordinating and controlling the operations of the components [10]. It uses various ports, including analog ports (ADC) for sensors such as humidity (e.g., A0) and pH (e.g., A1). Digital ports (GPIO), like ports 2, 3, 4, are used to connect devices like relays, servo motors, and digital sensors. Serial communication (TX) is used to interact with other devices and transmit data.
- **Lolin D1 Mini Pro:** Microcontroller ESP8266, capable of Wi-Fi communication. Digital ports on Lolin D1 Mini Pro (D0 to D8) operate in a voltage range of 0–3.3V.
- **HC-SR04** [11]**:** Ultrasonic sensor that measures distances using sound waves, capable of measuring distances from 2 cm to 4 m with excellent accuracy. This module emits an ultrasonic pulse and measures the time it takes for the pulse to return after reflecting off an object. Uses two digital ports, and is powered by 5V.
- **Rain Sensor:** Determines the presence of water on the reservoir surface. Connected to an analog port. Can be powered by 3.3 V or 5 V, depending on the model.
- **Temperature Sensor (DS18B20):** An analog temperature sensor providing accurate readings. Connected to an analog port. Powered by 3.3 V to 5 V.
- **pH Sensor (E-201C-Blue)** [12]**:** Measures hydrogen ion activity in water-based solutions, commonly used to gauge liquid pH. It plays a pivotal role in various fields,

including the chemical, pharmaceutical, dye, and scientific research industries, requiring acidity and alkalinity analyses. The integrated unit board in this kit supports both 3.3 V and 5 V systems. Connected to the Arduino's analogy port. Generally powered by 5 V.

- **Servo Motor:** Actuators are used to open and close the water reservoir lid and to control the chlorine dosing. This is made up of three components: 1) Actuator System. This is a electric motor, usually DC, which contains a set of gears to amplify torque, forming a reduction box. 2) Sensor. Typically, a potentiometer is associated with the servo's shaft. The potentiometer's electrical resistance indicates the shaft's angular position. 3) Control Circuit. Consisting of electronic components like an oscillator and a PID controller. Receives signals from the sensor (shaft position) and control signal, adjusting the motor to position the shaft as desired.
- **Relay** [13]**:** Allows controlling devices with varying power using logic signals of 0–5 v (5 mA). Has switching capacity up to 250VAC@10A / 30VDC@10A, and can control AC and DC devices like solenoids, motors, lights, fans, etc.
- **Solenoid Valve:** Used to control water replenishment in the purifier reservoir. Typically, they are designed to operate at different voltage ranges, such as 12 V, 24 V, or 110–120 V, depending on the application and design. In this case, it will be 12 V.

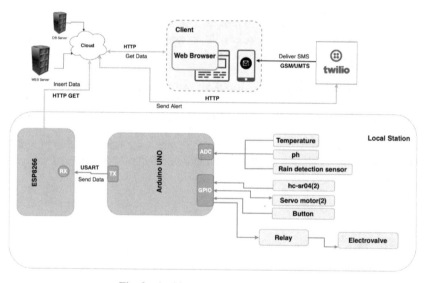

Fig. 2. Architecture of the APPT System

The system is based on an Apache web server with PHP and a MySql server. To send alerts, the Twilio platform is used. This cloud platform for communications provides developers with the ability to seamlessly integrate voice, messaging, and video services into their applications. Used to send SMS messages containing sensor values, it enables effective and real-time communication about the monitored parameter.

The Arduino UNO plays a vital role in collecting data from sensors and controlling actuators. Additionally, it is responsible for transmitting information to the Lolin D1

Mini Pro through serial communication. The Lolin D1 Mini Pro receives data from the Arduino UNO and facilitates communication with the server, as will be explained in the ESP32 to MySQL using HTTP (Indirect Method).

Regarding sensors, a temperature sensor, a pH sensor, and two HC-SR04 sensors were used, along with a rain detection sensor, as seen in Fig. 2. In the context of the two servo motors, one is responsible for the operation of opening and closing the lid on the first reservoir, while the other controls the chlorine dosing system, in addition to the temperature, pH, and water level sensors responsible for water purification. The sensors are distributed between both reservoirs (Fig. 3), where the first includes one of the HC-SR04 sensors and a button for manual activation of the lid servo motor, while the second encompasses the remaining sensors and actuators. The system incorporates a relay and a solenoid valve, enabling the replenishment of water in the purifier reservoir (second reservoir), as clearly indicated. All these actuators operate based on information collected by the sensors, ensuring precise and efficient control of the system.

To calculate the amount of chlorine needed, it is necessary to compare the volume of water entering the second reservoir with the water quality standards established in Portugal, which is 0.5 mg of chlorine per liter. To meet this standard, a chlorine dosing system (Fig. 4), was developed, capable of delivering exactly 0.5 mg of chlorine per liter of water. According to the guidelines established by the World Health Organization (WHO), a concentration of 0.5 mg per liter (mg/l) of free residual chlorine in water, maintained for a contact period of 30 min, is considered sufficient for effective water treatment. This means that when these conditions are met, the chlorine in the water can eliminate microorganisms such as bacteria and viruses, making the water safe for human consumption [14].

The calculation of the amount of chlorine in the water is performed using the formula:

$$\text{Chlorine Quantity} = \text{Volume of Water Entering the reservoir} \times \text{Chlorine Dosage}$$

Based on this calculation, the chlorine dosing system was configured to provide the appropriate amount of chlorine whenever an additional liter of water entered the system.

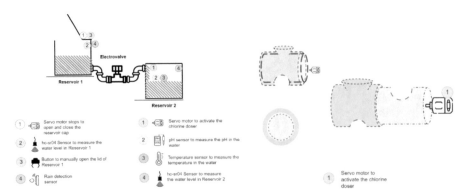

Fig. 3. The Two Reservoirs

Fig. 4. Simplified Scheme of the Chlorine Doser

ESP32 to MySQL using HTTP (Indirect Method): In the context of communication between the ESP8266 and a MySQL server, the indirect approach proves to be a secure and efficient strategy. In this method, the ESP8266 initially establishes a connection with a web server using the HTTP protocol, and subsequently connects to the MySQL server.

The functioning of this approach is described as follows:

HTTP Request Sending: The ESP8266 initiates by sending an HTTP request to the web server, incorporating sensor data into the request. Once received, the request is processed by a PHP script running on the web server. The PHP script extracts the information from the HTTP request, performs the necessary processing, and establishes communication with the MySQL database.

Communication with MySQL: The PHP script analyzes the received data and, after processing, communicates with the MySQL database. This indirect approach offers several advantages compared to the direct method.

Enhanced Security: Data security is strengthened as the MySQL user account credentials (username and password) are stored on the server, providing an additional layer of protection. Access to the MySQL account is restricted to localhost, further bolstering the system's security.

Simplicity and Efficiency: To mitigate complexity on both the ESP8266 and the MySQL server, data processing is centralized in the PHP script. This significantly simplifies both the ESP8266 program and MySQL management.

Efficient Memory Management: To prevent memory issues on the ESP8266, the PHP script sends only relevant data after processing, optimizing resource usage.

5   Simulation and Testing

During development, tests were conducted simultaneously with code implementation, as sensor value calibration played a crucial role in achieving optimized accuracy. This approach not only ensured the consistent functionality of the system but also allowed precise sensor calibration, ensuring the accuracy of the collected data.

Tests were conducted to read and visualize pH, water level, and temperature values. Additionally, testing was performed for SMS and website notifications regarding water level, pH, and chlorine values. Testing was conducted on the SMS notification functionality in the event of the water level falling below the established limit, indicating the need to replenish the reservoir: "Critical water level. Reservoir replenishment is required". The notification was verified if the pH and temperature values exceeded the defined limits, with the system sending the message: "pH, chlorine, or temperature values are outside the limits. Take corrective measures". In both tests executed, the expected results were obtained. Figure 5 displays the received notifications.

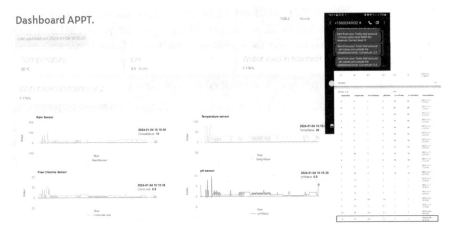

Fig. 5. System Dashboard and Notifications Received via SMS

Regarding the Opening and Closing of the Reservoir, the ability of the servo motor to control the opening and closing of the reservoir lid was tested. The system demonstrated successful results concerning rain presence as well as the manual button functionality to open and close the reservoir lid.

The chlorine doser was tested and it was observed that it is activated whenever the water level in the tank increases by approximately 1 L. These tests were conducted to ensure the proper functioning of each component of the system, ensuring effectiveness and reliability in various situations. The positive results thus far validate the system's efficiency, providing a solid foundation for the project's advancement. This ongoing process of testing and refinements represents a continual commitment to excellence and delivering a quality, low-cost solution to end users.

6   Conclusions and Future Work

The APPT system was designed to monitor water quality in households using rainwater or river water as a source of potable water supply. Distinguishing itself from conventional systems that merely present water quality values, the APPT system transcends this functionality by taking actions in response to critical water quality situations based on information provided by the sensors. Additionally, it stands out for its ability to disinfect water in an economically efficient manner, utilizing an automatic chlorine dosing system. Looking ahead, the perspective is to enhance the accuracy of the APPT system to optimize its performance in real-world environments. More specifically with regard to the physical installation of the entire system in a home and the carrying out of acceptance tests. This includes the implementation of more efficient methods for the exhaustive analysis of all parameters to optimize the monitoring process.

References

1. Araújo, A.: Contribuição para o estudo da viabilidade/sustentabilidade da dessalinização enquanto técnica de tratamento de água. Faculdade de Ciências e Tecnologia, Lisboa (2013)
2. Saraiva, S.: Avaliação da eficiência da desinfeção de efluentes urbanos com recurso a cloro. Universidade Nova de Lisboa, Lisboa (2015)
3. Vartak, S., Nimbalgundi, N., Poojary, K., Rodrigues, B.: The monitoring of water quality in an IOT environment using MQTT protocol. IoT-based Water Quality Monitoring System for Soft-Shell Crab Farming, p. 4 (2018)
4. Huang, A., Huang, M., Shao, Z., Zhang, X., Wu, D., Cao, C.: A practical marine wireless sensor network monitoring system based on LoRa and MQTT. Wireless Sensor Networks for Water Quality Monitoring using LoRa and MQTT, p. 7, 10–13 May 2019
5. Gibbsa, M.S., Morgana, N., Maier, H.R., Dandy, G.C., Holmes, M., Nixon, J.: Use of artificial neural networks for modelling chlorine residuals in water distribution systems. Monitoring and Modeling Chlorine Residual Concentration in Water Distribution Systems Using Artificial Neural Networks, p. 06 (2003)
6. Bowden, G.J., Maier, H.R., Dandy, G.C.: Optimal division of data for neural network models in water resources applications. Water Resour. Res. 2–38 (2002)
7. Paz, M., et al.: Wireless sensors for measuring drinking water quality in building plumbing. ACS ES&T Engineering, pp. 423–433, 18 December 2021
8. Zukeram, E.S.J., Provensi, L.L., Oliveira, M.V.D., Ruiz, L.B., Lima, O.C.D.M., Andrade, C.M.G.: In situ IoT development and application for continuous water monitoring in a lentic ecosystem in South Brazil. Water 20 (2023)
9. Paz, E.F.M., et al.: Wireless sensors for measuring drinking water quality in building plumbing: deployments and insights from continuous and intermittent water supply systems. Real-Time Monitoring of Chlorine Residuals in a Drinking Water Distribution System Using Wireless Sensor Networks, pp. 423–433, 24 October 2021
10. arduino. https://store.arduino.cc/products/arduino-uno-rev3. Accessed 21 Oct 2023
11. Robot Electronics: SRF04 - Ultra-Sonic Ranger Technical Specification (2003). https://www.robot-electronics.co.uk/htm/srf04tech.htm. Accessed 08 Mar 2023
12. Mauser: Grove - Sensor de medição PH (E-201C-Blue) – Seeed. https://mauser.pt/catalog/product_info.php?products_id=096-8594. Accessed 08 Nov 2023
13. Rajguruelectronics: Relay Module 2 Channel with Optocoupler 12V. https://rajguruelectronics.com/ProductView?tokDatRef=MTMyODE=&tokenId=NjE=&product=ADIY%202%20CHANNEL%20RELAY%20BOARD%2012v. Accessed 08 Nov 2023
14. Instituto Regulador de Águas e Resíduos, Recomendação IRAR n.º 05/2007 DESINFECÇÃO DA ÁGUA DESTINADA AO CONSUMO HUMANO (2007)

A Unified Approach to Real-Time Public Transport Data Processing

Juraj Lazúr[(✉)] [iD], Jiří Hynek[iD], and Tomáš Hruška[iD]

Faculty of Information Technology, Brno University of Technology,
Božetěchova 1/2, 612 66 Brno, Czech Republic
{ilazur,hynek,hruska}@fit.vut.cz

Abstract. The use of real operations data is essential for the planning
and management of modern public transport systems. With the expansion of universal formats for describing the structure of public transport systems, such as GTFS or Transmodel, the use of these data has
expanded far beyond the public transport domain. On the other hand,
the effort to use these data encounters the problem of its processing,
storage and integration with the structure of the transport system due
to the volume and speed of data generation from real operations. These
problems are even more evident in the case of further use of these data
as inputs for machine learning, or data mining, where integration of data
from different systems into a single model is necessary. The purpose of
this paper was to design a method by the which big data from real operations could be integrated with the changing structure of the transport
system so that this data could be stored long term without loss of granularity, or entropy value. As a result, we proposed a data model with big
data transformation algorithm, whose functionality has been verified in
testing over the public transport system of the second largest city in the
Czech Republic.

Keywords: Public Transport · Big Data Processing · Big Data
Visualisation · GTFS

1 Introduction

Public transport has been undergoing a gradual digitisation process for more
than 50 years [1]. On the side of transport companies and authorities, there are
two interlinked areas. The first involves the planning of operations, while the
second provides tools for real-time traffic management [2]. It is the data from
real operations that serves as the necessary basis for development planning and
problem solving in the system [3]. As a result, there is a closed loop of continuous
improvement of the system based on its behaviour in the real environment.

The importance of processing and analysing data from the real operation has
increased significantly in the last decades [4,5]. This trend is partly due to the
growth of urban agglomerations and the associated increase in the number of

© The Author(s), under exclusive license to Springer Nature Switzerland AG 2024
Á. Rocha et al. (Eds.): WorldCIST 2024, LNNS 989, pp. 86–95, 2024.
https://doi.org/10.1007/978-3-031-60227-6_8

passengers and the size of the area to be operated. This is also linked to the expansion of transport infrastructure, such as dedicated lanes, the funding of which is strictly based on data analysis. On the other hand, the de-carbonisation of public transport and the use of alternative powered vehicles is putting pressure on the efficiency of planning the use of individual resources [6,7].

While in the past, the analysis of data from public transport systems was based on targeted data collection [8], current tools or algorithms use mainly standardized formats (e.g. GTFS) for describing transport systems as their input [9]. The use of these formats is partly based on the widespread use of these formats and the associated availability of data in a unified format [11], but also on the potential for replication of useful analyses over systems described by the same format [12]. In addition, the availability of these formats is used in areas other than the optimization of public transport systems [11].

On the other hand, the internal data models of the published tools, or algorithms, are different from each other in contrast to the unified inputs. This inconsistency then causes duplication of tools with the same functionalities. But more importantly, it makes difficult to compare different transport systems with each other [9]. The method of obtaining the necessary amount of raw real operations data also varies. In many cases, the studies themselves must be preceded by the data collection and storage [11,12]. Some simplification is to retrieve this data via various APIs, but even in this case further annotation of the data is necessary [9]. In addition, when collecting real operations data, it is crucial to address how to process, compress and store this big data, due to its quantity and speed of generation.

One of the systems over which it has been necessary to address these issues is the Brno public transport system. As a part of the expansion of open datasets, the Data Department of the Municipality of Brno has obtained access to real operations data of this transport system. In order to extract new knowledge from these data, an analytical tool has been designed. Due to the amount of data that had to be stored, the use of an appropriate compression method seemed necessary. Although efficient compression methods exist not only in the field of Big Data [10], the requirement to integrate the real operation data with the transport system structure necessitated the design of a transformation algorithm, that would also be able to reduce the real operation data. The proposed solution, combines the transport system structure with real operations data. While the proposed data model allows to store the changing structure of the public transport system, the real operations data integration and compression requirements are solved by the proposed transformation algorithm. Thus the proposed solution would be able to store both the expected and at the same time the actual state of the transport system over a longer time horizon.

To verify the proposed solution, this concept was implemented and tested over the Brno public transport system. As a result, the proposed data model based on GTFS format with transformation algorithm was able to reduce the raw data volume by 75% on average while maintaining the same granularity and provides a data model for analytical tools. The generality of the resulting data model should also make easier to re-use this concept for another transport systems. Also, the proposed solution could further simplify the subsequent use of data from real operations in analyses, machine learning, or data mining. These benefits should bring efficiency gains, cost reductions and, most importantly, increased passenger satisfaction.

2   Transport Systems Structure Modeling

The complexity of public transport systems is reflected in the internal data models of the tools that operate them. The basic structure of transport systems consists of physical infrastructure, such as stops, and a timetable that determines from where, when and to where each connection will be made. Thus these data models have to deal both with inputs coming from different sources [13] and with continuous changes in transport systems. Modularity thus becomes an important feature of these models, resulting from the fact that some elements, such as stop position, change less frequently than, for example, line schedules [13]. Linking these data models, which implement these requirements differently, not only in the context of public transport system integration thus becomes a complex problem. Furthermore, these systems work with a very similar set of entities, but the ambiguous ontology causes some entities to represent different things in different systems [14]. These problems have been largely eliminated with the introduction of standardised open formats for describing the structure of transport systems.

These formats, such as GTFS[1], shown in the Fig. 1, or Transmodel[2], were originally developed both to share information between transport systems [18] and to simplify the development of various third-party applications [15], such as trip planning application. The use of these formats as inputs for various analyses comes later, while the use of these formats is related both to the widespread availability of data in these formats by transport authorities [15] and to the unified ontology, as the different parts of the formats represent the same elements in different systems. Together with real operation data, they have become the most common basis for advanced transport systems analysis.

The majority of studies is generally based on two data sources. The first one is the structure of the transport system and the second one is the data from real operation [17]. While the structure is mostly available from standardized formats, the extraction of real operation data is often based on collecting data during the progress of the study from various APIs [16]. A certain simplification that more recent studies have been working with is standardisation of formats for sharing data from real operation. Formats such as GTFS Realtime or SIRI, contains real-time information on the current state of the system in standardized form. However, the integration as well as the real-time processing of these two data sources still needs to be addressed separately in almost every analysis. An exception is the Transmodel family sub-format, OpRa[3]. This standard should

---

[1] https://gtfs.org/
[2] https://www.transmodel-cen.eu/
[3] https://www.opra-cen.eu/

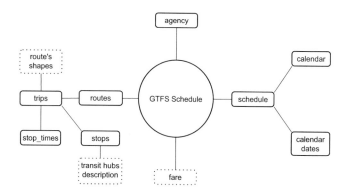

Fig. 1. A simplified GTFS structure diagram for describing the structure of a particular public transport system at a particular moment in time. Each element represents a single file. Elements with a dotted border are optional in the standard. By combining these data, it is thus possible to get an overview of when, when and which trip is served in the system.

primarily be used for sharing and enable storing raw information about the behaviour of the transport system, in order to study and optimise it. However, this format is still under development.

Expansion of open formats for the description of public transport systems has provided a new and highly efficient entry point for a wide variety of studies [16] whose purpose has long been beyond public transport [11]. On the other hand, there still remains the problem of collecting real operations data over a longer time interval. Despite efforts at unification of analyses input in the form of GTFS Realtime or OpRa standards, transformation, processing and storing the real operations data still represents a repeated part of the studies. In addition, the characteristics of the real operations big data create the need to address the volume as well as the velocity of the creation of this data.

3   Proposed Solution

Goal of our solution is to improve the availability of data describing the real behaviour of transport systems. It connects the public transport systems structure and real operations data. Designed in this way, the solution enables to store the real behaviour over a longer time horizon. The proposed solution can be divided into two main parts. The first part is the proposed data model, based on the GTFS format, which enables to store transport system's expected structure. The second part is an transformation algorithm, which enables compression and integration of real operations data into the proposed data model. By integrating real operations data, the actual behaviour of the transport system is stored in the proposed data model. The structure of the solution itself is then shown in the Fig. 2.

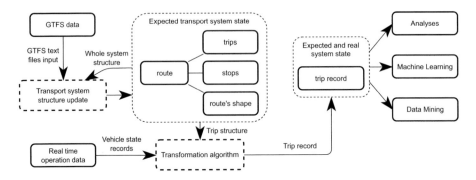

Fig. 2. The model of usage. The expected state of the system is periodically updated based on the current GTFS data. Afterwards, the global expected system state is stored for each day. The real operation data is integrated and stored intro the data model using a proposed transformation algorithm. Trip records, which include the expected as well as the actual state of the system, serve as the basis for subsequent use of the data.

The proposed data model uses GTFS format data as its input. This is due to the fact that the GTFS format provides an unambiguous ontology. Also, it is provided by a high number of transport authorities and is fully adapted for automated processing. The basic element of the data model is a route. Each route consists of a list of stops, the definition of the route's shape over a road or rail network and trips. A trip is the realization of a specific route on a specific time. By connecting these basic elements, the proposed model produces the expected system structure for the current day. The proposed data model is then daily updated with the current GTFS data. As a result, a list of trip records that should have been made is stored for each day. This not only stores the expected structure of the system, but it is possible to track individual changes in the structure by comparing the stored states.

The proposed transformation algorithm connects the data from the real operation with the trip records. While the structure of the transport system is described by available standardised formats, the real operation data have generally different record structures, and come from various sources. In our design, we divided the real operation data into two groups. The first group includes the data pertaining to an entire trip record or group of trip records. Such data are, for example, the records of cancellation of a trip, replacement of a trip, or maintenance of an entire line. This data can easily be assigned as an additional attribute to a trip record. In contrast, the second group consists of data that changes during the course of the trip, such as vehicle delays or occupancy. By simply aggregation the second data group into a single value, there would be a significant loss of granularity. However, storing all the raw data would be unsustainable in the long term, both in terms of speed of creation and quantity. Our solution to these conflicting requirements is the proposed transformation and compression algorithm shown in Fig. 3.

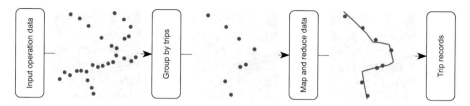

Fig. 3. Three main steps of the transformation algorithm for integrating and compression of real operation big data into the proposed data model. Organizing and storing raw real operations data based on the trips to which that data belongs allows compression, but also preserves all of its external relationships.

The essence of this proposed algorithm is the use of the geographical component of the real operations data and the assumption that the data values changes only at certain route points. The first step of the algorithm is to group the input records from the real operation according to the individual trips. The second step is to assign each record, based on its geographical part, to the corresponding logical segment of the trip. The third step is the compression, which is graphically illustrated in Fig. 4.

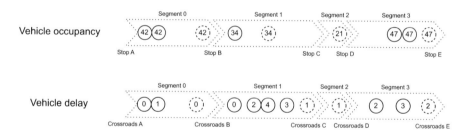

Fig. 4. Only one record is stored for each segment, indicated by a dotted line. The records are sorted from oldest to newest. For clearly given logical sections, such as vehicle occupancy, the assignment of the most recent record is unambiguous. However, for example vehicle delays, the reduction is dependent on the chosen length of the logical segments. Thus, in some cases, information may be lost. An example is segment 1, where the sharp increase and decrease in delay is lost. However, an overall delay increase of delay value 1 over segment 0 is recorded.

Record reduction is based on the assumption that data values from real operations change only at certain route points. Therefore, the route is divided into logical segments, with only one value assigned to each segment and stored. While in some cases, such as vehicle occupancy, the determination of these points is unambiguous, for example in the case of actual vehicle delay, the location of the points depends on the desired level of granularity. This is due to the fact that the delay can vary even at a single route point in the case of traffic jam. Our proposed algorithm determines the logical length of the segments based on the route's shape. By merging the segments, it is then possible to obtain a higher compression rate at the cost of losing granularity, or vice versa.

The structure of a transportation system without real operation data cannot track the dynamics of the system behavior, while without knowledge of the structure, modeling the relationships between real operation data is difficult. This relationship becomes even deeper when we need to store real time data over a longer time horizon. Our proposed data model attempts to reflect these requirements in an attempt to achieve the best possible compression to granularity ratio.

4   Results and Evaluation

To validate our proposed concept, we implemented and tested the whole model on the Brno public transport system in cooperation with the Data Department of the Municipality of Brno. The chosen transport system represents the second largest city in the Czech Republic. It provides sufficient complexity, as well as the availability of suitable data sources. Testing the implemented model focused on two areas and included an average of 50 lines and 6500 trips per day during the 30 days period. The first area was the ability of the proposed solution to store the structure of the transport system and its changes. The second area of testing and evaluation was calculating the transformation algorithm compression rate.

The structure of the monitored Brno public transport system consists of 11 tram lines, 13 trolleybus lines and 37 bus lines. For a greater diversification of the results, 28 railway lines of the integrated system of the South Moravian Region were also included in the testing. The source of input data was a regularly updated GTFS dataset and a database of real operations data, which contained raw records for 1 previous day each time. A significant problem was caused by the absence of route's shapes in the input GTFS data, which we solved by interpolating these data using our own routing algorithm.

The implementation itself consisted of a full-stack application for processing and basic aggregation and visualization of the stored records. The NoSQL database MongoDB was used for data storage, while the server part, which implements the proposed data model and transformation algorithm, uses NodeJS technology and the Express framework. The user interface was implemented using the React library. The entire implementation that was used in the testing is then available in the public repository[4]. The live version of the system in the test run is then available on the Brno city website[5].

The first tested area was to verify the ability of the proposed solution to store the structure of the transport system. The stored data was compared with the reference data of the transport operator. The comparison then showed that the proposed solution is able to store the structure with an accuracy of 93%, while being able to deal with a wide set of anomalies. On the other hand, the bottlenecks were some inaccuracies in the input GTFS data, as well as the inability of

---

[4] https://github.com/Jorgen98/BPTSAT-Public
[5] https://kod.brno.cz/bptsat/

the system to deal with specific changes in stop identifiers and the duplication of some trips in the case of replacement services.

The second tested area was testing the compression rate of the data, which was calculated by simply comparing the raw and processed data sizes. The results are shown in Table 1. The size of the processed data also includes the proportional part of the whole transport system structure.

Table 1. Results of measuring the average size of input and processed records belonging to a single trip. The resulting compression rate depends on the length of the route and the number of input records, whereby the number of records depends on the quality of the signal between the vehicle and the dispatcher centre within the RIS system.

	Trams	Trolleybuses	Buses	Trains
Avarage raw data record size [kB]	40.35	20.04	71.47	98.71
Avarage stored trip record size [kB]	12.68	10.12	11.63	16.41
Compression ratio	3.18	1.98	6.15	6.02

5   Discussion

The result of our work is the solution that can connect and store the expected and the actual state of transport systems over time. We were able to validate this concept during a test implementation over the Brno public transport system. This resulted in a functional data model, transformation algorithm that can reduce the input data by an average of 75% while keeping the same information and in the analytical application shown in Fig. 5. The solution designed in this way can simplify the use of data from real operations, which should increase the efficiency of transport systems and the satisfaction of passengers. On the other hand, there still remains a small set of anomalies, e.g. duplication of trips during service replace, that the proposed model is not able to deal with yet.

Further development of the proposed solution could focus on its extension with other optional attributes of the GTFS format, such as fare. Also, the robustness of the proposed data model should be enhanced, e.g., by checking the input based on the interpolation of already stored records. In the case of the transformation algorithm, improving the determination of logical segments, e.g. by determining them in real time or based on a statistic computed from raw data, seems to be a very suitable area for further development.

Fig. 5. The implemented user application for the analysis of trip delays over time. The average delay for the selected trip and time period are displayed. Delay categories, separated by colour, can be selected by the user. More detailed data can be obtained by clicking on the relevant part of the trip's shape.

6   Conclusion

The purpose of this paper was to find a way in which the structure and behaviour of public transport systems could be connected and stored. For this purpose, we proposed the data model for storing the structure of transport systems based on the GTFS format and a transformation algorithm that aims to integrate data from real operations into this model. We have validated our concept by implementing and testing the proposed solution on the Brno public transport system. As a result, the proposed data model and transformation algorithm should further simplify the access and use of the behaviour data of public transport systems for advanced analyses, the use of which goes beyond the field of public transport.

Acknowledgments. This work was supported by project Smart information technology for a resilient society, FIT-S-23-8209, funded by Brno University of Technology.

References

1. Lampkin, B., Wren, A.: Computers in Transport Planning and Operation. Operational Research Quarterly (1970-1977). JSTOR **23**(3), 404 (1972). https://doi.org/10.2307/3007903
2. Zito, P., Amato, G., Amoroso, S., Berrittella, M.: The effect of advanced traveller information systems on public transport demand and its uncertainty. Transportmetrica **7**(1), 31–43 (2011). https://doi.org/10.1080/18128600903244727
3. Symes, D.J.: Automatic vehicle monitoring: a tool for vehicle fleet operations. IEEE Trans. Veh. Technol. **29**(2), 235–237 (1980). https://doi.org/10.1109/t-vt.1980.23846
4. Pelletier, M.-P., Trépanier, M., Morency, C.: Smart card data use in public transit: a literature review. Transp. Res. Part C Emerg. Technol. **19**(4), 557–568 (2011). https://doi.org/10.1016/j.trc.2010.12.003
5. Using Archived AVL-APC Data to Improve Transit Performance and Management (2006). https://doi.org/10.17226/13907

6. Gallet, M., Massier, T., Hamacher, T.: Estimation of the energy demand of electric buses based on real-world data for large-scale public transport networks. Appl. Energy **230**, 344–356 (2018). https://doi.org/10.1016/j.apenergy.2018.08.086
7. Li, J.-Q.: Battery-electric transit bus developments and operations: a review. Int. J. Sustain. Transp. **10**(3), 157–169 (2014). https://doi.org/10.1080/15568318.2013. 872737
8. van Egmond, P., Nijkamp, P., Vindigni, G.: A comparative analysis of the performance of urban public transport systems in Europe. Int. Soc. Sci. J. **55**(2), 174 (2003). https://doi.org/10.1111/1468-2451.55020144
9. Aemmer, Z., Ranjbari, A., MacKenzie, D.: Measurement and classification of transit delays using GTFS-RT data. Public Transport **14**(2), 263–285 (2022). https:// doi.org/10.1007/s12469-022-00291-7
10. Wiseman, Y., Schwan, K., Widener, P.: Efficient end to end data exchange using configurable compression. ACM SIGOPS Oper. Syst. Rev. **39**(3), 4–23 (2005). https://doi.org/10.1145/1075395.1075396
11. Nishino, A., Kodaka, A., Nakajima, M., Kohtake, N.: A model for calculating the spatial coverage of audible disaster warnings using GTFS realtime data. Sustainability **13**(23), 13471 (2021). https://doi.org/10.3390/su132313471
12. Chondrodima, E., Georgiou, H., Pelekis, N., Theodoridis, Y.: Particle swarm optimization and RBF neural networks for public transport arrival time prediction using GTFS data. Int. J. Inf. Manag. Data Insights **2**(2), 100086 (2022). https:// doi.org/10.1016/j.jjimei.2022.100086
13. Kizoom, N., Miller, P.: A Transmodel based XML schema for the Google Transit Feed Specification with a GTFS/Transmodel comparison. Kizoom Ltd., London (2008)
14. Ruckhaus, E., Anton-Bravo, A., Scrocca, M., Corcho, O.: Applying the LOT methodology to a public bus transport ontology aligned with transmodel: challenges and results. Semantic Web **14**(4), 639–657 (2023). https://doi.org/10.3233/ sw-210451
15. Antrim, A., et al.: The many uses of GTFS data-opening the door to transit and multimodal applications. Location-Aware Information Systems Laboratory at the University of South Florida, vol. 4 (2013)
16. Wessel, N., Allen, J., Farber, S.: Constructing a routable retrospective transit timetable from a real-time vehicle location feed and GTFS. J. Transp. Geogr. **62**, 92–97 (2017). https://doi.org/10.1016/j.jtrangeo.2017.04.012
17. Wessel, N., Widener, M.J.: Discovering the space-time dimensions of schedule padding and delay from GTFS and real-time transit data. J. Geogr. Syst. **19**(1), 93–107 (2016). https://doi.org/10.1007/s10109-016-0244-8
18. Knowles, N., Miller, P., Drummond, P.: Transmodel and GTFS-Comparison and Convergence. Briefing Paper for the Public Transport Coordination Group (PTIC), Version, vol. 4 (2009)

Proportional Integral and Derivative Auto Tuning of Industrial Controllers Using the Relay Feedback Method

Salazar-Jácome Elizabeth[1]([✉]) [iD], Sánchez-Ocaña Wilson[2] [iD], Tulcán-Pastas Ana[3] [iD], and Tustón-Castillo José[2] [iD]

[1] Departamento de Ciencias de la Ingeniería, Universidad Tecnológica Israel, Quito, Ecuador
msalazar@uisrael.edu.ec
[2] Departamento de Eléctrica y Electrónica, Universidad de las Fuerzas Armadas ESPE, Latacunga, Ecuador
wesanchez@espe.edu.ec
[3] Departamento de Ciencias Administrativas, Universidad Tecnológica Israel, Quito, Ecuador
atulcan@uisrael.edu.ec

Abstract. This research intends to validate the self-tuning test developed by Aström and Hägglund in 1984, known as the relay feedback method, with the self-tuning used by the industrial level Programmable Logic Controllers. The methodology used was experimental, analysis and diagnostic; a series of measurements of the process (monovariable level) was obtained in open loop, by means of a Datalog, in csv format, for the treatment and obtaining of the model of space - state and thus to obtain the function of transference. The results of the calculation of the values of the constants Proportional gain, Integration constant and Derivation constant, were modeled in the application of the relay method (Aström and Hägglund), to then load these parameters in the Simulink Proportional, Integral y Derivative_Controller and thus obtain as response the tuning curve of the process in simulated form; finally, these same constant values are entered manually as well as automatically in the Proportional, Integral y Derivative__Compact Version 2 of Tia Portal V16. The automatic tuning curve, as a response of the Proportional Integral and Derivate Optimizer, causes a much smoother behavior of the industrial process against set point changes as well as disturbances, with maximum values of over pulse, stabilization time and error within normalized ranges, because by Programmable logic controller design Proportional, Integral y Derivative_Compact Version 2 use weighting coefficients for both Proportional and Derivative Action, making the tuning resulting from the process much finer, generating the values of constants and weighting coefficients automatically.

Keywords: Aström and Hägglund · self-tuning · relay feedback method · PLC Optimizer

© The Author(s), under exclusive license to Springer Nature Switzerland AG 2024
Á. Rocha et al. (Eds.): WorldCIST 2024, LNNS 989, pp. 96–107, 2024.
https://doi.org/10.1007/978-3-031-60227-6_9

Nomenclature

PID Proportional, Integral and Derivative
Kp   Proportional gain
Ki   Integration constant
Ti   Integral time
Kd   Derivation constant
Td   Derivation time
PLC Programmable logic controller
P&ID Piping and instrumentation diagram
CPU Central Processing Unit
ku   Critical gain
tu   Critical time period
I/O Input/Output

1 Introduction

PID controllers are widely used in industry; their operation is based on the measurement of the error produced by the process variable and the set point [1]. This type of controller improves the quality of the output in the transient state and permanent state, it is formed by i) the derivative element, it acts on the transient state of the controlled signal since it prevents rapid changes in the process variable, ii) the integral action, which notably decreases the error in stable state and iii) the proportional action, corrects the error in stable state of the output variable [2] (Fig. 1).

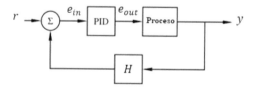

Fig. 1. Block diagram of a closed-loop control system with PID controller.

The PID controller is define by the following expression [3]:

$$e_{out}(t) = K\left(e_{in}(t) + \frac{1}{T_i}\int_o^t e(\theta)d\theta + T_d \frac{de_{in}(t)}{dt}\right) \tag{1}$$

Being the controller parameters: K proportional gain, T_i integration time, T_d derivation time. The transfer function of the controller is:

$$C(s) = K\left(1 + \frac{1}{sT_i} + sT_d\right) \tag{2}$$

The regulation of these parameters makes it possible to achieve stability of the variables in the system, quick responses to changes in the reference value and minimum error in the stable state, for which tuning methods are applied [4]. Among the tuning methods are: a) Ziegler and Nichols, uses the error to reach the adjustment of the controller b) Square Minims, is a method of estimation, applies Gaussian probability techniques or Bayesian estimation. c) Levenberg-Marquardt, applies a method of nonlinear approximation d) Cohen - Coon, is a method that analyzes the behavior of the response curve of the plant to an input unit step [5]. However, in industrial processes it is difficult to apply these methods, due to the difficulty in obtaining an analytical description of the processes, so self-tuning techniques of controllers have been developed to improve their performance in industry [6].

One of the auto tuning techniques is the one developed by Aström and Hägglund in 1984, where a self-tuning test of the parameters is performed to identify the critical point by means of a feedback relay; this technique is known as the relay feedback method [7]. This method is characterized by the speed in the tuning of the controller, can also be used in closed loop, which facilitates its application in industry as it does not significantly interfere in the daily development of processes, Fig. 2 shows the block diagram of relay method [8].

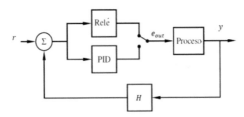

Fig. 2. Classic diagram of the relay method.

The relay method calculates the PID parameters of the controller, using the open circuit Nyquist curve [9], where the point of intersection of this curve with the real negative axis must be established, increasing the gain of a proportional controller so that it is possible to obtain a sustained oscillation of constant amplitude [10]. Aström and Hägglund determined that by replacing the proportional controller with a relay it is possible to establish the point of intersection, this is because the relay induces a limit cycle when the loop is closed and thus establishes the amplitude of the oscillation with the amplitude of the characteristic of the relay [11]. For the adjustment of the parameters the estimation of the critical gain ku is considered and the time period for the oscillation is called the critical time period tu [12].

Researchers from the University of Timisoara in Romania propose the use of fuzzy integral proportional controllers, dedicated to the control of servo systems, using symmetric optimal methods of feedback tuning [13], similar to the study proposed in this research, with the difference that the control used for servo positioning is based on a discretized control, and in this proposal the technique will be used for continuous processes.

Studies by researchers at the University of Craiova and Salerno, about Nonlinear Optimal Control of Oxygen and Carbon Dioxide Levels in Blood, use a type of control based on the linearization of the dynamic model of the oxygenator through the expansion of Taylor series and calculation of Jacobian matrices, since its principle is based on a mathematical modeling [14] similar to the basis of the optimal control of the auto synthesis using industrial programmable logic controllers based on PID compound controls.

2 Methodology

2.1 Data Collection

The present research was carried out using a work station of the Laboratory of Hydronic and Neutronics of the University of the Armed Forces ESPE Extension Latacunga, [15] the same one that contains a didactic module for the control of variables of control: flow, pressure, level and temperature in closed loop, counts on a PLC Siemens S7-1200 CPU 1214C ac/dc/rly, and allows the control of the frequency of the motor of a centrifugal pump; as it is shown in the Fig. 3. The diagram P&ID and each one of the components are represented in the Fig. 4.

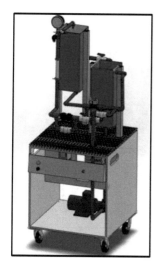

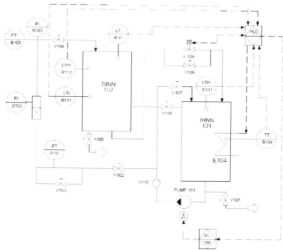

Fig. 3. Training module, multivariable process

Fig. 4. P&ID, Multivariable process

For the development of the tuning, first a series of process measurements is made, from which the input (ultrasonic sensor) and output (speed variator) values will be obtained in open loop [16]. A control algorithm is designed for the PLC that commands the process, the measurement values are stored in a Datalog, for which a period of operation of approximately 60 min is considered, where every 10 s there was a random variation of setpoint delivered to the variator between values of 0% to 100% to observe the variation in the level. These values are exported to a file (CSV), from which the data will be manipulated for later analysis. See Fig. 5.

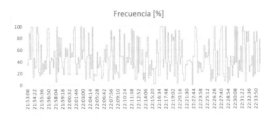

Fig. 5. Output data (frequency converter), single-variable level process

2.2 Generation of the Mathematical Model of the Level Process

The behavioral values of the process are exported to MatLab and through the Control System Toolbox, the identification of the plant model will be carried out [17]. With this tool and the measured input-output values of the system, a mathematical model can be created to emulate a physical system. Figure 6 shows the PID Tuner complement where the mathematical model is generated, for which three steps are considered: the import of the input-output test data, the identification of the plant model and the use of the model generated within the environment [18]. In the MatLab environment, the data generated by the Datalog is imported into the workspace as shown in Fig. 6.

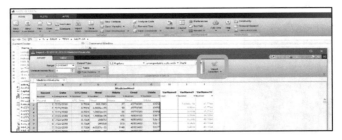

Fig. 6. Importing data into the MatLab model

With the data stored in separate variables, one for input values and one for output values, the PID Tuner plug-in is chosen, located in the APPS tab, for importing the data into the application, there are different forms of response, among which Arbitrary I/O Data is selected [19]. There is detailed the name of the variables that contain the values of the input and output data, as well as the start time of the measurement and the sampling time of the measurement as detailed in Fig. 7.

Once the import is done, two curves are generated, one that represents the entered data, and another that is the result of the identification process of the plant, which can be

Fig. 7. Configuring the import parameters

manipulated to fit a structure defined as: one pole, two poles or a model of state space. In this case a state space is selected, where an automatic estimation is made, in Fig. 11 the result is visualized [20]. In order for the generated curve to be more in line with the measured parameters, a modification can be made in the order of the plant and the estimation can be made again, until the result is acceptable, as shown in Fig. 12; after this, the final model is accepted. Finally, the generated plant model is exported to the workspace for its future applications.

3   Analysis and Results

As a result of the export, a variable of type 'idss' is obtained that represents a model of state spaces, to find the transfer function of the process, the MatLab instruction 'ss2tf' is used, which converts the representation of the state space into a transfer function; where b and a represent the poles and zeros of the transfer function [21].

From here we obtain the following transfer function that represents the physical plant:

$$G(s) = \frac{0,0003513s^8 + 0,003407s^7 + 0,01767s^6 + 0,1156s^5 + 0,2313s^4 + 0,5843s^3 + 0,7713s^2 + 0,7672s + 0,6144}{s^8 + 6,319s^7 + 44,99s^6 + 203,7s^5 + 475,2s^4 + 805,2s^3 + 1075s^2 + 605,3s + 4,744}$$

(3)

With the function already established (Ec. 3), the relay method is applied in the Simulink environment as shown in Fig. 8 and the results obtained are shown in Fig. 9.

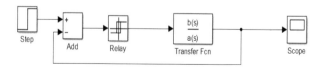

Fig. 8. Relay method application model

Through measurements, parameters a = 0.1121 and t_u = 19.69 are set. With these values the different parameters of the PID controller can be calculated according to the

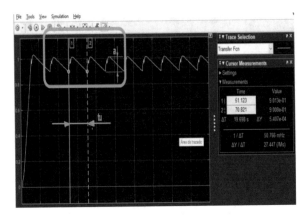

Fig. 9. Response to relay method

Table 1. PID Controller Parameters.

Controller	Kp	Ti	Td
P	0.5*Ku	-	-
PI	0.45*Ku	Tu/1.2	-
PID	0.6*Ku	Tu/2	Tu/8

following Table 1:

$$k_u = \frac{4d\,(relay\,amplitude)}{\pi\,a} = \frac{4 * 100}{\pi * 0.1121} = 1362 \tag{4}$$

$$K_p = 0.5 * ku = 0.5 * 1362 = 681 \tag{5}$$

$$T_i = \frac{tu}{2} = \frac{19.69}{2} = 9.8s \tag{6}$$

$$T_d = \frac{tu}{8} = \frac{19.69}{8} 2.45s \tag{7}$$

These values are loaded into the Simulink PID parameters (Fig. 10) to test their operation for step response.

To contrast with the real part, an s7-1200 PLC is configured to control the real process, so that it works with a PID block configured with the parameters obtained from the relay method. See Fig. 11.

The PID_Compact V2 block, the SIEMENS S7-1200 controllers are based on the following parameters (Table 2):

$$y = K_p\left[(b \cdot w - x) + \frac{1}{T_I \cdot s}(w - x) + \frac{T_D \cdot s}{a \cdot T_D \cdot s + 1}(c \cdot w - x)\right] \tag{8}$$

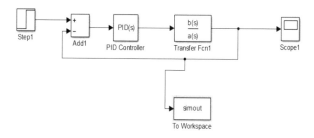

Fig. 10. Application model_PID Controller

Table 2. SIEMENS S7-1200 Controller Parameters.

Symbol	Description
y	Output value of the PID algorithm
Kp	Proportional gain
s	Laplacian operator
b	Weighting of the share P
w	Setpoint
x	Actual value
TI	Integration time
TD	Derivative time
a	Coefficient for the delay of the derived action (delay of the derived action T1 = a × TD)
c	Weighting of action D

Using the PLC Optimizer S/-1200 automatically, the constants shown in Fig. 12 and the respective response curve in Fig. 13 are generated, and if we compare them with the calculated and simulated values, it is established that they are not coincident, because in the calculation algorithm (Ec 8) of the constants used by the PLC PID_Compact V2 (for PLC's S7-1200 and S7-1500) they use weighting coefficients for both Proportional Action and Derivative Action, making the tuning resulting from the process much finer, generating the values of the constants and weighting coefficients automatically; allowing the process to stabilize either in a step change (Fig. 12) or in front of a disturbance (Fig. 13) in an adequate way.

Response curve of the automatic optimization of PID_Compact V2.

To calculate the values of the constants Kp, Ti and Td, the application of the relay method (Aström and Hägglund) is modeled, to then load these parameters into the Simulink PID_Controller and obtain as a response the tuning curve of the process in simulated form; Finally, these same constant values are entered manually in the PID_Compact V2 of Tia Portal V14, generating a similar tuning curve to the simulated one, with values of tss = 77 s and ess = 0 coincident, the simulated PO is 8% while in the PLC it is 0.2%.

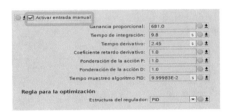

Fig. 11. Parameters manually assigned to PID_Compact V2

Fig. 12. Constants Gp, Ti and Td, in automatic Fine Optimizer

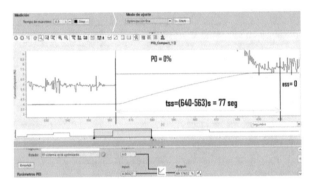

Fig. 13. Response curve of the PLC PID Fine_Optimization

When implementing the PID application models (Fig. 14) with the constants Kp, Ti and Td calculated as well as those obtained from the automatic optimization of the PLC, the behavior curves of the level process are obtained as a response (Fig. 15); from which it is established that the manual curve has a higher value has a PO = 8% and the automatic one a value of PO = 2%; the stabilization time has been improved in 11% to the algorithm used by the PLC, and with respect to the ess with the two methods a value of 0 is obtained.

For the development of the tuning, first a series of process measurements were made, both input and output in open loop, then they are stored in a Datalog, which is exported in a file (CSV) for treatment and obtaining the space state model, and finally find the transfer function of the process.

The type of control used for the auto tuning of the level variable of a process, is based on the use of an optimal control, in order to find the best way to control the process, based on the optimization of the objective function, as well as the speed of response of the variable, while the use of a non-optimal control would generate ambiguity and would depend on the experience and technical, practical and feasible skills of the optimizing technician.

Considering that the ultrasonic transmitter used for the analysis of the level variable has internal electronic compensation, it guarantees that the behavior of the same is linear

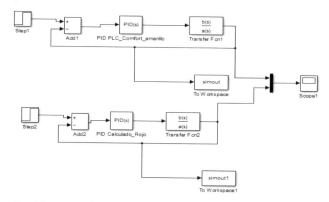

Fig. 14. Application models_PID Controller, PLC and calculated

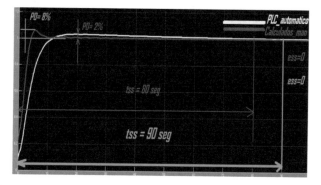

Fig. 15. Curves of the Automatic and Manual PID_application Models.

and its range of use is from 0 to 100%, it benefits that the generation of the constants product of the auto tuning guarantees the reliability in all its field of measurement.

It is important to note that the mathematical analysis of the PID control algorithm is in continuous time, but the principle used by the method in discussion Relay Feedback Method is performed in a discrete manner, due to the switching of its contacts, evidencing such behavior in the self-tuning graphs that facilitates the PID comfort, in the cyclic calculation of critical frequencies and determination of the constants Kp, Ti and Td.

4 Conclusions

The most widely used technique of self-tuning in Industrial Programmable Logic Controllers is the one developed by Aström and Hägglund in 1984, known as the relay feedback method, in which he tests a self-adjustment of the parameters to identify the critical point by means of a feedback relay.

Implemented the PID application models with the constants Kp, Ti and Td both calculated and those obtained from the automatic optimization of the PLC, the process behavior curves are obtained as a response, resulting in the ma-annual curve has a higher value with a PO = 8% and the automatic one with a value of PO = 2%; the stabilization time has been improved by 11% to the algorithm used by the PLC, and with respect to the ess with the two methods a value of 0 is achieved.

The automatic tuning curve, as a response of the Optimizer of the PLC S7–1200, provokes a behavior of the industrial process in front of changes of setpoint as well as perturbations, much softer with values of on impulse, times of stabilization and error in stable state within normalized ranges, because by design of PLC PID_Compact V2 (for PLC's S7-1200 and S7-1500) they use weighting coefficients for both Proportional Action and Derivative Action, making the tuning resulting from the process much finer, generating the values of constants and weighting coefficients automatically.

References

1. Berner, J.: Automatic tuning of PID controllers based on asymmetric relay feedback. Doctoral dissertation, Department of Automatic Control, Lund University (2018)
2. Serrano, F.E.: Comparative Analysis of PI/PID Tuning Methods in the Frequency Domain in Control Problems. IEEE Conescapan, Tegucigalpa – Honduras (2020)
3. Levy, S., Korotkin, S., Hadad, K., Ellenbogen, A., Arad, M., Kadmon, Y.: PID autotuning using relay feedback. In: IEEE 27th Convention of Electrical and Electronics Engineers in Israel, pp. 1–4 (2019)
4. Berner, J.: Automatic Controller Tuning using Relay-based Model Identification. Lund University (2017)
5. Hurtado, C.A.C., García, W.C.: Robust tuning of PID controllers using nonlinear estimation. Universidad Tecnológica de Pereira, Faculty of Electrical Engineering, Electronics, Physics, and Computer Science, Electronic Engineering (2016)
6. Júnior, A., Silva, M., Nascimento, A., Bispo, H., Silva, J.: Auto tuning using armax model. In: Blucher Chemical Engineering Proceedings, vol. 1, no. 2, pp. 12744–12751 (2015)
7. Díaz, J.A.: Analysis of the Astrom-Hagglud and Ziegler-Nichols methods for tuning PID controllers. ENGI Electron. J. School Eng. 1(1) (2018)
8. Björk, M., Levenhammar, R.: Relay Auto-tuners in Modelica. Lund University (2017)
9. De Frutos Bolzoni, L.: Development of advanced control algorithms for industrial programmable devices. University of Valladolid (2015)
10. Caulcrick, C., Huo, W., Hoult, V., Vaidyanathan, R.: Human joint torque modelling with MMG and EMG during lower limb Human-Exoskeleton interaction. IEEE Robot. Autom. Lett. 6(4), 7185–7192 (2021). https://doi.org/10.1109/LRA.2021.3097832
11. Guardiola, D.: MatLab modeling of transtibial prosthesis, Francisco José de Caldas District University, Revista Tekhnê, vol. 12, no. 2, pp. 13–22 (2015). ISSN 1692-8407
12. Kim, K.H., Bae, J.E., Chu, S.C., Sung, S.W.: Improved Continuous-Cycling Method for PID Autotuning (2021). https://doi.org/10.3390/pr9030509
13. Precup, E.-E., Preitl, S., Rudas, I.J., Tomescu, M.L., Tar, J.K.: Design and experiments for a class of fuzzy controlled servo systems. IEEE/ASME Trans. Mechatron. 13(1), 22–35 (2008). https://doi.org/10.1109/TMECH.2008.915816
14. Rigatos, G., et al.: Nonlinear optimal control of oxygen and carbon dioxide levels in blood. Intell. Ind. Syst. 3, 61–75 (2017)
15. Romero-Bustamante, J., Zurita-Herrera, B., Gutiérrez-Limón, M., Hernandez-Martinez, E.: Robust model-based control of a packed absorption column for the natural gas sweetening process. Int. J. Chem. Reactor Eng. 21(4), 461–471 (2023). https://doi.org/10.1515/ijcre-2022-0112

16. Kalpana, D., Chidambaram, M.: Auto-tuning of decentralized PI controllers for a non-square system. In: Komanapalli, V.L.N., Sivakumaran, N., Hampannavar, S. (eds.) Advances in Automation, Signal Processing, Instrumentation, and Control. LNEE, vol. 700, pp. 253–259. Springer, Singapore (2021). https://doi.org/10.1007/978-981-15-8221-9_23

17. Chidambaram, M., Saxena, N.: Relay Tuning of PID Controllers for Unstable MIMO Systems. Springer, New Delhi (2018). https://doi.org/10.1007/978-981-10-7727-2

18. Chandramohan, E., Seshagiri, A., Chidambaram, M.: Improved decentralized PID controller design for MIMO process. IFAC Papers On Line **53**(1), 153–158 (2020). https://doi.org/10.1016/j.ifacol.2020.06.026

19. Chandramohan, E., Seshagiri, A., Chidambaram, M.: Improved decentralized PID controller design for MIMO processes. IFAC-PapersOnLine **53**(1), 153–158 (2020). ISSN 2405-8963. https://doi.org/10.1016/j.ifacol.2020.06.026

20. Zeng, D., et al.: Research on improved auto-tuning of a PID controller based on phase angle margin. Energies **12**, 1704 (2019). https://doi.org/10.3390/en12091704

21. Salazar, E., Sánchez, W., Salazar, D., De la Torre, J., Aguas, L.: Biomechanical modeling of the knee for rehabilitation in patients of latacunga's patronato. Int. Rev. Autom. Control (IREACO) **15**(3) (2022)

Reinforcement Learning for Process Mining: Business Process Optimization

Ghada Soliman[1]($\boxtimes$) (iD), Kareem Mostafa[2] (iD), and Omar Younis[2] (iD)

[1] Lead Data Scientist, Orange Innovation Egypt, Giza, Egypt
ghada.soliman@orange.com
[2] Data Scientist, Orange Innovation Egypt, Giza, Egypt
{kareem.mostafa.ext,omar.younis.ext}@orange.com

Abstract. Process mining aims to extract knowledge from event data to understand, analyze, and improve Processes. Utilizing the benefits of Reinforcement Learning enables the automation of the business processes discovery, by systematically exploring the state space based on a certain environment interaction to achieve a certain goal, that reduces time and effort required for manual analysis and decision-making. This paper proposes an automated system capable of discovering certain organization's business processes and identifying the optimal sequence of transitions towards the ticket terminal state avoiding the bottleneck. To identify the structure of the business processes, the available source/target transitions are extracted, along with the computed transition probabilities between each possible transition. These probabilities are then incorporated into the reward design, which is needed to adjust the agent's behavior. By applying the Q-Learning algorithm, a Q-function learns the quality (q-value) of taking each possible action, given a certain state, encoded as a table that is updated iteratively during the training until reaching an optimal policy. Finally, after a certain number of episodes, the optimal sequence of transitions is identified. The Deep Q-Network (DQN) algorithm is applied to compare the results between the two approaches, and the Q-Learning agent tends to produce the optimal path that is aligned with the business processes.

Keywords: Process Mining · Business Process Optimization · Machine Learning · Reinforcement Learning · Q-Learning · Deep Q-Network

Abbreviations

DFG : Directed Flow Graph
MDP : Markov Decision Process
RL   : Reinforcement Learning
ANN : Artificial Neural Network
DQN : Deep Q-Network

Á. Rocha et al. (Eds.): WorldCIST 2024, LNNS 989, pp. 108–125, 2024.
https://doi.org/10.1007/978-3-031-60227-6_10

1 Introduction

Process mining empowers businesses to drive process excellence, enhance competitiveness, and achieve strategic objectives by providing businesses with a clear and objective understanding of how their processes are actually executed [1]. Process mining techniques utilize the generated event logs that keep track data about specific events or activities occurring, providing a trace of the activities performed within the system or process [2]. Analyzing business processes through process mining techniques is valuable to identify the optimal sequence of transitions based on real process data. Processes in real-life scenarios can be complex, involving numerous interactions, dependencies, and variations. Capturing and modeling these complexities accurately in process mining can be challenging, computationally intensive and time consuming [3]. It was intended to utilize the benefits of Reinforcement Learning to build a system capable of identifying the available business processes and their associated relationships and connections through an environment composed of the possible activities and a customized reward function to guide the agent through a specific behavior to reach an objective of interest. The proposed approach builds an intelligent agent, such that given a certain state, it can predict the next optimal state to apply the transition into, till reaching the target state with an objective to maximize the cumulative reward obtained for each taken action [4, 5]. According to this technique, the agent is able to identify the valid and realistic transitions between activities associated in the event log and it is possible to identify the optimal sequence of activity transitions, avoiding the bottleneck state towards the end of the activities trace [6]. The utilized event log is related to the ticketing system of a certain organization that keeps track the initiated tickets about the different issues that arise in the system along with the possible activities that take place for the ticket until the issue is resolved and the ticket is terminated. Figure 1 shows the diagram of the high-level components that constitute our system. For the following sections, we will go into the details of each of these components.

The paper is organized into the following: Sect. 1 introduces our paper, Sect. 2 is devoted to describing the event log used in the experiment associated with the preparatory steps needed for applying the experiment. In Sect. 3 the applied methodologies including the Q-Learning and Deep Q-Network algorithms are explained, Sect. 4 introduces the achieved results for the applied techniques. Finally, Sect. 5 concludes the paper.

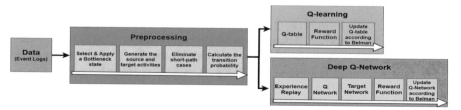

Fig. 1. The system main pipeline

2 Data Description

In this section, the event log structure is introduced along with the applied steps for preparation. These steps involve defining the available states and the possible actions that can be taken, to construct the needed environment for building an intelligent Reinforcement Learning agent. The agent interacts with the constructed environment and discovers the structure of the business processes. To effectively guide the agent towards the desired behavior, a reward/penalty function is designed considering realistic and valid transitions associated with their transition probabilities [7–9].

2.1 Event Log Structure

The proposed experiment primarily relies on an event log that tracks the activities occurring within the initiated tickets of a specific organization's ticketing system. The event log consists of (3414) tickets extracted from an internal ticketing system. The event log comprises the following fields:

- case ID: incident (ticket) identifier number
- activity: an event that occurs through the ticket life cycle
- timestamp: time of event occurrence

Figure 2 shows the plot of the Directed Flow Graph (DFG) for the events that take place during the life cycle of the initiated tickets, associated with the frequency for each possible transition. Table 1 introduces the description of the available events from the business perspective.

2.2 Select a Bottleneck State

The initial step involves identifying a certain state to serve as a bottleneck within the business process. The bottleneck state acts as an absorbing state that the agent should identify and avoid during the training. The criteria for selecting this state, includes:

- A state that once reached, it cannot be left according to the definition of the absorbing state [17]
- Having a high weight of outgoing edges indicating it has a high impact within the flow of the business processes

After conducting a visual analysis of the Directed Flow Graph depicted in Fig. 2, it was observed that the event *Pending Customer Action* is connected to (703) subsequent events distributed among 6 states with a frequency of transitions from that particular state towards the other states as shown in Fig. 3. After evaluating the position of the state within the normal flow of the business processes and considering the weights of transitions compared to other states, it was determined that this particular state meets the criteria for being a potential bottleneck state as explained further in Sect. 2.3. After a bottleneck state is chosen in the business process, the proposed system is designed to recognize it and predict the optimal sequence of transitions based on the business policy. The aim is to optimize the business operations by avoiding this specific state.

Definition of *Pending Customer Action* Event. From a business perspective, Pending Customer Action event indicates that a ticket or incident is awaiting attention due to the need for establishing a direct communication with the customer for resolving the ticket escalation by gathering essential information related to the issue however, that connection is not successful. There is a failure in the connection with Level 1 support team for establishing a connection with the customer causing a delay in the issue resolution. It is now clear that this activity is a bottleneck state indicating a delay in the continuation of the ticket resolution normal flow.

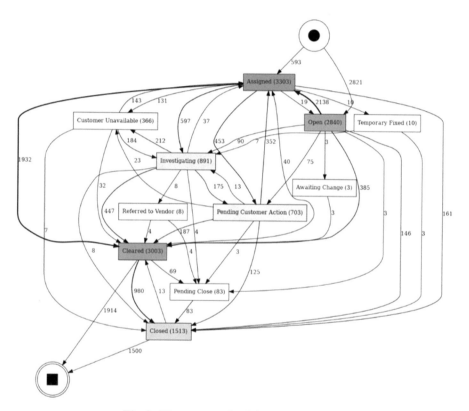

Fig. 2. The raw event log Directed Flow Graph

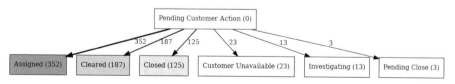

Fig. 3. Pending Customer Action outgoing edges DFG extraction

2.3   Apply the Absorption State Effect

In Markov Decision Process (MDP), an absorption state refers to a state from which it is impossible to transition to any other state. Once an agent reaches an absorption state, it remains in that state indefinitely [17]. Absorption states have a transition probability of (1) for staying in the same state and a probability of (0) for applying a transition to any other state. To simulate a realistic scenario where the proposed Reinforcement Learning agent encounters a specific node with significant penalties associated, a bottleneck state was designated as an absorption state within the event log. This absorption state serves as a node where the agent should not be selecting transitions. To achieve this effect, all subsequent events occurring after the bottleneck state are removed. In Fig. 4, the effect of the absorption state is illustrated by the red arrows direction, indicating that no events occur after that particular state.

2.4   Generate the Source and Target Activities from the Log Event

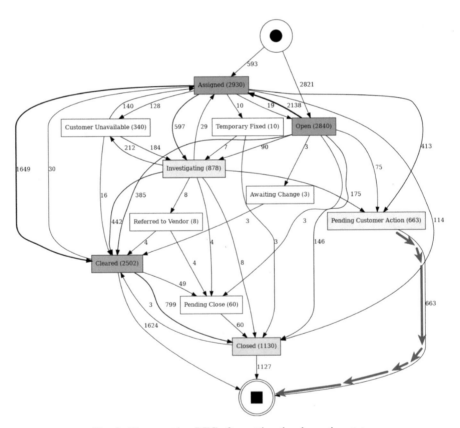

Fig. 4. The event log DFG after setting the absorption state

To calculate the transition probability for each unique source/target pair to be incorporated within the design of the reward function, it was necessary to prepare the event log in a source/target notation. This notation helps define the available states that the agent will interact with as a sequence of transitions from a source state to a target state. In this step, the business process structure is identified which will be used to define the environment states and actions with which the Reinforcement Learning agent will interact during training. Additionally, it was important to establish a common termination point (target) for each ticket, which the agent should aim to reach. As a result, a new state is defined with an 'end' notation to signify the termination of each case ID (ticket). Table 2 provides a summary of the aforementioned steps, taking into account the anonymization of the ticket (case) IDs for data confidentiality.

Table 1. The ticket events description from a business perspective

Event	Description
Assigned	Initial status for a manually opened ticket
Open	Initial status for an automatically opened ticket
Investigating	Analyzing ticket and adding internal comments
Cleared	When a task reaches its last activity in the life cycle and is successfully resolved
Closed	Last activity in a ticket life cycle
Awaiting Change	Waiting for a change request which is required to resolve the ticket
Pending Close	The business analysis stated that it is not used
Referred to Vendor	Add >>Vendor<< field appears
Temporary Fixed	Temporary solution implemented to address the issue
Customer Unavailable	Ticket is pending customer info if customer is responsible for car rier scope
Pending Customer Action	Ticket is pending customer info if customer is responsible for car rier scope

Table 2. Each source/target pair associated with the new terminal state (target) with anonymized case ID

case ID	source activity	source timestamp	target activity	target timestamp
beec43xa	Open	2022-10-06 03:11:03	Assigned	2022-10-06 03:20:24
beec43xa	Assigned	2022-10-06 03:20:24	Cleared	2022-10-06 04:06:19
beec43xa	Cleared	2022-10-06 04:06:19	Closed	2022-10-30 08:10:32
beec43xa	Closed	2022-10-30 08:10:32	**end**	2022-10-30 08:10:32
ceec44xb	Open	2023-01-23 02:16:08	Assigned	2023-01-23 02:19:03

Table 3. A sample of the eliminated tickets with short sequences

case ID	events
aa2947ax	(Open, Cleared)
aa2948ay	(Assigned, Closed)
aa2949az	(Open, Cleared, Closed)

2.5   Eliminate Short-Path Cases and Calculate the Transition Probability

Tickets with very short event sequences were excluded from the event log. This step was carried out to avoid misleading the agent with undesired sequences during the training process while maintaining realistic sequence of transitions aligned with the business logic. Table 3 demonstrates some samples of the eliminated event sequences. After eliminating the short event sequences, the transition probability for each unique source/target transition was computed after applying the above preparatory steps. For each source state, the count of the outgoing edges towards other states was identified to be divided by the total count of the outgoing edges associated with that particular source state. Table 4 presents the calculated transition probabilities after removing tickets with undesired short sequences. In the table, a bottleneck state transition is highlighted in red. Additionally, transitions towards the end state are highlighted in green representing the target of the agent.

Table 4. Transition probabilities for each source/target pair after the preparation steps

Source Activity	Target Activity	Transition Probability
Assigned	Investigating	0.362942
Assigned	Pending Customer Action	0.394460
Cleared	end	0.486804
Closed	end	1.000000
Investigating	Assigned	0.036683
Investigating	Cleared	0.338118
Open	Assigned	0.831021
Pending Customer Action	Pending Customer Action	1.000000

It can be observed that the bottleneck state can only be transitioned to itself as a self-loop with a transition probability of (1) to itself, according to the effect of the absorption state. As well there exist 2 states that can be transitioned to the target ('end') state which are: (Cleared, Closed).

3   Methodology

Reinforcement Learning is a branch of machine learning, alongside supervised and unsupervised learning. It introduces an intelligent agent capable of applying a sequence of actions based on a learned policy, which serves as a decision-making unit to achieve a specific goal. To develop an RL agent, it is essential to establish an environment where the agent can observe the consequences of its actions. By leveraging its learned policy, the agent aims to select actions that will maximize its total reward over time, based on the current state of the environment. According to the environment-agent interaction mechanism, the agent observes the current state of the environment, selects one of the available actions by referring to its decision making unit that can be a tabular method as the case in Q-Learning [10] or a parametric method as the case in the Deep Q-Network then as a consequence of its taken action, the agent receives a reward or penalty from the environment [9, 11, 12].

The reward or penalty system is predefined to align with the agent's objectives, dictating what behaviors should be encouraged to maximize gains and what should be avoided. This framework is designed to promote a strategy that prioritizes actions leading to the greatest increase in the agent's cumulative reward, directing it toward the desired outcome. After taking an action, the agent observes the subsequent state provided by the environment. Through training its decision-making unit, known as the policy, the agent learns to take appropriate actions that can help reach the expected objective. Figure 5 illustrates the reinforcement learning cycle and the agent's training process. In this iterative cycle, the agent refines its behavior to converge toward the optimal policy. The training episode terminates when the agent either achieves its objective or encounters a bottleneck state, at which point a new training episode begins.

In this section, we present the methodologies employed in the paper. These methodologies encompass Q-Learning and Deep Q-Network algorithms. Q-Learning is a conventional reinforcement learning algorithm that acquires the optimal policy through iterative updates to a Q-value table. On the other hand, Deep Q-Network is a more advanced approach that leverages artificial neural networks to approximate the Q-value function [11–13].

3.1 Q-Learning Algorithm

Q-learning algorithm is considered a model-free, value-based, and off-policy reinforcement learning algorithm [10].

- **Model-free:** the algorithm learns the optimal policy or value function directly from the experience of interacting with the environment, without explicitly modeling the transition probabilities or the reward function
- **Value-based:** the algorithm learns the optimal value function, such as the state-value function or the action-value function, and derives the optimal policy from it
- **Off-policy:** the policy used at training time (Greedy policy) is different from that at acting time (Epsilon greedy policy)

Q-Learning seeks to develop an agent that can achieve a specific goal by learning an action-value function, Q_π (s, a), which estimates the expected cumulative reward. This estimation is based on the agent starting from a given state, taking an action, and subsequently following the policy [9, 14]. Equation 1 presents the action-value function as the expected cumulative discounted reward value, which begins from state (s), proceeds with taking an action (a), and then follows the policy towards a goal (G)

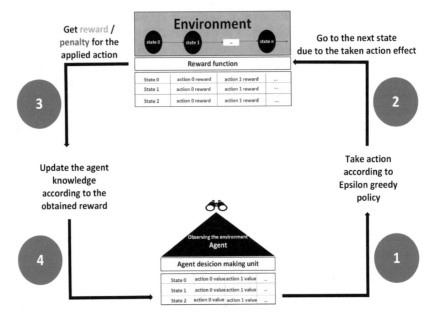

Fig. 5. Reinforcement Learning cycle paradigm

[12, 15]:

$$Q_\pi(s, a) = E_\pi[G_t|S_t = s, A_t = a] = E_\pi\left[\sum_{t=0} \gamma^t r_t|S_t = s, A_t = a\right] \quad (1)$$

where:

- **Policy** π: a greedy policy is followed to select the action that maximizes the agent's cumulative reward for reaching an optimal action-value function $Q(s, a)$
- **Discount Factor** γ: the discount factor that determines how far future rewards are taken into account in the return value

In Q-Learning, the agent learns an optimal action-value function, $Q(s, a)$, encoded as a Q-table, which determines the expected reward for taking a particular action in a given state. The Q-value function is updated iteratively, where the new Q-value is a combination of the old Q-value and the expected discounted reward that can be obtained from the next state according to the Bellman Equation taking into consideration the application of a greedy policy that aims to take actions that maximize the agent's cumulative reward [12, 15]. Equation 2 represents the formula used to update the Q-value during the training process of the agent. This update is based on the Bellman Equation, which is a simplified version of Eq. 1.

$$Q_{t+1}(s_t, a_t) \leftarrow Q_t(s_t, a_t) + \propto \left[r_t + \gamma max_{a_t} Q_t(s_{t+1}, a_{t+1}) - Q_t(s_t, a_t)\right] \quad (2)$$

The Q-table is continuously updated until an optimal table is achieved, enabling the identification of the best action for each state within the table.

The agent learns an optimal policy that maximizes its expected reward over time. The reward function assigns an immediate numerical value for each action the agent takes, and the learning rate determines the speed at which the agent updates its Q-value function. The state space includes all possible states in which the agent can exist, while the action space includes all possible actions the agent can take in each state. The Q-learning algorithm follows a straightforward principle: the agent selects an action in a given state, receives a reward, observes the new state, and updates its Q-value function accordingly utilizing Eq. 2.

Q-table Initialization. The Q-table for the proposed use case is composed of (12) states, according to the event log of the ticketing system for an organization. Each state represents a particular event that can take place within the ticket life cycle till its termination. The actions taken by the agent are defined as the transition from one state to another. The Q-table represents the available states in rows, the possible actions in columns, and the table cells refer to the learned Q-value for each possible state-action pair during the agent's training. At the beginning of the training, the Q-table is initially set with an arbitrary value, typically zero.

Reward Function Design. To effectively guide the agent towards the target of interest, it is crucial to formulate a specific reward function that maximizes desired outcomes and penalizes undesired ones [15]. The objective is to guide the agent to reach the terminal state *end*, while avoiding the bottleneck state through a valid and realistic transitions—a series of transitions that are aligned with the business logic. The reward function is designed to guide the agent to behave according to the following criteria as shown in Table 5:

Table 5. Q-Learning reward design

Invalid/Absorption transition	Valid transition	Target transition
-1	Transition Probability	10

Where:

- **Invalid/Absorption transition**: a transition between two states with no direct connection between them in the event log or a transition towards the bottleneck state, such transitions incur a penalty of (-1).
- **Valid transition**: a transition between two states with a direct connection between them, such transitions are rewarded with an immediate value as the transition probability representing the likelihood of transitioning between those states, which is the desired behavior
- **Target transition**: The transition towards the target (end) state, such transitions are rewarded a relatively significant reward to motivate the agent towards that objective

Training Process. Initially, the Q-table is populated with zeros. During training, which spans a specific number of episodes (e.g., 4000), a value (p) is randomly generated at each time step and compared to a threshold epsilon (ε), where $0 < \varepsilon < 1$. This comparison governs the action selection process, striking a balance between exploring potential actions at each time step and exploiting the knowledge acquired thus far according to the following criteria [9]:

$$\text{Taken action (a)} = \begin{cases} \text{Select a random action} & \text{if } p < \varepsilon \\ \text{Take the action with maximum Q - value} & \text{else } p \geq \varepsilon \end{cases}$$

The exploration factor ε initially has a value of (1.0), indicating a full exploration behavior. The exploration rate is gradually reduced over time using an exponential decay factor of (0.001). This gradual reduction allows the agent to rely more on its learned knowledge and follow a greedy policy to maximize its reward by exploiting its learned policy in the Q-table. When the agent encounters a certain state, it selects an action based on the exploration-exploitation criteria mentioned earlier, then it observes the resulting next state and the immediate reward. The q-value for the corresponding state-action pair is updated according to the mentioned Eq. 2. Throughout the training, outlined as a pseudocode in Algorithm 1, the agent learned to make sequential decisions with the goal of maximizing the reward over an episode, which is a sequence of states, actions, and rewards that concluded at a terminal state. After the completion of training, the updated Q matrix took the form of Table 6. It can be deduced that the Q-table shows the learned optimal policy that defines the best transition action to be taken according to the associated maximum Q-value per state. According to the resulted Q-table after the training, whenever the initial state is 'Open', the expected sequence of transitions is: **'Open'** −'Assigned' -'Cleared' -'end' based on the exploitation of the maximum computed Q-values given each state. It can be shown in Table 6 that both rows of the bottleneck state *Pending Customer Action* and the target state *end*, at the end of training, are still in their initial state with all zeros. The reason behind that is the applied policy during the training, in which the episode terminates whenever the agent transitions towards either the bottleneck or the target state that's why their Q-table values are not updated or changed.

```
Algorithm 1 Q-learning algorithm
 1: Input: env, episodes, max iter per episode, adaptive epsilon
 2: Initialize empty lists: rewards per episode, regrets per episode, epsilon per_episode
 3: for idx = 0 to episodes do
 4:     Initialize state by resetting the environment with initial state name" Assigned"
 5:     Initialize done as False
 6:     for step = 0 to max iter per episode do
 7:         action  ←choose action based on current state
 8:         next state, reward, done, info  ←execute action in environment
 9:         update agent state, action, reward, and next state
10:         state  ←next state
11:         if done is True then
12:             break the loop 13:
        end if
14:     end for
15:     Print 'Episode:', episode index + 1, 'Total reward:', total reward, 'Regret:', total regret
16:     Append total reward to rewards per episode list
17:     Append total regret to regrets per episode list
18:     Reset total reward and total regret attributes to 0
19:     if adaptive epsilon is True then
20:         Append current epsilon to epsilon per episode list
21:         Update epsilon using the exponential decay formula if it is greater than epsilon min
22:     end if
23: end for
    return Q, rewards per episode, regrets per episode, epsilon per episode
```

Table 6. Q-Table final result after training

	Assigned	Awaiting Change	Cleared	Closed	Customer Unavailable	Investigating	Open	Pending Close	Pending Customer Action	Referred to Vendor	Temporary Fixed	end
Assigned	0.510	0.394	2.527	1.358	0.437	1.007	0.423	0.165	-0.465	0.287	0.417	-0.442
Awaiting Change	0.064	-0.039	2.672	0.119	-0.110	-0.011	-0.064	-0.035	-0.145	-0.040	-0.089	-0.152
Cleared	0.904	0.273	1.307	1.886	-0.044	0.460	0.033	0.611	-0.397	0.347	-0.002	5.000
Closed	0.027	-0.041	0.390	0.190	-0.116	-0.035	-0.084	-0.054	-0.173	0.021	-0.058	4.956
Customer Unavailable	1.329	-0.042	0.276	0.145	-0.047	0.120	-0.074	-0.051	-0.138	-0.026	-0.121	-0.131
Investigating	0.334	-0.031	2.602	0.309	0.072	0.014	-0.074	-0.063	-0.198	0.126	-0.076	-0.179
Open	1.520	0.057	0.270	0.252	-0.096	0.084	-0.143	-0.053	-0.123	-0.017	-0.069	-0.100
Pending Close	0.071	-0.029	0.108	2.487	-0.069	-0.048	-0.072	-0.070	-0.166	-0.043	-0.094	-0.186
Pending Customer Action	0.000	0.000	0.000	0.000	0.000	0.000	0.000	0.000	0.000	0.000	0.000	0.000
Referred to Vendor	0.003	-0.040	2.530	0.107	-0.065	0.025	-0.051	0.144	-0.115	-0.065	-0.050	-0.123
Temporary Fixed	-0.051	-0.045	0.306	0.262	-0.090	1.536	-0.070	-0.034	-0.131	-0.047	-0.120	-0.138
end	0.000	0.000	0.000	0.000	0.000	0.000	0.000	0.000	0.000	0.000	0.000	0.000

3.2 Deep Q-Network Algorithm

To compare the results obtained from the Q-Learning algorithm, an experiment is implemented based on the Deep Q-Network algorithm. Instead of using a Q-function encoded as a table that is updated iteratively for each state-action pair directly until an optimal policy is reached, a parameterized Q-function in the form of an artificial neural network (ANN) is used to approximate the quality values (q-values) for each possible action at a given state (value-function estimation) [11–13].

Experience Replay. To enhance the learning process, the concept of experience replay is introduced [11, 12, 16]. At each time step, the agent's experience is represented as a tuple: $e_t = (s_t, a_t, r_{t+1}, s_{t+1}, , \text{done})$. The tuple consists of the current environment state s_t, the action taken at the given state a_t, the reward obtained r_{t+1} at time $t + 1$ as a result of the previous state-action pair (s_t, a_t), and the subsequent state of the environment s_{t+1} Additionally, a 'done' indicator is used to determine if an episode should be terminated in the subsequent step, which occurs if the agent reaches either the target or a bottleneck state. These experience tuples e_t are stored in a replay buffer of a predefined size, allowing the agent to reuse past experiences during the training and prevent the loss of valuable knowledge. At the start of training, the target network is initialized with the same parameters as the Q-network, and the replay buffer is initialized with a specific size. Using an epsilon factor probability, a uniformly distributed value between 0 and 1 is generated. If this random value < epsilon factor, a random action was chosen, enabling the agent to explore the action space of the environment (exploration). On the other hand, if the random value was greater than or equal to the epsilon factor, the action with the highest q-value was selected, allowing the agent to exploit the learned policy and take actions that maximize its reward (exploitation). Based on this action selection criteria, experience tuple samples were generated and stored in the replay buffer until the buffer reached its predefined threshold. Subsequently, a mini-batch of experience tuples is randomly selected from the buffer to train the Q-network and estimate the q-values.

Target Network Utilization. To train the neural network to approximate the q-values, a second network called the *target network* is defined to ensure a stable training, avoid oscillation and allow for computing the loss function between the predicted q-values for a certain state and the q-target values [10–12]. The loss function was optimized using the gradient descent algorithm to update the q-network parameters and improve q-value predictions. The target network is less frequently updated compared to the Q-network. The Q-network parameters are copied to the target network every certain number of steps to ensure a stable estimation of target values, leading ultimately to a stable and an optimal solution.

The following cases show the target network values [10, 11].

$$y^j = \begin{cases} r_j & \text{if episode terminates at step } j+1 \\ r_j + \gamma \ \max' Q(\phi, a'; \theta^-) & \text{otherwise} \end{cases}$$

- The q-target value is estimated as the immediate reward for taking an action at instant j, in case the episode terminates at the next step $(j + 1)$
- Otherwise, the q-target value is set as the maximum q-value for a certain action given the next state (φ_{j+1}) from the target network, discounted by the gamma factor (how much we value future rewards) plus the immediate reward for taking that action within the current state
- y^j: the target network Q-values estimation
- θ^-: the target network weights (parameters)
- γ: the discount factor that determines the importance of future rewards. It is a value between [0, 1], where 0 means only considering immediate rewards and 1 means considering all future rewards equally
- ϕ: the available state space
- j: the current step or time index in the episode

Function Design. To apply the experiment of the Deep Q-Network algorithm on the process mining event log, a specific environment was defined as a class called *Process-MiningEnvironment* that inherits from the Gym package. This class was used to define a space of the available states and actions, as well as the immediate rewards for taking those actions according to the business process structure identified in Sect. 2.4. The environment was designed in a way that helps the agent reach a specific target by maximizing the reward for certain actions given certain states. The criteria for designing the rewards can be referenced in Table 7.

Each case in Table 7 above, is defined as follows:

- **Invalid/Valid not of interest transitions**: an invalid transition occurs when the agent selects to apply a transition from one state to another that is not available or not applicable within the event log states, while a valid-and-not-of-interest transition

Table 7. DQN algorithm reward

Case	Immediate reward
Absorption state transition	-10
Invalid transitions	-10
Valid not of interest transitions	-10
Valid of interest transitions	0.1
Target transitions	100

occur when the agent selects to carry out an applicable transition towards any possible subsequent state other than the one with the highest transition probability

- **Valid state of interest**: a transition from a normal state to another state that is directly connected to it. Additionally, it includes transitions to the state with the highest transition probability among all other valid states. Such transitions are assigned a reward of (0.1)
- **Terminal state**: The transition from a normal state to the *end* state. This transition is assigned a relatively significant reward of (100), which motivates the agent to head towards the *end* state

The intuition for using a different reward function design for the Deep Q-Network compared to Q-Learning, is the unique manner by which each algorithm updates its policy. After conducting numerous experiments, we reached a conclusion: employing the same reward matrix as Q-learning leads to sequences that encompass invalid states that are not aligned with the business rules.

By utilizing the reward design shown in Table 7, the agent is able to traverse valid states until reaching a terminal state. Table 8 presents the artificial neural network architecture, consisting of three layers with dimensions (12, 32), (32, 32), and (32, 12) for the input, hidden, and output layers, respectively. These dimensions are determined by the presence of 12 states, including the bottleneck and terminal state, as well as 12 possible actions that correspond to transitioning towards those states.

Table 8. Artificial Neural Network (ANN) architecture

Total params: 1868
Trainable params: 1868 Non-Trainable params: 0
Input size (MB): 0.02 Forward/backward pass size (MB): 0.22 Estimated Total Size (MB): 0.25

Layer	Output Shape	Param
Linear-1	[-1,12,32,32]	416
Linear-2	[-1,12,32,32]	1056
Linear-3	[-1,12,32,12]	396

4   Results

The obtained rewards for the Q-Learning agent during the training can be shown in Fig. 7(a) which indicates that the agent is able to apply valid actions given each possible state to maximize its reward over time till reaching the target of interest during the training for about (2500) episodes, out of (4000) episodes until convergence.

In addition to the regret diagram that shows the regret by the agent for taking actions so the agent applies optimal actions to approach the target of interest and the loss decreases gradually over time as shown in Fig. 7(b). The reward and regret diagrams for the Deep Q-Network as well, can be shown in Figs. 7(c) and 7(d) respectively. According to the structure of the event log with respect to the available state transitions and their corresponding transition probabilities, in addition to the defined reward function, the agent was able to apply valid transitions aligned to the business. At the end of training, the final Q-table is optimized such that for each state, the optimal action can be identified with the maximum reward value as shown in Table 6. According to the Q-learning approach, the optimal sequence of transitions leading to the termination of the ticket (reaching the final state) and avoiding the bottleneck state, depicted in Fig. 6 heuristic miner petri net is: **'Open' –'Assigned' –'Cleared' –'end'.**

In comparison with the Q-Learning algorithm, the DQN experiment resulted an agent that was able to apply valid transitions adhering to the specified criteria within the proposed reward design that guided the agent to apply valid transitions such that for each possible state, the agent applies transitions towards the state with the maximum transition probability with an optimal sequence of transitions that is:

'Open' –'Assigned' –'Investigating' –'Cleared' –'end'. The Q-Learning agent result regarding the optimal path, was more business-aligned and appropriate for this business optimization objective.

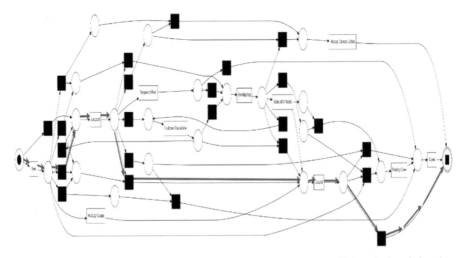

Fig. 6. The optimal path that is most aligned with the business rules utilizing the heuristic miner petri net

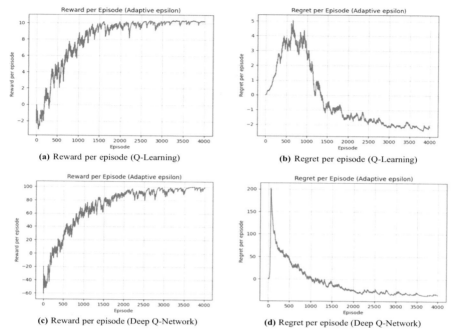

Fig. 7. A comparison between the Q-Learning and Deep Q-Network algorithms obtained (rewards – regrets) per episode (a) Reward per episode (Q-Learning), (b) Regret per episode (Q-Learning), (c) Reward per episode (Deep Q-Network) and (d) Regret per episode (Deep Q-Network)

5 Conclusion

In this study, we aimed to develop a proactive model capable of determining the optimal transitions among states in an organization's business processes. Our findings revealed that the optimal sequence of transitions towards the ticket termination state was **'Open' –'Assigned' –'Cleared' –'end'** that is most aligned with business optimization objective and rules. To achieve these results, we proposed an automated system that utilizes Reinforcement Learning with the Q-learning and Deep Q-Network approaches. Our intelligent agent demonstrated the ability to discover business processes by identifying the optimal sequence of transitions while considering transition probabilities for valid and business-wise transitions, thus avoiding bottlenecks. The Q-Learning agent, in particular, shows promise in identifying optimal transitions and avoiding bottlenecks, ultimately contributing to improved process performance and organizational outcomes. The construction of a customized environment that reflects the business process structure is essential for the success of our study. By investigating the dynamic interactions and relations between events, this approach enables our agent to exhibit the intended behavior, thus laying the foundation for groundbreaking advancements in the field. Our research contributes to the field of business processes optimization and discovery by showcasing the effectiveness of the Q-Learning and Deep Q-Network approaches in identifying the optimal sequence of transitions accurately, considering an alignment with the organization's business processes. Further exploration in this area could lead to advancements in automated systems that enhance business process efficiency and effectiveness. In conclusion, our research demonstrates the value of utilizing Reinforcement Learning techniques in creating intelligent agents that can navigate complex business processes. The Q-Learning agent shows promise in identifying optimal transitions and avoiding bottlenecks, ultimately contributing to improved process performance and organizational outcomes.

Fundings. This research received no external funding.

References

1. Van der Aalst, W., Weijters, T., Maruster, L.: Workflow mining: discovering process models from event logs. IEEE Trans. Knowl. Data Eng. **16**, 1128–1142 (2004)
2. Van Der Aalst, W.: Process mining: overview and opportunities. ACM Trans. Manag. Inf. Syst. **3**, 1–17 (2012)
3. Garcia, C.D.S.A., Junior, E.R.F., Dallagassa, M.R., Sato, D.M.V., Carvalho, D.R., Santos, E.A.P., Scalabrin, E.E.: Process mining techniques and applications—a systematic mapping study. Expert Syst. Appl. **133**, 260–295 (2019)
4. Chiorrini, A., Diamantini, C., Mircoli, A., Potena, D.: A preliminary study on the application of reinforcement learning for predictive process monitoring. In: Leemans, S., Leopold, H. (eds.) ICPM 2020. LNBIP, vol. 406, pp. 124–135. Springer, Cham (2021). https://doi.org/10.1007/978-3-030-72693-5_10
5. Di Francescomarino, C., Ghidini, C., Maggi, F.M., Milani, F.: Predictive process monitoring methods: which one suits me best? In: Weske, M., Montali, M., Weber, I., vom Brocke, J. (eds.) BPM 2018. LNCS, vol. 11080, pp. 462–479. Springer, Cham (2018). https://doi.org/10.1007/978-3-319-98648-7_27
6. Imran, M., Ismail, M.A., Hamid, S., Nasir, M.H.N.: Complex process modeling in process mining: a systematic review. IEEE Access **10**, 101515–101536 (2022)
7. Bousdekis, A., Kerasiotis, A., Kotsias, S., Theodoropoulou, G., Miaoulis, G., Ghazanfarpour, D.: Modelling and predictive monitoring of business processes under uncertainty with reinforcement learning. Sensors **23**(15), 6931 (2023). https://doi.org/10.3390/s23156931
8. Kotsias, S., Kerasiotis, A., Bousdekis, A., Theodoropoulou, G., Miaoulis, G.: Predictive and prescriptive business process monitoring with reinforcement learning. In: Krouska, A., Troussas, C., Caro, J. (eds.) NiDS 2022. LNNS, vol. 556, pp. 245–254. Springer, Cham (2022). https://doi.org/10.1007/978-3-031-17601-2_24
9. Sutton, R.S., Barto, A.G.: Reinforcement Learning, 2nd edn. An Introduction, MIT Press, Cambridge (2018)
10. Sharma, J., Andersen, P., Granmo, O., Goodwin, M.: Deep Q learning with Q-matrix transfer learning for novel evacuation environment. IEEE Trans. Syst. Man Cybern. Syst. **51**, 7363–7381 (2021)
11. Sunghyun, S., Liu, L., Bae, H.: Automatic Discovery of Multi-perspective Process Model using Reinforcement Learning. Preprint at https://doi.org/10.48550/arXiv.2211.16687 (2022)
12. Mnih, V., et al.: Human-level control through deep reinforcement learning. Nature **518**, 529–533 (2015)
13. Arulkumaran, K., Deisenroth, M.P., Brundage, M., Bharath, A.A.: Deep reinforcement learning: a brief survey. IEEE Signal Process. Mag. **34**, 26–38 (2017)
14. Wang, Y., Chen, W., Liu, Y., Ma, Z.-M., Liu, T.-Y.: Target transfer q-learning and its convergence analysis. Neurocomputing **392**, 11–22 (2020)

15. Watkins, C., Dayan, P.: Q-learning. Kluwer Academic Publishers, Boston (1992)
16. Lin, L.-J.: Reinforcement Learning for Robots Using Neural Networks. PhD thesis. Carnegie Mellon University, Schenley Park Pittsburgh, PA, USA (1993)
17. Voskoglou, M.Gr.: Applications of finite Markov chain models to management. Am. J. Comput. Appl. Math. **6**(1), 7–13 (2016). https://doi.org/10.48550/arXiv.1601.01304

SAMSEF: An Agile Software Maintenance Leveraging Scrum Framework for Improved Efficiency and Effectiveness

Muhammad Ali[1], Sehrish Munawar Cheema[2], Ammerha Naz[3],
and Ivan Miguel Pires[4(✉)]

[1] Department of Software Engineering, Superior University Lahore, Lahore, Pakistan
`msse-f21-008@superior.edu.pk`
[2] Department of Computer Science, University of Management and Technology,
Lahore, Pakistan
`sehrish.munawar@umt.edu.pk`
[3] Department of Computer Science, University of Sialkot, Sialkot, Pakistan
`22201009-001@uskt.edu.pk`
[4] Instituto de Telecomunicações, Escola Superior de Tecnologia e Gestão de Águeda,
Universidade de Aveiro, Águeda, Portugal
`impires@ua.pt`

Abstract. Agile methodologies have gained widespread recognition for their ability to produce high-quality software systems. Given that over half of all job, roles are associated with maintenance activities throughout a program system's entire lifecycle, researchers and engineers are actively exploring utilizing agile methods to ensure the continued robustness of the software. The agile software development lifecycle lacks a dedicated maintenance plan, prompting the need for a set of agile maintenance principles derived from both theoretical foundations and practical experience. These established standards encompass maintenance planning, customer involvement, incremental maintenance execution, documentation updates at each stage, and thorough testing of maintenance tasks. This study proposes an efficient framework named SAMSEF-based. It is a hybrid of Model-Based Software Engineering and Agile methodology, successfully used in a pilot study on health technology systems. It can improve project progression and system design documentation, enhancing agile capabilities.

Keywords: Software Agility · scrum efficiency · accelerated maintenance · software life cycle · MBSE

1 Introduction

Software development teams are increasingly adopting agile methodologies, which can be adapted to accommodate future changes in the software development process. Scrum, a lightweight approach influenced by Boehm's spiral model, is commonly employed to track and monitor software development progress [23]. According to Lehman's law of

Á. Rocha et al. (Eds.): WorldCIST 2024, LNNS 989, pp. 126–136, 2024.
https://doi.org/10.1007/978-3-031-60227-6_11

software evolution, software maintenance plays a crucial role in the long-term transformation of software [34]. With proper upkeep, the software would continuously change or retain its utility. Significant distinctions exist between software development and maintenance [42]. Most software projects leverage pre-existing code, interact with various organizational components, and are notoriously challenging to modify extensively.

Consequently, an iterative maintenance approach is necessary to adapt to future changes effectively [9]. Reports indicate that employing agile methods for software maintenance yields numerous benefits. Software maintenance can be grouped into four main types: adaptive maintenance, perfective maintenance, corrective maintenance, and preventive maintenance. Perfective and adaptive maintenance are typically perceived as alterations to the software design that enhance system functionality [27].

This research was based on two research questions: (RQ1) What are the distinguishing factors between software maintenance and software development? (RQ2) What is the most effective approach for managing urgent emergency customer requests within agile maintenance sprints?

The paper introduces a hybrid framework based on Model-Based Software Engineering (MBSE) and agile methodology, successfully used in a pilot study on health technology systems. The study found that SAMSEF significantly improved project progression and system design documentation, enhancing Agile's capabilities. This approach offers a pragmatic solution for harmonizing Agile and MBSE, addressing regulatory compliance, and mitigating technological debt challenges [44]. Future research will involve quasi-experimental investigation [14].

This research investigates how agile management tools can boost innovation in socio-economic contexts. Our main focus is understanding how economic and social structures can grow while sticking to sustainable development principles [22]. AGILE management technology was used to encourage creativity in line with these principles. Our analysis involves advanced decomposition modeling, resulting in a foundational decomposition, context diagram, and target tree. The ultimate aim is to ensure the long-term sustainability of the social system by effectively applying AGILE management principles. While this study offers valuable insights, we recognize the need for more research to understand how human resource management can drive innovation within the frameworks of sustainable development [19].

2   Literature Review

The literature review explores the impact of Agile management tools on innovation in socio-economic contexts, establishing the foundation for understanding AGILE principles in sustainable development and creativity. It also investigates the transformative potential of ICT in improving the quality of life for elderly individuals, highlighting the importance of these tools in fostering innovation [12]. The research aims to enhance well-being and decrease medical dependency by utilizing a co-creation strategy combining Design Thinking, Lean Startup, and SCRUM Agile techniques. The strategy was evaluated within the H2020 CAPTAIN initiative, resulting in increased partner engagement and high team performance.

The impact of organizational culture on Agile Project Management (APM), specifically Scrum in industries beyond software development, uses the Competing Values

Framework to assess corporate cultures of seven industrial and service organizations through in-depth interviews [6]. The research reveals that subcultures within agile teams have distinct characteristics centered on clan and market values. Scrum's fundamental principles emphasize audacity, transparency, and respect, which can be nurtured through openness and adaptability.

The origins and use of cultural concepts in Agile Project Management (APM), focusing on the Scrum framework [37], emphasizes the significance of acknowledging and promoting cultural characteristics within project teams, aiming to improve understanding of the complex network of cultural norms that form the foundation for agile approaches. The study also provides insights into the effectiveness of Agile Project Management in various business structures [25].

The software sector places great importance on maintaining high-quality software systems, leading to increased focus on enhancing software quality. This study explores using Bayesian Networks (BNs) in log monitoring to identify performance anomalies and suspicious usage patterns [4]. The primary objectives are to improve software quality and user experience for web application users. Three different BN models can be used to identify performance bottlenecks, malfunctions, and recurring user behavior patterns. The web application uses Scrum methodology, considering these outcomes in its process [39]. Data-driven, quality-oriented software maintenance holds significant potential, and the study process and replication kit can facilitate future research [31].

Agile software development is a systematic approach that uses a life cycle model with incremental and iterative processes. It involves diverse skill sets and autonomy, allowing teams to collaborate to address issues and generate resolutions [3]. This approach is gaining popularity among firms due to its ability to address challenges and ensure project success. Diverse tools in the software industry are increasingly used to facilitate effective communication and collaboration among teams.

Scrum Watch is designed to help project managers make informed decisions by providing comprehensive reports and visual representations using the cloud-based technology [29]. It addresses the challenges of large amounts of data generated by devices, focusing on openness and transparency in process and product measurements.

The "Agile Enterprise Architecture Development Method" (AEA-DM) is a tool that combines Enterprise Architecture (EA) and Agile methodologies, focusing on incremental development strategies. It was developed using the Design Science Research Methodology (DSRM) and is recommended for optimizing digital transformation initiatives [30]. The AEA-DM is a comprehensive guide that outlines team member responsibilities, facilitates agile demand management, and promotes the integration of agile software development practices with enterprise architecture approaches, thereby advocating for their use in digital transformations.

Agile development methodology outperforms traditional software development life cycles, highlighting the need for adaptation to evolving situations and circumstances. This approach demonstrates the collective impact of gradual enhancements [41]. However, challenges such as meeting deadlines, budgeting, and maintaining quality standards in agile project management persist due to the expansion of project scope and changing market needs [18]. The "Eclectic Agile Methodology" is a new approach to product development that enables organizations to experiment with various processes, procedures, tools, and validations before making a definitive choice. It addresses limitations in previous agile methods, such as shared environments, dependencies, multiple tracks, geographical dispersion, and seamless connection with upstream and downstream applications [16].

The Scaled Agile Framework (SAFe) is a key agile scaling methodology in software development, crucial for businesses to maintain competitiveness in the digital economy. It ensures visibility and alignment of projects with the company's objectives, particularly in departments like sales, finance, and marketing [7]. The use of SAFe is essential for developing scalable approaches for software delivery, as the importance of software in corporate success has increased significantly. This study explores the prerequisites for deploying SAFe. The case study showcases the implementation of SAFe in a prominent software organization, emphasizing its importance in enhancing software delivery and providing valuable insights for professionals and researchers in the field [10].

Pharmaceutical firms are utilizing digital platforms to enhance manufacturing and quality control processes, boosting client satisfaction and industry adherence while conducting periodic assessments to ensure accountability [40].

Table 1 contains a comparative analysis based on the research focus of different methodologies and their key findings and insights. The studies were selected based on their direct relevance to our research question and objectives, ensuring a focused and insightful examination. Emphasizing quality, recentness, and alignment with our theoretical framework, these studies represent a subset within the broader literature.

Table 1. Comparative Analysis Based on Research focus of Different Methodologies Along with their Key finds and Insights

Ref.	Research Focus	Methodology	Key Findings and Impact	Additional Insights
[38]	Co-creation strategy	Design Thinking, Lean Startup, SCRUM Agile	Increased partner engagement	Improved morale and productivity
[25]	Organizational culture's influence on APM	Competing Values Framework	Clan and market subcultures	Endorsing cultural characteristics
[14]	sMBSAP approach	Model-Based Software Engineering (MBSE), Agile	Positive impact on project progression	Addressing regulatory compliance
[31]	Bayesian Networks for log monitoring	Action research project, software logs	Identifying performance bottlenecks	Data-driven, quality-oriented maintenance

(continued)

Table 1. (*continued*)

Ref.	Research Focus	Methodology	Key Findings and Impact	Additional Insights
[19]	AGILE management tools for innovation	Decomposition modeling, sustainable development	Utilization of AGILE management principles	Human resource management's role
[45]	Scrum Watch	Cloud-based technology	Support for decision-making	Openness and transparency
[46]	Agile Enterprise Architecture Development Method	Design Science Research Methodology	Integration of agile practices	Expert opinions
[16]	Eclectic Agile Methodology	Inclusive methodology	Experimentation with processes	Addressing various challenges
[10]	Scaled Agile Framework (SAFe)	Agile principles in marketing, sales, and finance	Enhancing software delivery	Practical implementation case
[21]	Digital transformation in pharmaceutical firms	Design-based methodology, dynamic indicators	Monitoring digital transformation	Adaptable and dynamic indicators

3   Methodology

In the pursuit of comprehensive insights, our study meticulously interrogated research inquiries, theoretically and empirically. Our observations unveil the software development industry's commendable track record in delivering high-caliber software systems and widespread adoption of agile methodologies. This positive industry trend has prompted maintenance teams to embrace agile maintenance practices, a shift not without challenges arising from inherent disparities between software creation and maintenance processes [17]. Delving into this complex terrain, our primary focus centers on elucidating the nuanced distinctions between software development and maintenance, unraveling fundamental disparities that underscore the industry's dynamics. This exploration lays the groundwork for our subsequent research inquiry, strategically navigating the effective management of urgent client requests within agile maintenance sprints.

3.1   Empirical Study

Almashhadani et al. [1] studied agile maintenance and identified significant challenges from missed iteration goals. During the sprints, the team needed more planning for additional work requested by the customer. To address this issue, they proposed a solution of categorizing maintenance sprints into two types: short sprints for immediate tasks and longer sprints for tasks scheduled in advance. However, they observed that client requests were still made between these smaller sprints. To mitigate this uncertainty, it is advisable to allocate a buffer of unplanned time that can be utilized for unexpected events or tasks that arise later on [8].

3.2 Proposed Framework

Establishing a well-structured approach to handle urgent client requests and prevent interruptions to sprint goals is essential. We successfully addressed this issue by implementing the Scrum Software Maintenance Model, presented in Fig. 1, enabling us to manage critical client requests effectively while maintaining regular sprints [5]. The Scrum Software Maintenance Model initiates the planning phase by employing version control to track maintenance changes. During the planning stage, careful consideration is given to the required maintenance activities. The first step to start corrective maintenance involves creating a new code branch. Routine maintenance requests are scheduled regularly, with specific monitoring conducted to identify urgent and significant sprints. In the event of an urgent request, the ongoing sprint is stopped, and its progress is saved in the version control system [20]. A new type is then created to address the urgent request.

This model ensures that pending requests are not overlooked, as it allows us to resume work from where it was left off by restoring the source code from version control once the urgent demand is completed [26]. By incorporating the sprint state into the version control system, this approach prioritizes emergency sprints that are of greater importance to our client's businesses. The following factors are emphasized:

- *Planning for Maintenance:* During the sprint planning session, it is crucial to consider both the critical and regular emergency tasks. It is advisable to allow for flexible iterations based on the nature of the maintenance work. The length of these iterations need not be uniform. Employing a version control system, such as Git, for each iteration is highly recommended. This practice facilitates efficient management, enables auditing capabilities, and ensures traceability throughout the process [32];
- *Iterative Execution of Maintenance Tasks:* Regardless of the nature or significance of the task, it is essential to approach it methodically and integrate each successful step into the existing system [35];
- *On-Site Customer Presence*: Engaging customers in sprint planning is essential to solicit their feedback on planned and unplanned emergency work. By involving them, we can assess the significance of each task and accordingly make adjustments to the sprint plan [43]. This collaborative approach ensures that customer input is valued and incorporated into decision-making.
- *Documentation Update at Each Phase*: Ensure the system's accuracy by meticulously documenting each step. This can be accomplished using either a well-defined use case or a meticulously constructed test case [20];
- *Testability of Maintenance Processes*: You can make the maintenance process easier to test by doing each step in a different version and using version control [36]. It will be possible to do accurate integration and error testing and simultaneously iterate on urgent emergency work.

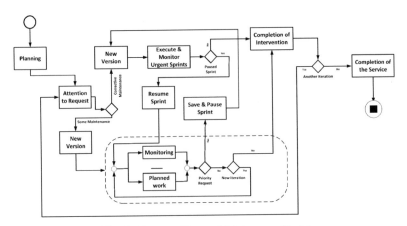

Fig. 1. Proposed Scrum Maintenance Model

3.3 Case Study

We used a case study approach to integrate an existing exercise system, allowing fitness instructors to offer online training and remote exercise plans. The customer was in the US, while the maintenance team was in South Asia. The Internet, email, and affordable international connections facilitated communication between the customer and the maintenance team. The advent of these digital resources has greatly facilitated the growth of remote work agencies, digital groups, and remote companies [24]. Our maintenance team followed agile scrum maintenance methodologies; however, during their work, they encountered several challenges:

- The objectives set for the current sprint were not achieved, as priority was given to urgent emergency sprints;
- Sprints that were substituted by emergency sprints had to be readdressed due to the inability to preserve the progress made in the original sprint;
- Conducting comprehensive system testing proved challenging due to the absence of distinct code versions for each update, making it difficult to pinpoint the source of the issues [33].

The Scrum Software Maintenance Model ensured the system's continuous operation, with maintenance tasks planned in iterative cycles. The client actively participated in meetings to assess the urgency of emergency work and ensure the sprint plan remained current. The report covers three months of maintenance work, including six planned and four unplanned sprints. The initial plan included six sprints with buffers. The maintenance engineers used version control software to save alternative versions of their work and continued emergency sprints [28]. After finalizing the disaster maintenance, they returned to the remaining sprints, allocating additional time and buffers. Use cases and test cases were used to track the process.

4   Findings

This study delves into the distinctive aspects of software maintenance and development and the optimal approach for managing urgent client requests within agile maintenance sprints. The findings emanate from the maintenance team's collaborative efforts and adoption of the Scrum Software Maintenance Model.

Regarding RQ1, "What are the distinguishing factors between software maintenance and soft-ware development?", the Scrum Software Maintenance Model reveals notable disparities between software maintenance and development. Incremental updates, client involvement in sprint planning, and prioritization contributions emerged as critical factors, underscoring the nuanced nature of maintenance tasks compared to the broader development scope.

Regarding RQ2, "What is the most effective approach for managing urgent emergency customer requests within agile maintenance sprints?", The study advocates an effective strategy for handling urgent client requests in agile maintenance sprints. Collaborative discussions during Scrum meetings, active participation of the author with the Scrum master and team, and adopting the Scrum Software Maintenance Model collectively contribute to heightened client satisfaction [15]. This approach streamlines emergency task integration, facilitates efficient versioning for integration testing [13], reduces paperwork through use cases and test case adoption[11], and ensures a swift, reliable, and accountable software maintenance process [2].

5   Conclusion

The Scrum Software Maintenance Model is a tool developed to address the challenges of agile maintenance in the software development industry. It focuses on handling urgent emergency sprints within planned sprints, involving client involvement, and managing iterations using version control, testing, and documentation. The model has been proven to increase client satisfaction by involving them in sprint planning, prioritizing, and addressing essential emergency assignments. Version control helps maintain the sprint state, reduces rework, and improves support engineers' happiness. Documentation in use and test cases makes the software system more manageable, meeting the maintenance and agile scrum goals.

Acknowledgments. This work is funded by FCT/MEC through national funds and co-funded by the FEDER – PT2020 partnership agreement under the project **UIDB/50008/2020.** This article is based upon work from COST Action CA21118 - Platform Work Inclusion Living Lab (P-WILL), supported by COST (European Cooperation in Science and Technology). More information on www.cost.eu.

References

1. Almashhadani, M., Mishra, A., Yazici, A., Younas, M.: Challenges in agile software maintenance for local and global development: an empirical assessment. Information **14**, 261 (2023)
2. Almeida, F.: Communication and coordination issues in managing distributed scrum teams. In: Perspectives on Workplace Communication and Well-Being in Hybrid Work Environments, pp. 193–212. IGI Global (2023)
3. Anand, A., Kaur, J., Singh, O., Ram, M.: Optimal resource allocation for software development under agile framework. Reliab Theory Appl **16**, 48–58 (2021)
4. Avila, D.T., Van Petegem, W., Snoeck, M.: Improving teamwork in agile software engineering education: the ASEST+ framework. IEEE Trans. Educ. **65**, 18–29 (2021)
5. Bomström, H., et al.: Information needs and presentation in agile software development. Inf. Softw. Technol. 107265 (2023)
6. Bundhun, K., Sungkur, R.K.: Developing a framework to overcome communication challenges in agile distributed teams–case study of a mauritian-based IT service delivery centre. Glob. Transit. Proc. **2**, 315–322 (2021)
7. de Castro, R.O., Sanin, C., Levula, A., Szczerbicki, E.: The development of a conceptual framework for knowledge sharing in agile it projects. Cybern. Syst. **53**, 529–540 (2022)
8. Chang, S.-C., Wongwatkit, C.: Effects of a peer assessment-based scrum project learning system on computer programming's learning motivation, collaboration, communication, critical thinking, and cognitive load. Educ. Inf. Technol. 1–24 (2023)
9. Edison, H., Wang, X., Conboy, K.: Comparing methods for large-scale agile software development: a systematic literature review. IEEE Trans. Softw. Eng. **48**, 2709–2731 (2021)
10. Fagarasan, C., Cristea, C., Cristea, M., Popa, O., Mihele, C., Pisla, A.: The delivery of large-scale software products through the adoption of the SAFe framework, pp. 137–143. IEEE (2022)
11. Habib, B., Romli, R., Zulkifli, M.: Identifying components existing in Agile software development for achieving "light but sufficient" documentation. J. Eng. Appl. Sci. **70**, 75 (2023)
12. Heimicke, J., Dühr, K., Krüger, M., Ng, G.-L., Albers, A.: A framework for generating agile methods for product development. Procedia CIRP **100**, 786–791 (2021)
13. Hossain, A.: A Systematic Mapping Study on Scrum and Kanban in Software Development (2023)
14. Huss, M., Herber, D.R., Borky, J.M.: An agile model-based software engineering approach illustrated through the development of a health technology system. Software **2**, 234–257 (2023)
15. Huss, M., Herber, D.R., Borky, J.M.: Comparing measured agile software development metrics using an agile model-based software engineering approach versus scrum only. Software **2**, 310–331 (2023)
16. Jain, A., Porwal, R., Kansal, V.: Design of a new agile methodology: eclectic--an approach to overcome existing challenges. Int. J. Intell. Eng. Syst. **16** (2023)
17. Joskowski, A., Przybyłek, A., Marcinkowski, B.: Scaling scrum with a customized nexus framework: a report from a joint industry-academia research project. Softw. Pract. Exp. (2023)
18. Karar, A.N., Labib, A., Jones, D.F.: A conceptual framework for an agile asset performance management process. J. Qual. Maint. Eng. **28**, 689–716 (2021)
19. Kopytko, M., Lutay, L., Chornenka, O., Markiv, M., Grybyk, I., Dzyubina, A.: A sustainable socio-economic system model leveraging AGILE management technologies for fostering innovations. Int. J. Sustain. Dev. Plan. **18**, 1951–1956 (2023)

20. Mishra, A., Alzoubi, Y.I.: Structured software development versus agile software development: a comparative analysis. Int. J. Syst. Assur. Eng. Manag. **14**, 1–19 (2023)
21. Miura, K., Masuda, Y., Shirasaka, S.: Designing performance indicator in human-centered agile development. In: Zimmermann, A., Howlett, R., Jain, L.C. (eds.) KES-HCIS 2023, pp. 143–152. Springer, Singapore (2023). https://doi.org/10.1007/978-981-99-3424-9_14
22. Mora, M., Marx-Gomez, J., Wang, F., Diaz, O.: Agile it service management frameworks and standards: a review. In: Arabnia, H.R., Deligiannidis, L., Tinetti, F.G., Tran, Q.-N. (eds.) Advances in Software Engineering, Education, and e-Learning. TCSCI, pp. 921–936. Springer, Cham (2021). https://doi.org/10.1007/978-3-030-70873-3_66
23. Morandini, M., Coleti, T.A., Oliveira, E., Jr., Corrêa, P.L.P.: Considerations about the efficiency and sufficiency of the utilization of the Scrum methodology: a survey for analyzing results for development teams. Comput. Sci. Rev. **39**, 100314 (2021)
24. de Oliveira, E.R., Ribeiro, P.C.C., Méxas, M.P., de Oliveira, S.B.: Scrum method assessment in Federal Universities in Brazil: multiple case studies. Braz. J. Oper. Prod. Manag. **20**, 1496 (2023)
25. Patrucco, A.S., Canterino, F., Minelgaite, I.: How do scrum methodologies influence the team's cultural values? A multiple case study on agile teams in Nonsoftware industries. IEEE Trans. Eng. Manag. **69**, 3503–3513 (2022)
26. Petrescu, M., Motogna, S.: A Perspective from Large vs Small Companies Adoption of Agile Methodologies (2023)
27. Plant, O.H., van Hillegersberg, J., Aldea, A.: How DevOps capabilities leverage firm competitive advantage: a systematic review of empirical evidence, pp 141–150. IEEE (2021)
28. Raharjo, T., Purwandari, B., Budiardjo, E.K., Yuniarti, R.: The essence of software engineering framework-based model for an agile software development method. Int. J. Adv. Comput. Sci. Appl. **14** (2023)
29. Ranawana, R., Karunananda, A.S.: An agile software development life cycle model for machine learning application development, pp. 1–6. IEEE (2021)
30. Rashid, N., Khan, S.U., Khan, H.U., Ilyas, M.: Green-agile maturity model: an evaluation framework for global software development vendors. IEEE Access **9**, 71868–71886 (2021)
31. del Rey, S., Martínez-Fernández, S., Salmerón, A.: Bayesian Network analysis of software logs for data-driven software maintenance. IET Softw. **17**(3), 268–286 (2023)
32. Saarikallio, M., Tyrväinen, P.: Quality culture boosts agile transformation—action research in a business-to-business software business. J. Softw. Evol. Process **35**, e2504 (2023)
33. Shafiee, S., Wautelet, Y., Poelmans, S., Heng, S.: An empirical evaluation of scrum training's suitability for the model-driven development of knowledge-intensive software systems. Data Knowl. Eng. **146**, 102195 (2023)
34. da Silva, E.F., Magalhães, A.P.F., Maciel, R.S.P.: Software evolution and maintenance using an agile and MDD hybrid processes. In: Filipe, J., Śmiałek, M., Brodsky, A., Hammoudi, S. (eds.) ICEIS 2020. LNBIP, vol. 417, pp. 437–457. Springer, Cham (2021). https://doi.org/10.1007/978-3-030-75418-1_20
35. Singh, A., Kukreja, V., Kumar, M.: An empirical study to design an effective agile knowledge management framework. Multimed. Tools Appl. **82**, 12191–12209 (2023)
36. Suárez-Gómez, E.D., Hoyos-Vallejo, C.A.: Scalable agile frameworks in large enterprise project portfolio management. IEEE Access (2023)
37. Subramanian, N., Suresh, M.: Assessment framework for agile HRM practices. Glob. J. Flex. Syst. Manag. **23**, 135–149 (2022)
38. Tessarolo, F., et al.: Developing ambient assisted living technologies exploiting potential of user-centred co-creation and agile methodology: the CAPTAIN project experience. J. Ambient Intell. Humaniz. Comput. 1–16 (2022)
39. Tona, C., Juárez-Ramírez, R., Jiménez, S., Quezada, Á., Guerra-García, C., López, R.G.P.: Scrumlity: an agile framework based on quality assurance, pp. 88–96. IEEE (2021)

40. Traini, L.: Exploring performance assurance practices and challenges in agile software development: an ethnographic study. Empir. Softw. Eng. **27**, 74 (2022)

41. Tyagi, S., Sibal, R., Suri, B.: Empirically developed framework for building trust in distributed agile teams. Inf. Softw. Technol. **145**, 106828 (2022)

42. Udvaros, J., Forman, N., Avornicului, S.M.: Agile storyboard and software development leveraging smart contract technology in order to increase stakeholder confidence. Electronics **12**, 426 (2023)

43. Uraon, R.S., Chauhan, A., Bharati, R., Sahu, K.: Do agile work practices impact team performance through project commitment? Evidence from the information technology industry. Int. J. Product. Perform. Manag. (2023)

44. Van Wessel, R.M., Kroon, P., De Vries, H.J.: Scaling agile company-wide: the organizational challenge of combining agile-scaling frameworks and enterprise architecture in service companies. IEEE Trans. Eng. Manag. **69**, 3489–3502 (2021)

45. Vega, F., Rodríguez, G., Rocha, F., dos Santos, R.P.: Scrum Watch: a tool for monitoring the performance of Scrum-based work teams. J. Univers. Comput. Sci. **28**, 98 (2022)

46. Visweswara, S.: An agile enterprise architecture methodology for digital transformation (2023)

Software and Systems Modeling

Distribution of Invalid Users on an SSH Server

Kai Rasmus[1]([✉]), Tero Kokkonen[2], and Timo Hämäläinen[3]

[1] Luode Consulting Oy, Lutakonaukio 7, 40100 Jyväskylä, Finland
`kai.rasmus@luode.fi`
[2] Institute of Information Technology, Jamk University of Applied Sciences, Jyväskylä, Finland
`tero.kokkonen@jamk.fi`
[3] Software and Communications Engineering, Faculty of Information Technology, University of Jyväskylä, Jyväskylä, Finland
`timo.t.hamalainen@jyu.fi`

Abstract. The Secure Shell (SSH) server on a Unix-like system is a viable way for users to login and execute programs on the system remotely. Remote access is something that hackers also want to achieve, making SSH servers a target for attack. A quantitative study was made of the distribution of usernames and IP addresses in failed login usernames on a publicly available SSH server. The failed logins and IP addresses were ranked according to the number of occurrences producing a distribution. The results indicated that the elements followed approximately a distribution with an inverse relationship with the rank of the element similar to what is known as the Zipf's Law. An important consequence of the Zipf's law is that 20% of elements are responsible for 80% of consequences, which means that by blocking 20% of the failed login usernames or IP addresses, 80% or more of the failed logins are also blocked. This was found to be true for a real-world scenario. Some topics were identified for further research.

Keywords: SSH · invalid usernames · Zipf's Law · brute-force attack

1 Introduction

The Secure Shell Protocol (SSH) is a cryptographic network protocol based on a client-server architecture (that is) used for remote login and command-line execution of programs in Unix-like operating systems, such as Linux. SSH was developed already in 1995 [1], and the most common implementation of it, OpenSSH, was released in 1999 for the OpenBSD operating system, even though it has now been ported to many different operating systems.

SSH operates on a 3-layered protocol suite: the transport layer, the user authentication protocol, and the connection protocol. The user authentication protocol is able to use, among other certificate-based methods, usernames and passwords for authenticating users. Certificate-based methods are more secure but sometimes it is still necessary to use a username and password-based authentication for login to the system.

When an SSH server is open to the public internet, it is subject to abuse that comes in the form of unauthorized login attempts. Remote access to a server is a sought-after

© The Author(s), under exclusive license to Springer Nature Switzerland AG 2024
Á. Rocha et al. (Eds.): WorldCIST 2024, LNNS 989, pp. 139–151, 2024.
https://doi.org/10.1007/978-3-031-60227-6_12

asset amongst attackers. SSH has the capability of tunneling TCP connections creating a challenge for firewalls to secure the traffic. If an attacker gains SSH access to a server, they can then tunnel almost any type of connection through the tunnel. A proxy server can be placed at the firewall that decrypts the data in the tunnel thus making it easier for the firewall to spot malicious traffic. Statistical methods can also be used to look at the encrypted packets to spot what protocol is being sent through the tunnel [2].

Brute-force attacks, in which username and password pairs are generated and tried repeatedly until a correct pair is found, have always been a problem. In a 2008 study they were discovered to be the most common form of attack facing SSH, FTP and Telnet servers [3]. SSH is used in many Internet-of-Things (IoT) applications, i.e. brute-force attacks are still a problem in the modern era [4] and mitigation methods are still necessary. Brute-force attacks mimic benign network traffic so they do not have a clear signal that could be used to block the traffic. Artificial intelligence-based machine learning systems have been used to help in finding this attack traffic [5].

To keep an SSH server secure from attacks, it is important to know what usernames are being used to try to log in to the server. The system usernames used in Linux systems are well known and are usually not allowed to log in remotely. Work on identifying the usernames used in brute-force attacks is still ongoing with increasing knowledge on the types of usernames being used [6]. The passwords are not stored in the SSH log but can be studied using an SSH honeypot or some other system. Studies on the passwords show that they are not random either, but attackers have used precompiled lists of username-password-pairs, which are widely distributed over the internet [3].

The purpose of this work is to use a log of failed login attempts from a real online public-facing SSH server to study the distribution of the usernames and IP addresses used. The distribution will be compared to Zipf's Law, which states that there exists an inverse relation between the count of an element, i.e., the number of times it appears in a distribution, and the rank within the distribution. The distribution of passwords is outside the scope of this study, because they were not stored in the log of the server.

A numerical model was then used to show how throttling effects affect the distribution. Temporarily blocked IPs caused a lower number of attacks from the most common IP addresses after a certain number of failed logins in a specific amount of time. The model was also used to make the case for a permanent ban on constantly abusive addresses. The results could then be used in helping classification of anomalous traffic in intrusion detection systems of SSH servers such as that proposed by González et al. [7].

This study starts with some background to SSH and attacks against SSH servers, after which the methods and modelling work are described together with the gathered dataset. The results of the statistical analysis and modelling work are then described, after which a test of the 80–20 rule is made with a public-facing SSH server. The final part of the study is a discussion of the results together with some conclusions.

2 Theory

Zipf's Law states that there exists an inverse relation between the count of an element in a distribution and its rank [8–11]. In its simplest form it is:

$$D(R) = \frac{D_1}{R^\alpha} \tag{1}$$

where D_1 is the count of the first element and R is the rank of the element. The parameter α is usually close to 1, and that value will be used in this study.

A consequence of Zipf's Law is the so-called 20–80 rule, which states that 20% of the elements are responsible for 80% of the effects. In this SSH login context that would mean that 20% of the usernames would be responsible for 80% of the failed login attempts. Additionally, 80% of the failed login attempts would come from 20% of the IP addresses. The term 'law' will be used in place of Zipf's Law distribution.

This law was originally observed in the field of linguistics [10] by counting words in a text and sorting them in the order of occurrences. The inverse relationship between rank and count was found in all languages.

The law has then been applied to a large variety of fields: investment [12], census studies in urban environments [13], music classification [14], and industrial safety [15]. Studies of interconnected computers have shown that failure rates are related to the number of interconnections in a relation similar to Zipf's Law [16]. Military expenditure by countries has also been studied using Zipf's Law [17]. Looking at the flow of money from companies of similar rank has given a theoretical background to Zipf's Law [11].

A critique of Zipf's Law and especially the way in which it is often tested has been presented by Urzúa [18], and a good formal test is documented by Urzúa [19]. The study of military expenditure also found that the number 1 ranked country (the USA) spent more than double than the second ranked country in contrast to Zipf's Law [17]. The distribution of cities in Turkey [20] was also found to not follow the law. These findings underline the fact that Zipf's Law is not a deterministic law in the strictest sense.

3 Methods

A quantitative methodology was chosen for the research in this work. A dataset of usernames and IP addresses from failed SSH logins was taken from a public-facing server. These were then sorted according to the number of occurrences. The IP addresses were then deleted, and only the ranks were used afterwards to preserve privacy. Additionally, the IP addresses and usernames were from different periods in the logs and were not studied together.

Model datasets were then produced using Eq. (1) with the number of occurrences of the most common failed login as the parameter value for D_1. One dataset scenario, called VE1 was produced with all the failed logins and another, called VE2, including only invalid usernames. VE1 included login attempts using the built-in Linux usernames, such as the root-user. The third model dataset was made for the IP addresses used in the failed logins.

Statistical characteristics were then calculated for all scenarios separately. In these calculations the observed dataset is x_i, and the modelled dataset is y_i. These parameters have been used by [21] to compare numerical models of the Baltic Sea. Means were calculated for both types of datasets:

$$\bar{x} = \frac{1}{N} \sum_{i=1}^{N} x_i, \tag{2}$$

$$\bar{y} = \frac{1}{N} \sum_{i=1}^{N} y_i, \tag{3}$$

where N is the number of ranks, i.e., not the number of observations. Standard deviations were calculated for both types of datasets:

$$\sigma_{xx} = \sqrt{\frac{\sum (x_i - \bar{x})}{N - 1}}, \tag{4}$$

$$\sigma_{yy} = \sqrt{\frac{\sum (y_i - \bar{y})}{N - 1}}. \tag{5}$$

A correlation coefficient, R_{xy}, was then calculated using:

$$R_{xy} = \frac{\sum (x_i - \bar{x})(y_i - \bar{y})}{(N - 1)\sigma_{xx}\sigma_{yy}}. \tag{6}$$

The mean absolute error, *MAE*, was then calculated using:

$$MAE = \frac{\sum |x_i - y_i|}{N}. \tag{7}$$

The root mean square error, *RMSE*, was then calculated using:

$$RMSE = \sqrt{\frac{\sum (x_i - y_i)^2}{N}}. \tag{8}$$

The mean error, also called the bias *B*, was then calculated using:

$$B = \bar{y} - \bar{x}. \tag{9}$$

All of the statistical characteristics were calculated for the IP addresses and both scenarios of usernames: VE1 and VE2.

4   Dataset

Over 250,000 failed login attempts were observed over a period of several months. The failed login attempts are shown in Table 1 for the first 25 most common failed logins, most of which were invalid usernames. Most of the failed login attempts were made with the root-user, which is an existing username on this Linux server. If the existing usernames are excluded, then the most common failed login attempt on this server is.

Table 1. The 25 most commonly used invalid usernames.

Invalid username	Rank	Count	Exisiting
Root	1	167,791	Yes
Admin	2	6,194	No
Ubuntu	3	5,616	No
User	4	5,420	No
Postgres	5	5,148	Yes
Test	6	1,869	No
Oracle	7	1,127	No
Ftpuser	8	814	No
Git	9	577	No
Guest	10	564	No
Ubnt	11	430	No
Pi	12	425	No
Debian	13	386	No
Mysql	14	344	Yes
user1	15	343	No
Hadoop	16	332	No
test1	17	305	No
test2	18	302	No
Usuario	19	295	No
Support	20	293	No
Testuser	21	290	No
ftp	22	254	No
Jenkins	23	248	No
Dev	24	234	No
Deploy	25	229	No

with the admin-username. The least common failed logins were misspellings of real usernames or usernames consisting of only one letter.

Some of the attempts used usernames that are related to software ubiquitous on Linux-systems, such as the mysql user, which is installed when the MySQL database package is installed. As the database was installed on the system, the user was also installed. Another example is the Hadoop user, which refers to the Apache Hadoop software library used to analyze large datasets [22]. 'Usuario', on the other hand, is 'user' in Spanish. These failed usernames are similar to those found in the failed login analysis of [6].

144   K. Rasmus et al.

The dataset for the VE1-scenario together with the theoretical Zipf distribution is illustrated in Fig. 1. The same for the VE2-scenario is shown in Fig. 2. A striking feature of VE1 is that the theoretical distribution grossly overestimates the distribution. However, the slope of the theoretical distribution is almost correct after the rank of 10. This means that a better fit would not be found even with a higher value than 1 for the exponent α. The theoretical distribution line can be made to cross the observed distribution by choosing an α-value of 1.5 (Fig. 1) The initial value D_1 is too large to produce the correct theoretical distribution. The total number of usernames for VE1 was 13,091 and for VE2 13,064 shows that only a few existing usernames were attempted for the system.

The distribution of IP-addresses is shown in Fig. 3 and this distribution does not follow the theoretical distribution very well, except for the first 9 ranks.

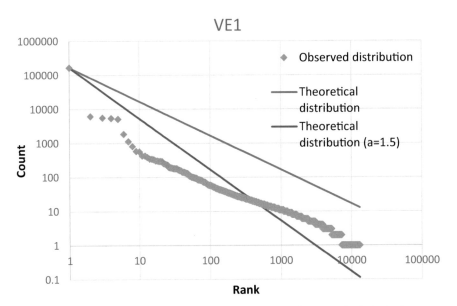

Fig. 1. Distributions of all failed login attempts for the scenario VE1. This includes login attempts using the 'root'-user. Note base 10 logarithms on both axes. A second theoretical distribution using an α-value of 1.5 is also shown.

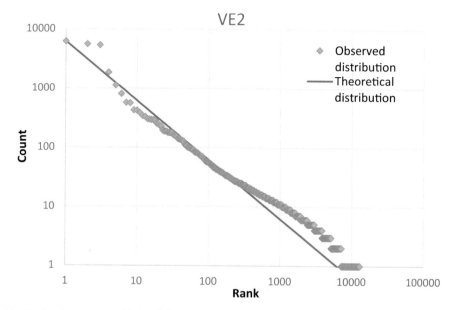

Fig. 2. Login attempts with invalid usernames only, scenario VE2. Note base 10 logarithms on both axes.

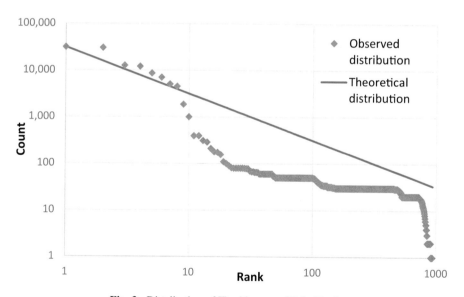

Fig. 3. Distribution of IP addresses of failed logins.

5 Model Development

A model was developed to study how the distribution changes when additional constraints are added. A set of addresses was selected randomly assuming an ideal Zipf distribution of the probability of picking an address. Therefore, the address with rank 1 had the highest probability of being chosen. An address was selected once per cycle, and a total of N cycles were made so that the total number of addresses was the same as the total number of observed addresses. Initially no constraints were added.

In the second case, an address was banned for an additional M number of cycles if it was picked K times within a period of M consecutive cycles. The third case was similar to the second case but the ban was permanent. For this study M was set to 100 and K was set to 10.

6 Results

The statistical characteristics for the two scenarios are shown in Table 2. The results for VE1 show that with the root-user, the distribution does not follow the Zipf distribution. Using the total number of failed logins, the number of failed logins for the second ranks should be approximately 128,000 but in reality, it is just over 6,000. The root-user has a special significance in Linux systems and is overrepresented by several orders of magnitude. Regardless of this, the correlation coefficient is over 0.8, even though the RMSE and B are high. The correlation coefficient for VE2, which does not include the root-user, is much higher (over 0.92), and the correlation can be seen visually in Fig. 2. The RMSE is two orders of magnitude smaller than for VE1, and the bias (B) is almost non-existent. In both cases the bias is negative i.e., the theoretical Zipf distribution is an overestimation.

The correlation coefficient for the IP addresses is the highest, and the bias is positive, meaning that Zipf's Law underestimates the IP addresses.

Table 2. Statistical characteristics for the two scenarios.

Characteristic	VE1	VE2	IP
N	13,091	13,064	936
D_1	167,791	6,194	31,413
$\bar{x}$	19.6	6.3	149.8
$\bar{y}$	128.9	73.9	249.0
σ_{xx}	1,469.9	90.8	1,592.7
σ_{yy}	1,876.5	73.9	1,292.7
R_{xy}	0.819	0.923	0.940
MAE	109.3	2.3	156.2
RMSE	1,085.0	36.3	592.5
B	−109.3	−1.4	99.2

When testing the 20-80rule, the results (Table 3) show that the predicted number of failed logins for VE2 has an error of 4% related to the observed number of failed logins for 20% of the usernames. When the root-user is included, the error is already 15%. This means that the 20-80rules can be used to block out 80% of failed logins when the top 20% most common usernames that do not include the root-user are blocked. Another consequence is that even more failed logins can be blocked when the root-user is included. The error for the IP addresses is 9% meaning that the 20–80 rule would block slightly less than 80% of the invalid login attempts.

Table 3. Values related to the test of the 20–80 rule. The values needed for comparison are in bold type.

Characteristic	VE1	VE2	IP
Total sum of failed logins or IPs	256,705	82,838	140,190
80% of total	**205,364**	**66,270**	**112,152**
Total distinct usernames or Ips	13,091	13,064	936
20% of distinct usernames or IPs	2,618	2,612	187
Sum of failed logins using only top 20% of distinct usernames or Ips	**237,285**	**63,586**	**124,673**
Difference theoretical - observed	31,921	−2,684	12,521
Difference %	16%	4%	9%

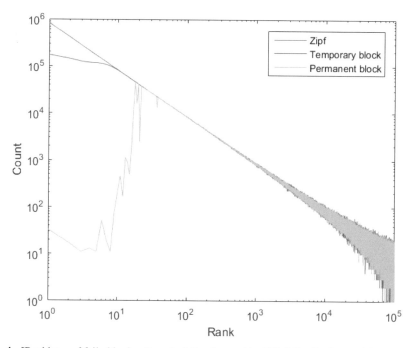

Fig. 4. IP address of failed login attempts following an ideal Zipf distribution and then with two levels of blocking: a temporary ban and a permanent ban. The probability that an IP address with a rank of 1,000 is picked is small, leading to a spread in the results.

The model results (Fig. 4) show that as time goes on, the most common addresses are picked less if they are subjected to bans, but after the first 20 most common addresses the ban has no effect. The permanent ban clearly removes the most common addresses but again has no effect after 20 of the most common addresses.

The shape of the curve with a temporary ban is similar to the observed IP address and username curves indicating that the security feature of the SSH server in use that throttles the number of login attempts in case of abuse is working.

7 Security Solution

Regardless of whether the invalid login attempts follow the Zipf distribution or not, the results need to be built into a real security solution for the results of this work to be useful. To this end, one day of failed logins was analyzed, and the top 20% of IP addresses with the most login entries were blocked at the firewall.

The log items for the SSH server per hour for failed attempts (Fig. 5) show a steady average of 6700 items per hour before the 28th of December, after which the average increases to 9600 items per hour. Coincidentally the day chosen for the collection of the IP information used for deciding which IP addresses to block was the 28th.

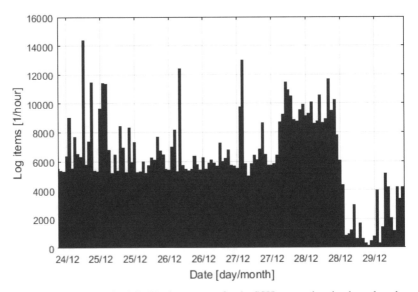

Fig. 5. Log items per hour for failed login attempts for the SSH server showing how they decrease after the implementation of the security control. The number of log items is never a constant with daily variations occurring constantly,

After the implementation of the security control the log items decrease to an average of 1600 per hour. This is slightly less than 20% of the average of the 28th, showing that this simple use of the 20–80 rule produces significant results without endangering the legitimate use of the server.

8 Discussion and Conclusions

The results seem to indicate that Zipf's Law holds well for the VE2 case, which does not include the root-user. However, it does not entirely work for the VE1 case, which does include the root user. This shows how much of an importance the root-user has in Linux systems, and why it is never a good idea to allow the root-user to login remotely, via SSH or any other means, even though extensions built on top of the original SSH protocol proposed to make this safer [23]. Modern Ubuntu versions of the Linux operating system do not have the root-user activated at all, and administrators use their own accounts and the sudo program to administer the system. This also helps in auditing events, because a log can be kept of which account is responsible for the event. Since the 20–80-rule is applicable to the SSH failed login cases not including the root user, it would help system administrators if a system were devised that would block the top 20% of invalid usernames. That way they would need to deal with 80% less attempts overall.

Future work on this topic could include trying different values for the exponent α in Eq. 1. Different values would produce different correlation coefficients, with an α-value of more than one producing a steeper curve as the rank increases and an α-value less than one would produce a less steep curve. A steeper curve would be necessary in the VE1 case that included the root-user and for the IP addresses.

A critical viewing of Fig. 2 shows that the Zip law only actually starts to work after a rank of 10 or so and after that, the dependence on rank is not strictly linear in the log-log plot. This means that due to the nature of some of the usernames being system users, they are more common in the failed logins than they should be according to the theory. The relatively simplistic nature of Zipf's Law means that it does not work in all cases. Even though it has been applied to many cases in a large variety of fields, it has been found not to work in some cases. One such case is the distribution of cities in Turkey [20].

The theoretical framework developed by Kawamura and Hatano [11] showed that a distribution converges to the Zipf distribution relatively slowly. In their work, the convergence had not happened even after the number of samples was n $= 10^7$. The differences found in this study and the ideal distribution can therefore be attributed to the small sample size used, and this could be improved by looking at a larger number of invalid logins.

A quantitative study was made of failed logins on a Secure Shell (SSH) server, which was connected to the public internet. Three cases were studied: one which included all failed logins, one which included failed logins of usernames that did not exist on the system, and one which included the IP address of the failed login attempt. The first case included the root-user, which is the administrator username on Linux systems. The failed logins were ranked according to the number of occurrences, and this distribution was then compared to the theoretical Zipf distribution, which states that there exists an inverse relationship between an element and its rank. A consequence of Zipf's Law is that 20% of elements are responsible for 80% of consequences.

The results show that the scenario including the root-user did not follow Zipf's Law, and this was found to be because the root-user was over-represented in the observed distribution by several orders of magnitude. A better correlation between the observed and theoretical distributions, with a correlation coefficient of over 0.9, was found when only failed logins with nonexistent usernames were used.

The second case also followed the 20–80 rule with an error that was 4% of the expected result. This means that the 20–80 rule can be used to block out over 80% of failed logins. If the root-user is also blocked, then this value becomes even higher. The root-user can be blocked from establishing a remote connection because it is often not allowed to remotely connect via SSH in any case. The 20–80 rule produced an error of 9% for the IP addresses.

The implementation of these results into a practical security solution was outside the scope of this study and could be done in the future. The validity of the results with more failed login data is also left for future work.

Acknowledgements. This research was partially funded by the Resilience of Modern Value Chains in a Sustainable Energy System project, co-funded by the European Union and the Regional Council of Central Finland (grant number J10052).

The authors would like to thank Ms. Tuula Kotikoski for proofreading the manuscript.

References

1. Ylönen, T.: SSH-secure login connections over the Internet. In: Proceedings of the 6th USENIX Security Symposium (1996)
2. Dusi, M., Gringoli, F., Salgarelli, L.: A preliminary look at the privacy of SSH tunnels. In: Proceedings of the 17th International Conference on Computer Communications and Networks, August 2008
3. Owens, J., Mathews, J.: A study of passwords and methods in brute-force SSH attacks. In: USENIX Workshop on Large-Scale Exploits and Emergent Threats (LEET), (2008)
4. Raikar, M.M., Meena, S.M.: SSH brute force attack mitigation in internet of things (IoT) network: an edge device security measure. In: 2021 2nd International Conference on Secure Cyber Computing and Communications (ICSCCC) (2021)
5. Wanjau, S.K., Wambugu, G.M., Gabriel, N.K.: SSH-brute force attack detection model based on deep learning. Int. J. Comput. Appl. Technol. Res. **10**, 42–50 (2021)
6. Khandait, P., Tiwari N., Hubbali, N.: Who is trying to compromise your SSH server? An analysis of authentication logs and detection of bruteforce attacks. In: Adjunct Proceedings of the 2021 International Conference on Distributed Computing and Networking (2021)
7. González, S., Herrero, A., Sedano, J., Zurutuza, U., Corchado, E.: Different approaches for the detection of SSH anomalous connections. Logic J. IGPL **24**(1), 104–114 (2016)
8. Zipf, G.K.: Human Behavior and the Principle of Least Effort. Hafner, New York (1965)
9. Li, W.: Zipf's law everywhere. Glottometrics **2002**(5), 14–21 (2002)
10. Zipf, G.K.: Human Behavior and the Principle of Least Effor: an Introduction to Human Ecology. Ravenio Books (2016)
11. Kawamura, N., Hatano, N.: Universality of Zipf's law. J. Phys. Soc. Jpn. **71**(5), 1211–1213 (2002)
12. Ausloos, M., Bronlet, P.: Strategy for investments from Zipf law (s). Physica A **1–2**(324), 30–37 (2003)
13. Moura, N.J., Jr., Ribeiro, M.B.: Zipf law for Brazilian cities. Physica A **367**, 441–448 (2006)
14. Manaris, B., et al.: Zipf's law, music classification, and aesthetics. Comput. Music. J. **1**(29), 55–69 (2005)
15. Wang, Z., Ren, M., Gao, D., Li, Z.: Exploring industrial safety knowledge via Zipf law, arXiv preprint arXiv:2205.12636. (2022)

16. Adamic, L.A., Huberman, B.A.: Zipf's law and the Internet. Glottometrics **1**(3), 143–150 (2002)
17. Arvantidis, P., Kollias, C.: Zipf's law and world military expenditures, peace economics, peace. Sci. Public Policy **22**(1), 41–71 (2016)
18. Urzúa, C.M.: Testing for Zipf's law: a common pitfall. Econ. Lett. **112**(3), 254–255 (2011)
19. Urzúa, C.M.: A simple and efficient test for Zipf's law. Econ. Lett. **66**, 257–260 (2000)
20. Duran, H.E., Cieslik, A.: The distribution of city sizes in Turkey: a failure of Zipf's law due to concavity. Reg. Sci. Policy Pract. **5**(13), 1702–1719 (2021)
21. Myrberg, K., et al.: Validation of three-dimensional hydrodynamic models of the Gulf of Finland. Boreal Env. Res. **15**, 453–479 (2010)
22. White, T.: Hadoop: The Definitive Guide. O'Reilly Media Inc. Sebastopol (2012)
23. Thorpe, C.: SSU: Extending SSH for Secure Root Administration. In: LISA (1998)

BERT Transformers Performance Comparison for Sentiment Analysis: A Case Study in Spanish

Gerardo Bárcena Ruiz[1,2]([✉]) and Richard de Jesús Gil[3]

[1] Universidad Americana de Europa, Av. Bonampak Sm. 6, Mz. 1, Lt. 1,
77500 Cancún, QR, México
gbarcena@up.edu.mx
[2] Universidad Panamericana, Augusto Rodin 498, 03920 Ciudad de México, México
[3] Universidad Internacional De La Rioja (UNIR), Av. de La Paz 137,
26006 Logroño, La Rioja, Spain
richard.dejesus@unir.net

Abstract. Despite the fact that the Spanish language is the second most spoken language in the world, research about AI and sentiment analysis is few compared with English language as research target. This paper refers to some works about sentiment analysis for the Spanish language that use BERT transformers and other technologies for this kind of sentiment analysis task; but the quality model based on indicators such as accuracy level could be different according to the tool or BERT version used. In addition, about the BERT family, it is challenging to determine which versions or subversions could perform better for sentiment analysis and also comply with the Spanish language when they are used on common platforms such as Colab. Therefore, the present study seeks to address this issue by establishing objectives, such as identifying relevant datasets based on the quality of Spanish used and having balanced subsets; also, locating different Spanish trained models; and proposing a method of comparison that involves relevant variables. We propose a weighted index that combines the F1-Score and the retraining time in different scenarios to help making better decisions. The results of this study indicate that the DistilBERT, RoBERTa, and ALBERT models have highest performances, but BERT remains in top positions as a consistent model.

Keywords: BERT · transformer · performance · sentiment analysis · Spanish language

1 Introduction

According to Cervantes Institute, in the world there are almost 500 million native Spanish speakers and close to 600 million if we add students and those that have limited competence to speak it. This represents between 6% and 7% of the world's population. Spanish is the second most spoken language after Mandarin Chinese as a native language and it is the 4th most spoken language overall after English, Mandarin, and Hindi [17].

Spanish-speaking countries are a big marketplace, above all if services are available to assist users and customers. Sentiment analysis is a very useful tool for product analysis, market research, sentiment of customers reviews, stock market trend predictions, reputation or user experience analysis [18], that can help users in services with Spanish language.

But if we observe some surveys about the use of Spanish in research about AI in general and sentiment analysis in particular, the research papers including Spanish as a target language are few. Table 1 shows some works that include Spanish and other languages for research in AI.

Table 1. Research works about sentiment analysis and used languages.

Reference	English Tests	Spanish Tests	Other Tests	Percentage for Spanish
[18]	18	1	0	5.26%
[19], adding all tests	201	7	17	3.11%
[20]	61	3	12	3.94%

Likewise, from a total of 83 works about emotion detection in texts only three of them were using Spanish as target language [21]. This lack of research about tools for Spanish provokes an uncertainty about which tool could fit better for a particular task.

As expressed previously, sentiment analysis is a very good tool, but the interest of this work is using sentiment analysis with Spanish, which assumes comprehensive understanding of the language. Currently, transformers are novelty tools that can perform natural language processing (NLP) tasks with acceptable performances, this is the reason to focus our research on them.

The original or vanilla transformer [2], also known as sequence-to-sequence, has an active encoder and decoder that usually performs neural machine translation tasks [1]. An example of this type of transformer is the Bidirectional and Auto-Regressive Transformer (BART) or Text-To-Text Transfer Transformer (T5).

Nevertheless, it is possible to turn off the encoder block and use only the decoder. This use case is typically for sequence generation or language modeling [1]. This tool is also known as an autoregressive model. Generative Pre-trained Transformers (GPT) is the now-well-known example of this kind of transformer.

As a complementary case, there are encoder-centered or automatic coding transformers, which seek to deliver a representation of the input sequence. These models commonly perform natural language understanding tasks, such as text classification and sequence labeling. Moreover, currently transformers mainly based on AI deep learning models, are tools used in areas such as computer vision, audio or voice processing, and even chemistry as well as NLP [1], among many others. The basic example of this type of transformer is BERT, which is the interest of this paper.

BERT model is a pertinent tool for sentiment analysis because of its language understanding and text classification capabilities. This type of transformer performs sentiment analysis with text classifying, with positive or negative qualifications, and retraining gives the possibility of classifying texts not previously entered to the transformer.

There are different versions of BERT transformers, such as RoBERTa, ALBERT, DistilBERT, DeBERTa, or Longformer [3], and many other subversions. However, this work focuses on the above list because they are the first ones of their kind.

But which of them has a better performance for sentiment analysis in Spanish language? Different models will have different retraining times and classification accuracy, the aim of this work is to determine, as far as possible, these two variables to get a benchmark when choosing one of these models.

1.1 General Objective

The general objective of the work is to compare the performance of different BERT transformers for sentiment analysis in Spanish with a common reference dataset.

1.2 Specific Objectives

Consequently, some of the specific objectives that the study seeks to achieve are: i) identifying the different BERT models available in Spanish. ii) Determine the available datasets in Spanish that have acceptable quality to be the core element for the comparison of the different models. iii) Establish the comparison scheme (benchmarking) of the models against a common dataset. iv) Make the comparison using the proposed index.

2 Background

The attention mechanism proposed for the first transformer [2], also called self-attention or intra-attention, relates different positions of a sequence to compute its representation. The attention mechanism can perform reading comprehension, synthesis, and linking between texts, among other tasks. The self-attention mechanism has helped transformer models to compute input and output representations without using sequence-aligned recurrent neural networks or convolutional neural networks. Figure 1 shows a diagram of the first transformer.

2.1 Encoder

The encoder comprises a stack of six identical layers with two sublayers each. First, there are multiple self-attention heads, followed by a simple feedback network connected to the position function. There is also a residual connection around each sublayer and one more normalization layer [2].

2.2 Decoder

This component also comprises a stack of six identical layers, as well as two sublayers, and a third sublayer with multiple attention heads on the output. Likewise, there are residual connections around each sublayer and a normalization layer. A modified self-attention sublayer of the decoder implements a different masking to make the current positions to only pay attention to previous positions [2].

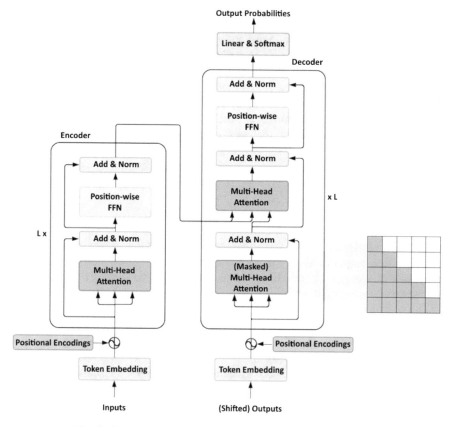

Fig. 1. Overview of vanilla transformer architecture. Source [1].

2.3 Attention

The attention function relates a query and a set of key-value pairs in the output, also known as the Query-Key-Value mechanism [1], all of them being vectors, as well as the output. The output is a weighted sum of the values, and a compatibility function calculates weights between the query and the corresponding key. The process calls the attention function in parallel (h times) with different linear projections and then concatenates and aligns several values into the output. This multi-head attention allows the model to pay attention to the information from different representation subspaces at different positions [2].

Additionally, one peculiarity of these transformers is that they make no assumptions about data structure. This gives them the potential to identify dependencies between different ranges, but it can also cause overfitting when there is a limited volume of data. One way to ease this problem is to introduce an inductive bias into the model. Also, a subsequent benefit of pre-training transformers on a large dataset is that they can learn universal language representations, and then retrain on particular datasets, which greatly improve performance [1].

2.4 BERT Versions

BERT. Its name comes from the pre-training of Deep Bidirectional Transformers for Language Understanding. There are two strategies for model pre-training: a) feature-based and b) fine-tuning. The first uses an architecture such as pre-trained representations. The second, as would be the case with GPTs, introduces minimal parameters and fine-tuning on previously trained parameters. Both approaches seek to learn general language representations with unidirectional models. This may limit the choice of the architecture used for pre-training, such as in question-answer oriented tasks, where it is important to maintain context in both directions [4].

Likewise, the BERT transformer seeks to ease unidirectionality by using the Masked Language Model (MLM), which randomly masks input tokens to predict the original element based solely on the context [4].

Furthermore, pre-training of BERT model has two parts:

a) First stage or application of the Masked Language Model. Also known as the Cloze task, training applies a random mask to the input tokens and then predicts them. *Softmax* function in the output receives the final hidden vectors from the masking. Typically, the process applies a mark over 15% of all input tokens. The process only predicts the masked words instead of the complete input, but that generates a mismatch between pre-training and fine-tuning. To avoid this, the training data generator, when the generator chooses 15% of random positions, then a [MASK] token is replaced 80% of the time, 10% of the time replaces it with another random token, and the remaining 10% of the time does not make any change [4].

b) Second stage or Next Sentence Prediction (NSP). Many times, transformers perform tasks like Question Answering (QA) or Natural Language Inference (NLI). Both tasks depend on the understanding of the relationship between two sentences, which the MLM stage did not get. For the NLP stage, the process uses a monolingual corpus and takes two sentences, 50% of the time the second one assures to follow, and the other 50% is a random sentence from the entire corpus [4].

RoBERTa. This version of BERT comes from Robustly Optimized BERT Pretraining Approach, aims to solve BERT undertraining, a deficiency identified by the RoBERTa authors. To achieve this, the changes relative to the original are: i) to perform more training with larger batches and more data; ii) to eliminate the NSP training stage; iii) to train with longer sequences; and iv) to change static masking for a dynamic one. Regarding masking, to avoid using the same mask in each training cycle or epoch, RoBERTa duplicates training data 10 times to mask in 10 different ways during the 40 cycles [5].

ALBERT. The name comes from A Lite BERT for Self-supervised Learning of Language Representations. This model incorporates two techniques to decrease the number of parameters and improve the model's pre-trained scalability capacity without severely impacting performance. The first decomposes the large, embedded vocabulary matrix into two smaller ones; this makes it easier to increase the size of the hidden layers without increasing the size of the embedded vocabulary parameters. The second technique allows the sharing of parameters between layers, which prevents the number of parameters from growing with the depth of the network. In equivalent configurations of

ALBERT and BERT, the former could have 18 times fewer parameters and the training process could be 1.7 times faster [6].

DistilBERT. The name comes from "a distilled version of BERT". AI models are growing in size and mobile industry concerns about this situation because: a) the environmental cost due to the energy consumption required to train the models; and b) the need to incorporate them into mobile devices that allow natural language processing but could not process large models [7]. For this reason, the industry prefers smaller models.

Moreover, knowledge distillation is a compression technique that a compact model, named "student", trains to reproduce the behavior of the "teacher", which is one or several larger models. In this model, DistilBERT is the "student" that has the same BERT model architecture, but without the token type embeddings and the grouper, and half of the layers, as well as the alignment and normalization layers, uses modern and optimized libraries. Training process takes exceptionally large batches (up to 4,000 examples per batch), using dynamic masking and without NSP (as in RoBERTa). The model has a pledge to be 40% smaller than BERT as well as 60% faster, with 97% language understanding [7].

DeBERTa. Decoding-enhanced BERT with Disentangled Attention, which uses these two techniques [8]:

a) Disentangled attention. In the BERT model, a vector that contains the sum of the content encoding and its position encoding represents each input word, but DeBERTa uses two vectors that encode the content and the position separately. Disentangled matrices based on their content and relative positions calculate attention weights.
b) Improved mask decoder. This model uses content and position encoding to perform the MLM stage, but the disentangled mechanism considers relative positions, which improves the prediction of tokens.

LongFormer. Although the self-attention mechanism is powerful, its computational and memory requirements grow quadratically with the length of the sequence, making it expensive to process long sequences. The LongFormer (Long-Document) transformer model aims to solve this problem by using self-attention that scales linearly according to the sequence length, allowing it to process long documents for tasks such as long document classification, question answering (QA), and coreference resolution. LongFormer has a combination of self-attention focused on the local context and global attention. These two types of attention allow the model to build contextual representations and full sequence representations for predictions [9].

3 Related Works

There are interesting research works about sentiment analysis that includes Spanish as a target language that let us observe different techniques and accuracies about text processing. Table 2 lists some works about Spanish sentiment analysis.

Indeed, researchers about text processing for language understanding are taking BERT transformers for this kind of task. As seen in Table 2, almost 62% of the works are using this kind of tool, and some of them pretend to use transformers later.

Table 2. Research works about sentiment analysis in Spanish.

About	Technology	Dataset	Accuracy
Hate speech (HS) detection [22]	MTL model using the monolingual Transformer-based model BETO[1] that integrates polarity and emotion knowledge for HS detection, named MTLsent + emo	HatEval and MEX-A3T	Macro-F1: 78.47 for HatEval and 86.58 MEX-A3T
Google Play Store comments [23]	BERT	Information, extracted by authors, about 15 top downloaded apps from Google Play Store	F1-Score: 0.80
Twitter comments [24]	Spanish adaptation of Affective Norms for English Words (ANEW), multi-label classifier based on k-nearest neighbors	Query to Twitter platform to extract 3000 tweets posted	Average Hamming Loss obtained by 10-fold cross validation process: 0.3192, best value
TripAdvisor reviews [25]	BETO and BERT fine-tuned with TF-IDF vectors	7413 reviews extracted by authors from TripAdvisor	F1-Score: 0.4280 for BETO and 0.2428 for BERT TF-IDF
Restaurant reviews [26]	Convolutional neural networks configured in two arrangements	At least 627 reviews provided by the 2016 Edition of SemEval competition	F1-Score: Among 0.9 and 0.93
Tweets about scientific dissemination [27]	Supervised learning with NLTK3 and ScikitLearn named OpScience	Authors query the Twitter API with hashtags PintofScience, PoS, PoS18, CienciaenRedes and CnR18, to get over 200,000 tweets	F1-Score: 0.7413
Instagram posts and Twitter tweets about tourism in Granada Spain [28]	BERT	For Twitter, they used the scraping tool named Twint to get 235,755 tweets, and for Instagram they used a java program with a job-scheduling process to deal with the daily data access restrictions to get 90,725 posts	F1-score: 0.5875
Financial data [29]	MarIA[2] and BETO	1,563 Spanish financial tweets got from 3,829 original tweets and 2,266 headlines from 92 selected newspapers	F1-score: 0.668261 for MarIA and 0.697432 for BETO

(continued)

[1] Spanish pre-trained BERT model.

[2] Spanish LLM model based on RoBERTa.

Table 2. (*continued*)

About	Technology	Dataset	Accuracy
Twitter comments with multilingual approach [30]	Multilingual transformer model XLM-RoBERTa with a data augmentation technique	TASS-2019 and TASS-2018 to get 3220 tweets for Spanish testing	F1-score: 0.7820, best value
RoBERTuito model for social media text analysis in Spanish [31]	Model based on RoBERTa	For training: 500 million tweets in Spanish from Spritzer collection. For evaluation: TASS 2020 Task A	Macro F1 score of 10 runs of the classification: 70.7, best value
Twitter comments [32]. In future works, the authors pretend to use BERT transformers	Universal Language Model Fine-Tuning (ULMFiT) based on weight-dropped LSTM	InterTASS Competition 2017 and InterTASS-PE Competition 2018	Macro F1-score: 0.567 for InterTASS 2017 and 0.463 for InterTASS-PE 2018
Multilingual negation detection [33]	Multilingual BERT and XLM-RoBERTa	SFU ReviewSP-NEG corpus from Simon Fraser University, for Spanish	F1-score: 0.8686, best value for Spanish and Russian
FIAVL tweets analysis [34]. In future works, the authors pretend to use BERT transformers	Support Vector Machines (SVM) and Naive Bayes (NB)	FIAVL official account with 18,000 tweets in Spanish language	F1-score: 0.987 for SVM and 0.878 for NB

Those works that use BERT or a subversion based on BERT have a F1-Score value of 0.597 on average and those that use RoBERTa reach a F1-Score value 0.756. Then, RoBERTa has a better performance indicator, according to values above.

4 Experimentation

First, the present study compares the performance and time dedicated to retraining one thousand sentences rated as positive or negative, coming from the IMDB Dataset of 50K Movie Reviews (Spanish) [10] dataset, extracted using a Pandas sampler. The main criteria to choose the dataset was having an equal number of positive and negative ratings and a high quality of Spanish language used in the texts.

On the one hand, the pre-trained models in Spanish, which are available in the HuggingFace vast collection, used for the comparison were: BERT [11], RoBERTa [12], ALBERT [13], DistilBERT [14], DeBERTa [15], and Longformer [16]. The selection criteria were the votes or downloads of the models.

Furthermore, we used Google Colaboratory Pro with T4 GPU. The retraining parameters were the following: for the tokenizer, the encoder used strings with a maximum length of 2048 (except ALBERT with 1024), the batch size was thirty-two (32), and there were ten (10) epochs. Although, for certain models, we changed diverse parameters due to GPU memory allocation issues.

In addition, for models benchmarking, the model performance criteria help to compare the classification process. Specifically, we take the F1-Score, based on recall, which is the value that shows the amount that the model can identify correctly; otherwise, it is the number of items correctly identified as positive out of the total number of true positives. The F1-Score combines accuracy and completeness measures into a single value. This helps to make it easier to compare the combined performance of accuracy and completeness among models.

5 Analysis of the Results

Certainly, the model retraining presented some drawbacks, in particular for ALBERT and LongFormer, since they consumed the entire memory of the GPU (trying to allocate 80 GB when only 15 are available), exceeding by far the assigned memory in Colab, so one way to solve the problem was to reduce the number of elements per batch and string sizes.

Table 3 shows the values obtained from the retraining. It is worth noticing ALBERT's better performance, but the retraining time was much longer than the rest of the models studied. It is also important to note that DistilBERT, according to the promise, was approximately 40% smaller than BERT and 60% faster.

Table 3. Accuracy score, training time, and other data of the retrained models.

Transformer	F1-Score	T(s)	T(m)	Batch	Model (MB)
BERT	0.80	216.21	3.60	32	419.10
RoBERTa	0.74	236.30	3.94	32	475.60
ALBERT	0.82	2777.00	46.28	16	224.50
DistillBERT	0.66	95.30	1.59	32	254.60
DeBERTa	0.80	334.94	5.58	32	1030.00
LogFormer	0.78	1723.97	28.73	8	567.20

Since the purpose of the study is to use the two variables, F1-Score and retraining time, as comparative elements, we propose a weighted index between 0 and 1, assigning the highest score to the best performer in the respective variable and a value proportional to the best performance to the others. There are 5 different scenarios: 1) taking F1-Score as the most important (F100); 2) taking F1-Score with 80% of the qualification weight and 20% of the time (F80 T20); 3) taking the same weight or importance for both (F50 T50); 4) taking that time matters more, but also taking the F1-Score into account (F20 T80); and finally, 5) time has the most important weight (T100).

Table 4 shows the models in descending order, according to the proposed index. The colors do not suggest a rating; they show the evolution of the different models in a simpler way. If we focus on accuracy as the main component, ALBERT, BERT, and DeBERTa are the best choices. But if we take the training time as the main component of the index, then DistilBERT is the undisputed winner, followed by BERT and RoBERTa.

As a result, and due to the drawbacks of memory management experimented with ALBERT and LongFormer, Table 5 shows the other four models. As seen, if retraining

Table 4. Position of the models according to their weighted rating.

F100	Idx	F80 T20	Idx	F50 T50	Idx	F20 T80	Idx	T100	Idx
ALBERT	1.00	BERT	0.87	DistilBERT	0.90	DistilBERT	0.96	DistilBERT	1.00
BERT	0.98	DistilBERT	0.84	BERT	0.71	BERT	0.55	BERT	0.44
DeBERTa	0.98	DeBERTa	0.84	RoBERTa	0.65	RoBERTa	0.50	RoBERTa	0.40
L.F	0.95	ALBERT	0.81	DeBERTa	0.63	DeBERTa	0.42	DeBERTa	0.28
RoBERTa	0.90	RoBERTa	0.80	ALBERT	0.52	L.F	0.23	L.F	0.06
DistilBERT	0.80	L.F	0.77	L.F	0.50	ALBERT	0.23	ALBERT	0.03

accuracy is the most important value to focus on, in this case, the winners would be BERT and DeBERTa, followed by RoBERTa. But, if retraining time is the most important value, DistilBERT takes the first position, followed by BERT and RoBERTa.

Table 5. Position of the filtered models according to their weighted rating.

F100	Idx	F80 T20	Idx	F50 T50	Idx	F20 T80	Idx	T100	Idx
BERT	1.00	BERT	0.89	DistilBERT	0.91	DistilBERT	0.97	DistilBERT	1.00
DeBERTa	1.00	DistilBERT	0.86	BERT	0.72	BERT	0.55	BERT	0.44
RoBERTa	0.93	DeBERTa	0.86	RoBERTa	0.66	RoBERTa	0.51	RoBERTa	0.40
DistilBERT	0.83	RoBERTa	0.82	DeBERTa	0.64	DeBERTa	0.43	DeBERTa	0.28

Additionally, for the case of DistilBERT, we performed an extra test changing the batch size of elements to process from 32 to 4. Then, F1-Score increased to 0.72, having a retraining time of 159.29 s, or 2.65 min, being the fastest of all models.

Figure 2 shows the confusion matrices of each of the retrained models. The T-axis corresponds to the elements originally classified as negative in the dataset, and the T + ones classified as positive. The P- and P + were those predicted as negative and positive, respectively.

It is an important note that RoBERTa is one of the most unbalanced, tending towards "pessimism". This model classifies initially positive elements as negative. The Distil-BERT is a quite balanced model, and the BERT model tends towards "optimism", which qualifies as positive elements that were originally negative.

Moreover, if we compare accuracy values of BERT and RoBERTa models with works previously listed in Related Works section, in this work BERT has a better performance. A reason for that could be the original training stage of the BERT model used in this paper. RoBERTa has equivalent values of accuracy to related works.

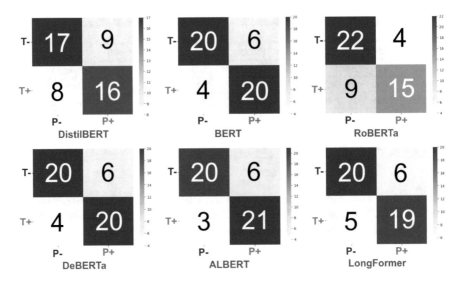

Fig. 2. Confusion matrices for each of the models under study. Note: Own source.

6 Conclusions

In conclusion, as seen in the study, most models operate with an acceptable F1-Score, but expected, some will take more time or resources, such as the ALBERT or LongFormer models, or may have lower performance classification as DistilBERT.

Moreover, according to values of previous tables, the DistilBERT model could be useful to implement, during the development stage, a proof of concept, and once the product is mature, BERT or RoBERTa could be a good option for the production stage. The original BERT model seems to still be a very good option due to its accuracy and shorter retraining time.

As a result of our benchmark, BERT and RoBERTa could be a good option to implement, and they are models that related works use widely. This work confirms, in a way, previous researchers' selections.

DeBERTa and DistilBERT models have interesting characteristics that could be a good option for future works.

References

1. Lin, T., Wang, Y., Liu, X., Qiu, X.: A survey of transformers. AI open **3**, 111–132 (2022)
2. Vaswani, A., et al.: Attention is all you need. In: 31st Conference on Neural Information Processing Systems (NIPS 2017), Long Beach, CA, USA
3. Mohammed, A.H., Ali, A.H.: Survey of BERT (Bidirectional encoder representation transformer) types. J. Phys. Conf. Ser. **1963**(1), 012173 (2021)
4. Devlin, J., Chang, M.-W., Lee, K., Toutanova, K.: BERT: pre-training of deep bidirectional transformers for language understanding (2019). arXiv:1810.04805
5. Liu, Y., et al.: RoBERTa: a robustly optimized BERT pretraining approach (2019). arXiv: 1907.11692

6. Lan, Z., Chen, M., Goodman, S., Gimpel, K., Sharma, P., Soricut, R.: ALBERT: a lite BERT for self-supervised learning of language representations (2020). arXiv:1909.11942
7. Sanh, V., Debut, L., Chaumond, J., Wolf, T.: DistilBERT, a distilled version of BERT: smaller, faster, cheaper and lighter (2020). arXiv:1910.01108
8. He, P., Liu, X., Gao, J., Chen, W.: DeBERTa: decoding-enhanced BERT with disentangled attention (2021). arXiv:2006.03654
9. Beltagy, I., Peters, M.E., Cohan, A.: Longformer: the long-document transformer (2020). arXiv:2004.05150
10. Fernandez, L.: IMDB Dataset of 50K movie reviews (Spanish). Kaggle (2021). Accessed Aug 2023. https://www.kaggle.com/datasets/luisdiegofv97/imdb-dataset-of-50k-movie-reviews-spanish
11. Romero, M.: BETO (Spanish BERT) + Spanish SQuAD2.0. Hugging Face, 11 Feb 2020. https://huggingface.co/mrm8488/bert-base-spanish-wwm-cased-finetuned-spa-squad2-es. Accessed Aug 2023
12. IIC - Institute of knowledge engineering, Autonomous university of Madrid. IIC/roberta-base-spanish-squades. Hugging Face, 17 Mar 2022. https://huggingface.co/IIC/roberta-base-spanish-squades. Accessed Aug 2023
13. DCCUChile - Department of Computer Sciences, University of Chile. dccuchile/albert-xlarge-spanish-finetuned-mldoc. Hugging Face, 11 Jan 2022 b. https://huggingface.co/dccuchile/albert-xlarge-spanish-finetuned-mldoc. Accessed Aug 2023
14. DCCUChile - Department of Computer Sciences, University of Chile. dccuchile/distilbert-base-spanish-uncased-finetuned-mldoc. Hugging Face, 11 Jan 2022. https://huggingface.co/dccuchile/distilbert-base-spanish-uncased-finetuned-mldoc. Accessed Aug 2023
15. PLN@CMM - Natural Language Processing Group of the Center for Mathematical Modeling, University of Chile, mdeberta-cowese-base-es. Hugging Face, 04 Jul 2022. https://huggingface.co/plncmm/mdeberta-cowese-base-es. Accessed Aug 2023
16. Heras, J.: joheras/longformer-base-4096-bne-es-finetuned-v2. Hugging Face, 03 May 2023. Accessed Aug 2023. https://huggingface.co/joheras/longformer-base-4096-bne-es-finetuned-v2
17. Sierra, J.A.: Spanish continues to grow and has almost 500 million native speakers, according to the Cervantes Institute's yearbook 2022. Atalayar, 27 Oct 2022. https://www.atalayar.com/en/articulo/culture/spanish-continues-grow-and-has-almost-500-million-native-speakers-according-cervantes/20221026154937158810.html. Accessed Dec 2023
18. Wankhade, M., Rao, A.C.S., Kulkarni, C.: A survey on sentiment analysis methods, applications, and challenges. Artif. Intell. Rev. **55**(7), 5731–5780 (2022)
19. Nazir, A., Rao, Y., Wu, L., Sun, L.: Issues and challenges of aspect-based sentiment analysis: a comprehensive survey. IEEE Trans. Affect. Comput. **13**(2), 845–863 (2022)
20. Shi, Y., Zhu, L., Li, W., Guo, K., Zheng, Y.: Survey on classic and latest textual sentiment analysis articles and techniques. Int. J. Inf. Tech. Dec. Mak. **18**(04), 1243–1287 (2019)
21. Alswaidan, N., Menai, M.E.B.: A survey of state-of-the-art approaches for emotion recognition in text. Knowl. Inf. Syst. **62**(8), 2937–2987 (2020)
22. Plaza-Del-Arco, F.M., Molina-Gonzalez, M.D., Urena-Lopez, L.A., Martin-Valdivia, M.T.: A multi-task learning approach to hate speech detection leveraging sentiment analysis. IEEE Access. **9**, 112478–112489 (2021)
23. López Condori, J.J., Gonzales Saji, F.O.: Análisis de sentimiento de comentarios en español en Google Play Store usando BERT. Ingeniare Rev chil ing. **29**(3), 557–563 (2021)
24. Palomino, R., Meléndez, C., Mauricio, D., Valverde-Rebaza, J.: ANEW for Spanish Twitter sentiment analysis using instance-based multi-label learning algorithms. In: Lossio-Ventura, J., Muñante, D., Alatrista-Salas, H. (eds.) Information Management and Big Data. SIMBig 2018, CCIS, vol. 898, pp. 46–53. Springer, Cham (2019). https://doi.org/10.1007/978-3-030-11680-4_6

25. Vasquez, J., Gomez-Adorno, H., Bel-Enguix, G.: Bert-based approach for sentiment analysis of Spanish reviews from Tripadvisor (2021)
26. Martínez-Seis, B.C., Pichardo-Lagunas, O., Miranda, S., Perez-Cazares, I.J., Rodriguez-Gonzalez, J.A.: Deep learning approach for aspect-based sentiment analysis of restaurants reviews in Spanish. CyS **26**(2), 899–908 (2022)
27. Sánchez-Holgado, P., Martín-Merino Acera, M., Blanco Herrero, D.: Del data-driven al data-feeling: análisis de sentimiento en tiempo real de mensajes en español sobre divulgación científica usando técnicas de aprendizaje automático. Disertaciones (Internet). 17 Jan 2020, vol. 13, no. 1, Accessed 28 Dec 2023
28. Viñán-Ludeña, M.S., De Campos, L.M.: Discovering a tourism destination with social media data: BERT-based sentiment analysis. JHTT. **13**(5), 907–921 (2022)
29. Pan, R., García-Díaz, J.A., Garcia-Sanchez, F., Valencia-García, R.: Evaluation of transformer models for financial targeted sentiment analysis in Spanish. PeerJ. Comput. Sci. **9**(9), e1377 (2023)
30. Barriere, V., Balahur, A.: Improving sentiment analysis over Non-English tweets using multilingual transformers and automatic translation for data-augmentation. In: Proceedings of the 28th International Conference on Computational Linguistics (Internet). Barcelona, Spain (Online): International Committee on Computational Linguistics, pp. 266–271 (2020). Accessed 28 Dec 2023
31. Pérez, J.M., Furman, D.A., Alemany, L.A., Luque, F., RoBERTuito: a pre-trained language model for social media text in Spanish (Internet). arXiv (2022). Accessed 28 Dec 2023
32. Palomino, D., Ochoa-Luna, J.: Advanced transfer learning approach for improving spanish sentiment analysis. In: Martínez-Villaseñor, L., Batyrshin, I., Marín-Hernández, A. (eds.) Advances in Soft Computing. MICAI 2019, LNCS, vol. 11835, pp. 112–123. Springer, Cham (2019). https://doi.org/10.1007/978-3-030-33749-0_10
33. Shaitarova A, Rinaldi F. Negation typology and general representation models for cross-lingual zero-shot negation scope resolution in Russian, French, and Spanish. In: Proceedings of the 2021 Conference of the North American Chapter of the Association for Computational Linguistics: Student Research Workshop (Internet). Online: Association for Computational Linguistics, pp. 15–23 (2021). Accessed 28 Dec 2023
34. Rivera-Guamán, R.R., Cumbicus-Pineda, O.M., López-Lapo, R.A., Neyra-Romero, L.A.: Sentiment analysis related of international festival of living arts Loja-Ecuador employing knowledge discovery in text. In: Botto-Tobar, M., Montes León, S., Camacho, O., Chávez, D., Torres-Carrión, P., Zambrano Vizuete, M. (eds.) Applied Technologies. ICAT 2020, Communications in Computer and Information Science, vol. 1388, pp. 327–339. Springer, Cham (2021). https://doi.org/10.1007/978-3-030-71503-8_25

Open Software Catalogue – Supporting the Management of Research Software

Marcin Wolski$^{(\boxtimes)}$, Jan Todek, Maciej Łabędzki, and Bartosz Walter

Poznan Supercomputing and Networking Center, Poznań, Poland
`marcin.wolski@man.poznan.pl`

Abstract. Software used for conducting research is one of the key assets that promote transparency and enable reproducibility. While several tools exist to help researchers use and share publications and datasets, the support for Research Software is still insufficient. In this vision paper, we outline expectations and requirements for managing Research Software and propose the Open Software Catalogue, a tool that relies on selected best practices in software engineering.

Keywords: Open Science · Research Software · Software Catalogue · FAIR data

1 Introduction

The process of scientific research usually involves a number of typical, repetitive tasks: literature review, collecting data, or analysis of results [12]. All those activities require transparency and reproducibility, which are indispensable for making a reliable and significant contribution to the body of knowledge. A number of dedicated tools, instrumentation, or even entire research environments support the operation of some of the tasks; for example, published papers are indexed by easy-to-search e-libraries, and data is stored and offered through public repositories.

Research Software (RS), which has been widely used for collecting, processing, and analysing data, is another category of tools that deserves proper curating and support [10]. RS refers to a variety of custom-made computer programs, usually developed and used by researchers, to conduct specific scientific experiments, analyse data, simulate processes, or perform other tasks related to their research.

Several initiatives attempt to transform RS into more Findable, Accessible, Interoperable, and Reusable digital resources (FAIR) [9]. The Software Heritage is an outstanding example of a repository ensuring software availability and traceability [4]. Despite these efforts, there is still a gap between the current functions provided by numerous solutions (infrastructures and corresponding services) supporting the management of RS and the expected maturity of archiving, referencing, describing, and crediting software source code in research [3]. The

Á. Rocha et al. (Eds.): WorldCIST 2024, LNNS 989, pp. 165–171, 2024.
https://doi.org/10.1007/978-3-031-60227-6_14

EOSC Task Force on Quality Research Software[1] formulated recommendations against this gap and in particular indicated governance, credit, sustainability, and metadata as these areas that require significant work for adoption.

Our goal is to address these open issues and the community's needs by providing the Open Software Catalogue, a solution dedicated to managing, sharing, and reusing RS. Additionally, we aim to provide means for crediting the creators and maintainers of RS to foster its development and sharing.

The remainder of the article is structured as follows: In Sect. 2, we write about the background of our subject. In Sect. 3, we outline the process of scientific research and the role of software therein. Next, in Sect. 4, we provide an overview of existing solutions. In Sect. 5, we propose the Open Software Catalogue, a dedicated tool for managing RS. Section 6 includes a summary and concluding remarks.

2 Background

The FAIR principles identify key features that make digital resources easier to find, accessible, interoperable, and reusable. Implementing FAIR helps researchers determine the impact of their work by reusing and citing the data they create, and it fosters collaboration between them [7].

Open Science, also commonly referred to as Open Research, represents a set of practices for collecting, managing, and disseminating research in the most open and transparent way possible. The FAIR Data Principles were developed to emphasise the importance of research data management throughout the research lifecycle, which is crucial to the success of Open Science.

Open Science (OS) is transparent and accessible knowledge that is shared and developed through collaborative networks [5].

In the context of FAIR and OS, software and data are regarded as different types of digital research objects [8]. As such, they share common characteristics that allow them to be treated the same with regard to some aspects of FAIR, such as being able to assign a digital object identifier (DOI) or having a licence. However, there are also several important differences between data and software as digital research objects [11]: Data are facts or observations that provide evidence. Software, on the other hand, is the result of a creative process that provides a data processing tool. As such, software is executable, whereas data is not. The lifespan of software, understood as versioning, is generally shorter than that of data. Consequently, dependencies as well as dependent software packages are subject to frequent changes.

3 Research Software

Software has become one of the essential assets used in any research-related procedure. The diagram presented in Fig. 1 depicts a typical research process and outlines steps that usually require software-based support.

---

[1] https://eosc.eu/advisory-groups/infrastructures-quality-research-software/.

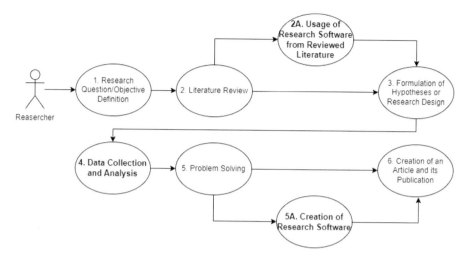

Fig. 1. Simplified research workflow

In Fig. 1, we can see several steps during which RS is used or created. Those steps are:

2A – When, during review, a researcher finds an interesting paper with some RS, it is possible they will try to run it by themselves to test it or see how it works,

4 – When gathering and analysing data, researchers may use someone else's tools or create their own,

5A – When conducting experiments, researchers may create some RS or use one that someone else created.

Two key actors in RS interaction are the **creator** and the **user**, each with distinct goals. The **creator** aims for efficient software development as part of a broader project, often with minimal documentation, seeking maximum value for minimal effort. On the other hand, the **user** desires reproducibility, ease of access, and user-friendliness to achieve RS's intended results. Balancing these divergent perspectives is crucial to creating an effective and accessible RS management tool.

3.1 Issues and Problems

RS is often a neglected scientific artefact. The software has become a more and more important part of research, but at the same time, the resulted software has not achieved enough quality to play a considerable role in research. It is due to several factors, e.g., time constraints, lack of peer review, focus on working prototypes, changing requirements, and the skill set of researchers.

Consequently, RS is non-transferable, non-reusable, and non-transparent. Using RS outside of its primary context as a tool linked with a particular research effort is a challenging task and, in some cases, even impossible. The fact that it is difficult to find RS by itself, not as an attachment to an article but as a fully-fledged software tool, creates a situation in which there is little incentive to create high-quality software.

This lack of places where RS can be found by itself can create a situation where the RS is stored only in one place – with the article. This is a dangerous situation: nothing should be stored in only one place and risks being lost.

Problems we have identified so far include the following:

1. RS is often of low quality (from an engineering point of view) – creators develop software that only meets their needs ignoring other possible users,
2. Users of RS have problems with running RS – creators do not provide their configuration and information on how to run their RS,
3. RS is not stored in persistent archives with appropriate redundancy,
4. Lack of exposure – RS is often only provided as an addition to the article itself, and it is hard to find it on its own by searching for it.

4   Overview of Existing Solutions

A large number of domain-specific RS registries and repositories have emerged for different scientific disciplines to ensure dissemination and reuse among their communities [13,14].

Services related to research can be divided into three categories [3]:

- *Publishers* – publishers are organisations responsible for preparing submitted research texts. Examples of publishers are Elsevier or Springer,
- *Aggregators* – aggregators are services focused on collecting information from diverse sources to increase the discoverability of digital content. The examples are RS Directory or Software Catalogue,
- *Archives* – archives are services with a primary goal of ensuring the long-term preservation of digital content. Examples of archives include Zenodo, Software Heritage, or Figshare.

Some existing archives allow for the storage of multiple types of research artefacts. However, they do not provide solutions that would help with solving the problems with RS we have identified – usually, they just allow for the upload of the RS but not much more. The RS Directory provides dedicated functionalities related to RS. This tool, however, is more focused on already-working and properly managed research tools than on all RS.

5   Open Software Catalogue

The Open Software Catalogue (OSC) would be a tool aimed at individuals creating RS and intending, most likely as part of the open science movement, to share the software developed during research experiments and development. At the same time, it should be the first go-to thing for any researcher who wants to find and run RS related to some article he found somewhere else.

The OSC would be an aggregator [3] that should answer the needs of the open community as well as be compliant with FAIR principles and best practices that have been identified by researchers and practitioners. The main features expected from OSC are the following: encourage best practices within RS [6]; enable the automated deployment and run of RS to speed up its evaluation; integrate with third-party solutions to facilitate the archiving of RS; expose researcher's achievements in software development and increase their visibility within the community; improved metadata for a better searching experience and interoperability.

5.1   Solution

The Open Software Catalogue will be built upon core functions delivered by the existing tool called Software Catalogue (SC). SC is a tool aimed at supporting the consistent exploration and reporting of software within an organisation, a community, or a service provider domain. SC is also a framework that enables the aggregation of many portions of data that are spread over multiple different locations (sources) into one coherent and well-structured piece of information[2].

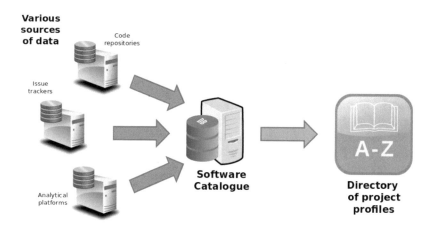

Fig. 2. Software Catalogue data flow

---

[2] https://bitbucket.software.geant.org/projects/SC/repos/softwarecataloguesuite.

SC provides a global view on software development from three main perspectives: a single software contributor, a single software initiative, and an institution having multiple such initiatives. SC constructs a big picture for all of them through the automatic collection and aggregation of information coming from separated data sources storing software artefacts such as source code repositories, issue tracking systems, and analytical platforms [2]. Data flow in SC is depicted in Fig. 2.

A software initiative may consist of multiple source code repositories (e.g., in the case of multi-modular software) and issue trackers while still being reflected by a single record in SC. The record is continuously updated by polling data sources against updates. Collected data are correlated and deduplicated. As a result, the constructed record contains a comprehensive set of metadata and statistics, such as community members, programming languages, libraries, lines of code, development time, and contributors' activity.

SC is currently deployed in the GEANT federated organisation as a supporting tool for the networking community [1]. The tool has successfully accomplished the transition to the production phase, and it currently stores consistent information about 40+ software projects and 3000+ individuals, including 400+ developers from GEANT, who have contributed to the software development[3].

6   Conclusions

In the course of discussion about RS, it is crucial to understand that its real value comes from peer researchers if they are able to run it, replicate the results of one experiment, or use it in other experiments. Even a tool developed for the sake of a single study should be reliably stored and made available to potential readers or users.

We propose the Open Software Catalogue, built upon the existing and mature framework, to bring the concept of FAIR RS closer to reality. By having the software metadata aggregated in one place, properly described, and referenced, we make searching for RS easier, both at the level of browsing through our aggregator and information being properly positioned on the search engines. The cooperation with third-party archives that store research tools could largely reduce the risk of RS being lost. Additionally, helping with making RS easier to run could foster reusability.

In this vision paper, we explored the relationship of Open Science to the RS. We identified some gaps that need to be bridged to make the RS more open-science-friendly. Based on that, we proposed how an existing tool could become the Open Software Catalogue and discussed the problems it would try to solve or address.

---

[3] https://sc.geant.org.

References

1. Wolski, M., Rodwell, T.: Software Governance in a large european project - GÉANT case study. In: Franch, X., Männistö, T., Martínez-Fernández, S. (eds.) PROFES 2019. LNCS, vol. 11915, pp. 620–625. Springer, Cham (2019). https://doi.org/10.1007/978-3-030-35333-9_48

2. Łabedzki, M., Wolski, M.: GEANT Software Catalogue As A Code Portfolio. GEANT Connect, p. 58 (2018). https://www.geant.org/News and Events/CONNECT/Documents/CONNECT 29 FINAL (1).pdf

3. Scholarly Infrastructures for RS (report from the EOSC task force) European Commission. Directorate General for Research and Innovation., 2020, ISBN: 978-92-76-25568-0. https://dx.doi.org/10.2777/28598

4. Di Cosmo, R., Zacchiroli, S. Software heritage: Why and how to preserve software source code. In: iPRES 2017-14th International Conference on Digital Preservation, pp. 1–10

5. Open Science now: A systematic literature review for an integrated definition, Vicente-Saez, Ruben and Martinez-Fuentes, Clara. J. Bus. Res. **88**, 428–436 (2018). https://doi.org/10.1016/j.jbusres.2017.12.043

6. Garijo, D., Ménager, H., Hwang, L., Trisovic, A., Hucka, M., Morrell, T., Allen, A.: Task Force on Best Practices for Software Registries, SciCodes Consortium. 2022. Nine best practices for RS registries and repositories. PeerJ Computer Science 8:e1023. https://doi.org/10.7717/peerj-cs.1023

7. Wilkinson, M., Dumontier, M., Aalbersberg, I. et al.: The FAIR guiding principles for scientific data management and stewardship. Sci. Data **3**, 160018 (2016). https://doi.org/10.1038/sdata.2016.18

8. Lamprecht, A.-L., et al.: Towards FAIR principles for RS. Data Sci. **3** 1–23 (2019). https://doi.org/10.3233/DS-190026

9. Groth, P., Dumontier, M.: Introduction - FAIR data, systems and analysis. Data Sci. **3**, 1–2 (2020). https://doi.org/10.3233/DS-200029

10. Hettrick, S., et al., UK RS Survey 2014, Zenodo (2014). https://doi.org/10.5281/zenodo.14809.

11. Katz, D.S., et al.: Software vs. data in the context of citation, PeerJ Preprints, e2630v1 (2016). https://doi.org/10.7287/peerj.preprints.2630v1

12. Atkinson, M.G., Montagnat, S., Johan Taylor, I.: Scientific workflows: past, present and future, Future Gener. Comput. Syst. **75**, 216–227 (2017). https://doi.org/10.1016/j.future.2017.05.041

13. Greuel, G.-M., Sperber, W.: swMATH – an information service for mathematical software. In: Hong, H., Yap, C. (eds.) ICMS 2014. LNCS, vol. 8592, pp. 691–701. Springer, Heidelberg (2014). https://doi.org/10.1007/978-3-662-44199-2_103

14. Gentleman, R.C., Carey, V.J., Bates, D.M., et al.: Bioconductor: open software development for computational biology and bioinformatics. Genome Biol **5**, R80 (2004). https://doi.org/10.1186/gb-2004-5-10-r80

Semantic Analysis of API Blueprint and OpenAPI Specification

Robert Pergl and Nikolas Jíša[✉]

Faculty of Information Technology, Czech Technical University in Prague,
Prague, Czech Republic
jisaniko@fit.cvut.cz

Abstract. This paper presents a comparative semantic analysis of API Blueprint and OpenAPI Specification, two widely used API description languages. It explores their structural and semantic differences, providing insights into their unique features and capabilities. A significant focus is placed on developing a robust method to convert API blueprint documents into OpenAPI format. This conversion blueprint is designed to maintain the semantic integrity of the original documents, and its referential implementation is provided as well. The study offers contributions to the field of REST API development, particularly for developers and organizations seeking to transition between these specifications without losing the essence of their REST API documentation.

Keywords: REST API · API Blueprint · OpenAPI Specification · Semantic Description

1 Introduction

In recent decades, the world has witnessed remarkable technological advancements, with the widespread integration of the Internet into various aspects of our daily lives, including social networks, online shopping and banking [8]. Central to these functionalities are the Application Programming Interface (API)s, the underlying mechanisms that facilitate numerous Internet activities. This emphasizes the relevance of the questions about how to optimize API development. One particular aspect of API development is semantic description, which can help developers rapidly find suitable APIs [10].

This paper delves into the examination of semantic description formats for Representational State Transfer (REST) APIs, specifically API Blueprint (APIB) and OpenAPI Specification (OAS), with the aim of conducting a comprehensive semantic analysis and establishing methodologies for mapping one format to the other. APIB and OAS are standards for REST API description and so far, their relationship has not been described in detail nor the conversion from one to the other.

The rest of this paper is structured as follows. Section 2 describes the research methodology, the research goal with research questions, the technological background and related work. The core part of this paper is Sect. 3 which describes

© The Author(s), under exclusive license to Springer Nature Switzerland AG 2024
Á. Rocha et al. (Eds.): WorldCIST 2024, LNNS 989, pp. 172–181, 2024.
https://doi.org/10.1007/978-3-031-60227-6_15

semantic analysis, as well as the mapping mechanism from APIB to OAS. Research questions are addressed in Sect. 4 along with some ideas on follow-up research. Finally, Sect. 5 offers concluding remarks.

2 Research Overview

The research in this paper was conducted according to Design Science Research Methodology (DSRM) [7] which consists of three closely related cycles:

- Relevance Cycle, which serves as an initiation of research with an application context providing research requirements and also defines criteria for evaluation of the results.
- Rigor Cycle, which ensures innovation of the research by putting together existing knowledge to be used as a basis for the research.
- Design Cycle, which iterates between the construction of an artifact, its evaluation, and subsequent feedback. This is the central cycle, as it depends on the other two cycles.

The relevance cycle for this paper is described in the next part of this section with research questions, which are answered in Sect. 4. The rigor cycle is described in Sect. 2.1 and Sect. 2.2. The design cycle is described in Sect. 3.

The goal of this research is to put together a conversion mechanism from APIB to OAS (or vice versa), which implies research questions as follows:

RQ1: What are the differences in semantic-description capabilities between OAS and APIB?
RQ2: What would be the mechanism for mapping APIB to OAS or OAS to APIB respectively?

2.1 Technological Background

Application Programming Interface (API). Application Programming Interface (API) abstracts problems and dictates client interaction with software solutions, typically via a library for use in various applications. It offers modular functionality for end-user programs, ranging from single functions to extensive, possibly open-source, components [9].

API Blueprint (APIB). [1] defines API Blueprint (APIB) as follows:

"API Blueprint (APIB) is a Web documentation-oriented API description language. The API Blueprint is essentially a set of semantic assumptions laid on top of the Markdown syntax used to describe a Web API."

APIB documents are structured into sections that are described in detail in [1]. The basic structure of APIB is that APIB contains Resource Sections which contain Action Sections with parameterized Uniform Resource Locator (URL) Templates. Action Sections contain Request Sections, which define supported requests for given action, and Response Sections defining supported responses for given action.

OpenAPI Specification (OAS). [5] defines OpenAPI Specification (OAS) as follows:

"The OpenAPI Specification (OAS) defines a standard programming language-agnostic interface description for Hypertext Transfer Protocol (HTTP) APIs, which allows both humans and computers to discover and understand the capabilities of a service without requiring access to source code, additional documentation, or inspection of network traffic. When properly defined via OpenAPI, a consumer can understand and interact with the remote service with a minimal amount of implementation logic. Similar to what interface descriptions have done for lower-level programming, the OpenAPI specification removes guesswork in calling a service."

OAS documents are in `JSON` or `YAML` format and consist of multiple types of objects specified in detail in [5]. Essentially, in OAS document, there is `Paths Object` containing parameterized URLs with `Path Item Objects`. `Path Item Object` can contain `Operation Object` for every HTTP method (for example, `get` or `post`). One of the implementations of the OAS standard is `Swagger` [6].

2.2   Related Work

In [4] there is implemented functionality to convert APIB to OAS. Our approach is, however, different in that, except for the functionality itself, it puts an emphasis on formal transformation beyond the implementation and conceptual comparison. For that matter, we performed multiple searches in https:// scholar.google.com/, https://www.scopus.com, https://www.webofscience.com/ like: "API Blueprint to OpenAPI", "Apiary to OpenAPI", "API Description Translation", etc. but we were unable to find relevant existing sources. In addition to that, we also conducted a few https://www.google.com searches and found multiple blogs and discussions comparing APIB and OAS such as [3] suggesting differences in semantic capabilities of APIB and OAS.

3   Implementation

3.1   Semantic Analysis

For the most part, APIB can be considered, in its semantic description capabilities, a subset of OAS, for that most of APIB description constructs can be expressed in OAS as well. We have identified one exception, which is that in APIB Header Sections, which are within the same `Action Section` and which have the same `HttpStatusCode`, have shared Dictionary of Headers in their OAS equivalent, which means that if `Header Sections` contain different values for the same key, then this situation cannot be represented in OAS. For the other mapping direction, OAS supports multiple constructs which are not supported in APIB, some of these constructs are: `Security Requirement Object`, `Security Scheme Object`, `Terms of Service`, `Contact Object`, `License Object`, `Server Object`, `Server Variable Object`, `External Documentation Object`,

Encoding Object, Callback Object, Reference Object, XML Object, OAuth Flows Object. Therefore, mapping from OAS to APIB is lossy. Because of this and due to the limited scope of this paper, we do not present this mapping here.

3.2   Mapping of APIB to OAS

Given that APIB documents have Markdown format and OAS documents have JSON or YAML format, which are all standard formats with multiple existing libraries for parsing, we have decided to not involve parsing as part of this paper and limited the scope to transformation from internal semantic model of APIB to internal semantic model of OAS.

The whole mapping procedure involves a lot of logic, and therefore we decided to mention only the most relevant parts in this section with emphasis put on graphical representation, which, although not as exact as complete formal specification, should serve illustration purposes much better. For the same reason, we excluded several implementation details, such as detailed information about used types. The complete formal specification is implemented in C# in a declarative way and is available in https://github.com/CCMi-FIT/MappingOfApibToOas.

The mapping from APIB to OAS has a APIB document as input and OAS OpenAPI Object, which is the root object of OAS document, as the output. At the highest level of abstraction, the mechanism is depicted in Fig. 1 showing that, at the top level, the mapping can be decomposed into separate parts:

1. *Getting Named Types.*
2. *Mapping of APIB Metadata Section to OAS Server Objects.*
3. *Mapping of APIB Api Name and Overview Section to OAS Info Object.*
4. *Mapping of APIB Resource Group Sections to OAS Tag Objects.*
5. *Mapping of APIB resources to OAS Paths Object.*
6. *Mapping to OAS Components Object.*

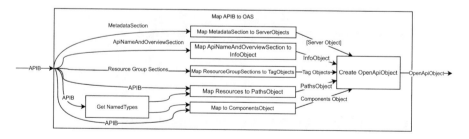

Fig. 1. Mapping of APIB to OAS

Getting Named Types. Step establishes a collection of Named Types, which are internal elements representing an APIB types put together from Attribute Sections of Resource Sections and Data Structure Sections. To elaborate, when an Attribute Section is in Resource Section or Data Structures Section, the type represented by this Attribute Section can be further referenced from other places in APIB by name. The mechanism of this step is depicted in fig. 2.

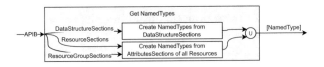

Fig. 2. Getting Named Types

Mapping of APIB Metadata Section to OAS Server Objects. Step converts the "HOST" value into an OAS Server Object. if it exists. In this case, the result is a one-item collection of OAS Server Objects.

Mapping of APIB Api Name and Overview Section to OAS Info Object. Step part takes 'Name' and 'Description' of APIB Api Name and Overview Section and stores them into 'Title' and 'Description' of OAS Info Object.

Mapping of APIB Resource Group Sections to OAS Tag Objects. Step takes in APIB Resource Group Sections and utilizes their identifiers to create OAS Tag Objects as depicted in fig. 3.

Fig. 3. Mapping of APIB Resource Group Sections to OAS Tag Objects.

Mapping of APIB resources to OAS Paths Object. Step takes as input whole APIB and collection of Named Types utilized to map APIB resources to URL paths with OAS PathItemObjects, according to *Mapping of APIB resources to OAS Paths Object*, which are then used to initialize the result OAS PathsObject.

Mapping of APIB resources to URL paths with OAS Path Item Objects. Step takes in the whole APIB and the collection of Named Types and works as follows:

1. APIB is utilized to get Resource Infos. Resource Info is a structure containing APIB Resource Section and it's parent APIB Resource Group Section (if any).
2. Then, APIB Action Sections of APIB resources in Resource Infos are mapped to the collection of Operations Infos according to section *Mapping of APIB Action Section to Operation Info*. Operation Info is a structure containing OAS Operation Object, URL path and HTTP Request Method.
3. Next, Operation Infos are grouped by the URL paths, which are then mapped to URL paths with OAS Path Item Objects according to *Mapping of APIB resources to URL paths with OAS Path Item Objects*.

This entire step is depicted in fig. 4.

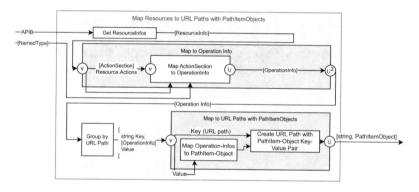

Fig. 4. Mapping of APIB resources to URL paths with OAS Path Item Objects

Mapping of APIB Action Section to Operation Info. Step takes in APIB Action Section, Resource Info and Named Types and its mechanism is as follows:

1. First, the URL path for that action is resolved. It is Uniform Resource Identifier (URI) Template Path for that APIB Action Section or, if not defined, parent APIB Resource Section URI Template Path. Furthermore, the query parameters are removed from this URI template.
2. Then, APIB Action Section is mapped into OAS Operation Object according to *Mapping of APIB Action Section to OAS Operation Object*
3. Lastly, Operation Info item is created from the URL path of the action, HTTP Request Method and from the OAS Operation Object.

The whole mapping step is shown in fig. 5.

Fig. 5. Mapping of APIB `Action Section` to `Operation Info`

Mapping of APIB Action Section to OAS Operation Object. Step takes in APIB `Action Section`, `Resource Info` and `Named Types`. The mechanism is as follows:

1. First, `Request Sections` of given APIB `Action Section` are mapped into pair of: OAS `Request Body Object` and collection of OAS `Parameter Objects`.
2. Next, the URL template of the given APIB `Action Section` is mapped to collection of OAS `Parameter Objects`.
3. Subsequently, the identifier of the parent APIB `Resource Group Section` is used to create OAS `Tag Object` (if any).
4. Then, APIB `Response Sections` of APIB `Action Section` are grouped by their HTTP Status codes and later mapped to OAS `Responses Object`.
5. Lastly, OAS `Operation Object` is created from APIB `Action Section`, OAS `Request Body Object`, `Parameter Objects`, `Tag Objects` and `Responses Object`.

The entire mapping step is depicted in fig. 6.

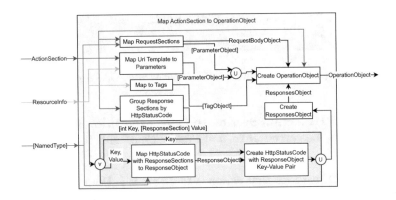

Fig. 6. Mapping of APIB `Action Section` to OAS `Operation Object`

Mapping of Operation Infos to OAS Path Item Object. Step takes in the collection of `Operation Infos`, which are expected to be meant for the same URL path. It creates OAS `Operation Object` for every `Operation Info` and sets it to the result OAS `Path Item Object` to the property corresponding to HTTP Request method (Get, Post, Put, etc.) as depicted in fig. 7.

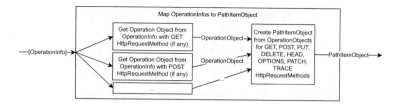

Fig. 7. Mapping of `Operation Infos` to OAS `Path Item Object`

Mapping to OAS Components Object. Step takes in whole APIB and collection of `Named Types` and produces `Schema Objects`, which are then utilized to create OAS `Components Object` as depicted in fig. 8.

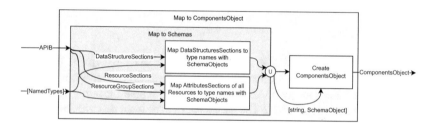

Fig. 8. Mapping to OAS `Components Object`

4   Evaluation and Discussion

We implemented the mechanism with C# following the functional programming paradigm to organize the code, so that it would match the mapping philosophy. The code is available in https://github.com/CCMi-FIT/MappingOfApibToOas. To validate the implementation, several automated tests were implemented as follows:

1. First, we downloaded examples of APIBs from [2].
2. Next, we put together the expected mapping result using [4] and manually verified the validity.
3. Then, we created an automated test for each of these examples:
   (a) Building manually domain objects according to the APIB,
   (b) Running our implemented mapping mechanism,
   (c) Comparing the result OAS document with the expected OAS result document.

The implementation utilizes 32-bit `integer`s in a few places associated with the number of items in collections, but other than that there are no arithmetic operations. This is a simplification, however, since it is unrealistic to expect any collection in the implementation to contain more than $2^{31} - 1$ items (that is, the maximum value of (signed) `integer` in .NET), there should not occur any overflow problem for real API descriptions.

The research question RQ1 is answered in Sect. 3.1. The mechanism of mapping from APIB to OAS is explained in Sect. 3.2, which answers the other research question RQ2.

There are multiple ways to conduct further research. One possible direction is to develop mapping the other way from OAS to APIB. Another direction is to establish a unified ontological model and to develop a mapping mechanism from APIB to this unified model, as well as a mapping mechanism from OAS to this unified model, which could later also involve other REST API description standards such as RESTful API Modeling Language (RAML) or Web Application Description Language (WADL). Another nice feature would be having a parser of APIB document to our internal representation, which would make this tool easily usable for any APIB document.

5   Conclusion

In this study, we conducted a semantic analysis of API Blueprint (APIB) and OpenAPI Specification (OAS), leading to the finding that, predominantly, APIB can be effectively mapped to OAS. However, the reverse mapping from OAS to APIB presents challenges due to the presence of semantic description constructs in OAS that do not have any corresponding equivalent in APIB. Furthermore, to facilitate practical application, a formal transformation from APIB to OAS was formulated, implemented and verified by automated tests.

Acknowledgement. This research was supported by the grant of Czech Technical University in Prague No. SGS23/206/OHK3/3T/18.

References

1. Api blueprint specification. https://apiblueprint.org/documentation/specification.html. Accessed 29 Oct 2023
2. Apiary api blueprint examples. https://github.com/apiaryio/api-blueprint/tree/master/examples. Accessed 29 Oct 2023
3. Apiary vs. swagger - what's the difference?. https://hellboundbloggers.com/apiary-vs-swagger-whats-the-difference/66113/. Accessed 04 Nov 2023
4. apib2swagger. https://github.com/kminami/apib2swagger. Accessed 29 Oct 2023
5. Openapi specification. https://spec.openapis.org/oas/latest.html. Accessed 29 Oct 2023
6. What is swagger? https://swagger.io/docs/specification/2-0/what-is-swagger/. Accessed 04 Nov 2023
7. Hevner, A.: A three cycle view of design science research. Scandinavian J. Inform. Syst. **19**, 4 (2007)
8. Rainie, L., Wellman, B.: 27The Internet in Daily Life: The Turn to Networked Individualism. In: Society and the Internet: How Networks of Information and Communication are Changing Our Lives, pp. 29–31. Oxford University Press (2019). https://doi.org/10.1093/oso/9780198843498.003.0002

9. Reddy, M.: API Design for C++. Elsevier Science (2011). https://books.google.cz/books?id=IY29LylT85wC

10. Wang, X., Sun, Q., Liang, J.: Json-ld based web api semantic annotation considering distributed knowledge.IEEE Access **8**, 197,203–197,221 (2020).https://doi.org/10.1109/ACCESS.2020.3034937

Classification Model for the Detection of Anxiety in University Students: A Case Study at UNMSM

Bryan Vera-Leon⬥, Laura Gozme-Avila⬥, and Yudi Guzmán-Monteza$^{(\boxtimes)}$⬥

Universidad Nacional Mayor de San Marcos, Avenue Germán Amézaga, Lima, Perú
{bryan.vera,laura.gozme,yudi.guzman}@unmsm.edu.pe

Abstract. Anxiety, a common mental disorder worldwide, affects millions of people, including university students. In Peru, in 2021, a high prevalence of anxiety of 69% was recorded among students at three institutions of the Consortium of Universities. Early detection and treatment of anxiety in academic settings are critical to ensure student well-being. This study proposes a classification model for which Machine Learning algorithms such as Random Forest (RF), Naive Bayes, and Support Vector Machine (SVM) were evaluated using the Universidad Nacional Mayor de San Marcos as a case study. Advanced techniques were implemented, including Adversarial Generative Networks data augmentation, and hyperparameter tuning was performed using GridSearch and Random Search. The results highlighted the SVM model, which achieved 85% accuracy and an F1-Score of 0.78, with a prediction time of 0.00100159 s. Notably, the data augmentation technique revealed a steady improvement in all models.

Keywords: Supervised Learning · Anxiety Detection · Classification Model · Data Augmentation Techniques

1 Introduction

According to the World Health Organization (WHO), in 2019, 1 in 8 people suffered from a mental disorder in the world, with anxiety being one of the most common problems, with 301 million people suffering from it [1]. In Peru, a mental health study conducted by the Consortium of Universities revealed that 69% of students had mild, moderate, severe, and highly severe anxiety [2]. In 2022, the negative impact of the COVID-19 quarantine on obstetrics students at the Universidad Nacional Mayor de San Marcos (UNMSM) was investigated [3]. Anxiety is a natural reaction to uncertain events and serves an adaptive function in the individual [4]. However, it can be considered a disorder when it interferes with the individual's daily activities, given its high intensity and duration. [5].

A university student is permanently in a competitive environment with persistent academic demands [6]. This can generate high levels of anxiety and trigger dropout and mental health complications, among other problems [7]. Therefore, at UNMSM, early detection and effective treatment of anxiety have become critical. In response to this

Á. Rocha et al. (Eds.): WorldCIST 2024, LNNS 989, pp. 182–190, 2024.
https://doi.org/10.1007/978-3-031-60227-6_16

issue, Machine Learning can be used to provide data-driven diagnoses and treatment suggestions. [8].

A classification model for anxiety detection at UNMSM based on Machine Learning (ML) has been proposed. A combination of techniques such as Relief for feature selection [9], SMOTE to counteract data imbalance [10], cross-validation as a standard method for model selection [11], and hyperparameter fitting to determine the optimal model configuration [12] were employed. The results obtained using these methodologies have demonstrated outstanding accuracy, thus consolidating a solid foundation for continuous improvement in prediction.

This research is structured in six main sections. The second section deals with a review of the literature on classification models. The third section details the methodological approach used. The fourth section presents the results, offering a detailed interpretation of the performance metrics obtained. The fifth section compares the proposed model with other relevant research. Finally, the sixth section provides key conclusions, limitations, and recommendations for future research in the field.

2   Background and Related Work

Many researchers have proposed using different machine learning models to address mental health issues, such as Random Forest (RF), Super Vector Machine (SVM), and Gradient Boosting, among others. In this research, we seek to look at approaches related to anxiety with the university population and with a nature of their data similar to that of those possessed.

Priya et al. (2020) aimed to predict anxiety, depression, and stress in modern life using machine learning algorithms. Data were collected from a total of 348 participants from India using the DASS-21 questionnaire, and Decision Tree (DT), Random Forest (RF), Naive Bayes, Support Vector Machine (SVM), and K-Nearest Neighbors (KNN) algorithms were applied. Where the Naive Bayes algorithm was found to be the best model; however, because the nature of the data produced unbalanced partition classes, the selection of the best model was based on the F1 metric with a value of 59.2% [13].

On the other hand, Bhatnagar et al. (2023), in their research, aimed to identify the extent and effects of anxiety in Indian college students. The data set was collected through an original questionnaire applied to 127 engineering students, in which SVM, Gradient Boosting Machine (GBM), Naive Bayes, RF, and KNN algorithms were used. Being the RF algorithm that yielded the best accuracy value, the author highlights the importance of considering in the future the categorizing of the disease according to its severity [14].

In turn, Verma, G. and Verma, H. (2020), in their research, aimed to identify the factors affecting the mental condition of students entering from traditional school education to technical education in college or university. The data were obtained through a proprietary questionnaire administered to 2029 students from colleges and universities in India. They applied various algorithms, such as logistic regression, SVM, KNN, Decision Tree, and Random Forest, using the Ensemble technique that combines them. The results showed an accuracy of 90.1%. This performance was more clearly visualized using the AUC-ROC curve metric, where the ensemble model showed a superior performance, approaching 1. This suggests that the combined model offers a more complete and visually improved explanation of the results obtained as opposed to the use of the algorithms individually. In proposing this technique, the author emphasizes the importance of considering the differences between stressors and how they affect the student population [15].

Constructing a predictive model involves selecting a machine learning algorithm and applying various techniques, from data processing to improving the final model. Among the feature selection techniques, the principal component analysis (PCA), which reduces the number of features by creating new ones that group the original ones [19], and the Relief algorithm, based on instance learning, which assigns weights to the features to select the most relevant ones, stand out [12].

When data sets are small, data augmentation techniques are used, such as Generative Adversarial Network (GAN), a model composed of a generator and a discriminator that compete to improve generation and discrimination by generating synthetic samples based on the existing data [20]. In the case of unbalanced datasets, the SMOTE technique is mainly employed, which randomly selects a point from the minority class, computes the K Nearest Neighbors, and generates synthetic points between the selected point and its Neighbors [13] to match the minority class with the majority class.

In addition, the technique of cross-validation, the most common method for model selection, is used. This process divides The data into "k" parts known as folds. One of these parts is reserved as a validation con-set, while the model is trained using the remaining "k-1" parts [14]. Finally, hyperparameter fitting is implemented to define the best configuration, significantly influencing the model performance [15].

The application of these techniques and algorithms reflects the approaches in the present study to ensure a precise and effective detection of anxiety in college students.

3   Methodology

The work has four main stages, as shown in Fig. 1

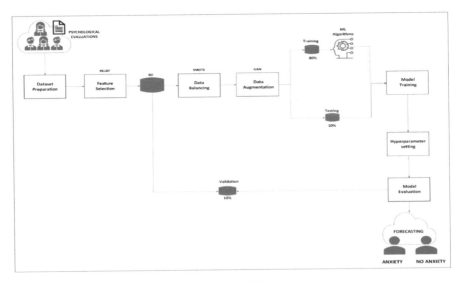

Fig. 1. Methodology

3.1   Data Preparation

The data come from a mental health screening test applied to 1100 San Marcos students from the first to tenth cycle from different faculties such as the Faculty of Industrial Engineering, Education, Accounting Sciences, Systems and Computer Engineering, among others, carried out in 2022, using the SRQ18 Mental Health Questionnaire, which is a binary option test and consists of 28 questions designed to detect the presence of mental illnesses such as depression, anxiety, among others. These questions were identified from SRQ1 to SRQ28, with answers coded as 1 or 0, and only the results for anxiety were taken, whose presence was assigned as "anxiety case" and "0" if it was not present.

The data were exported to a CSV file, and a binary categorization of the anxiety column was performed. Ninety percent of the data was used for preprocessing, training, and model testing, with the remaining 10% reserved for validation.

3.2   Preprocessing

Due to the nature of the data, three preprocessing techniques were applied, as detailed below.

Feature Selection.   The TURF method was selected as the only algorithm due to its iterative approach based on the RELIEF algorithm, which works with all types of data, thus improving the ability to accurately and efficiently identify the most relevant features that influence the detection of anxiety, obtaining seven features from the following Table 1.

Table 1. Dataset resulting from Feature Selection using RELIEF.

ID	Variable	Type	Options
SRQ1	Do you have frequent headaches?	BINARY	1–0
SRQ3	Do you have difficulty sleeping?	BINARY	1–0
SRQ6	Do you feel nervous, tense?	BINARY	1–0
SRQ8	Are you unable to think clearly?	BINARY	1–0
SRQ9	Do you feel sad?	BINARY	1–0
SRQ12	Do you have difficulty making decisions?	BINARY	1–0
SRQ18	Do you feel tired all the time?	BINARY	1–0
ANXIETY	ANXIETY CASE	BINARY	1–0

Data Balancing.   An imbalance was observed in the data set with 712 anxiety and 278 non-anxiety samples. The SMOTE technique was applied to address this, generating synthetic data and reaching 1424 samples.

Data Augmentation. To address the data limitation and improve the model's robustness, we implemented data augmentation in the balanced set by using Generative Adversarial Networks (GAN) with 1000 epochs and a batch of 128 as training parameters.

This technique generated 1000 new samples, enriching the diversity of our data set. This choice seeks not only to compensate for the data limitation but also to explore the effectiveness of data augmentation techniques in the context of our study, evaluating how they impact model performance.

3.3 Training

Random Forest, SVM, and Naive Bayes models were implemented due to the nature of our data and their effectiveness in dealing with similar problems according to the state of the art. These models worked with two data sets: one with balancing and the other with balancing and data augmentation. First, an instance was created for each model without hyperparameters, using a cross-validation of 10 folds to compare and select the best model to perform the hyperparameter adjustment subsequently. We also sought to evaluate whether the data augmentation technique improved model performance.

3.4 Hyperparameter Setting

First, a dictionary of hyperparameters was created, as shown in Table 2.

Table 2. Hyperparameters Dictionary.

Model	Hyperparameters	Values
Random Forest	n_estimators	[10, 100, 1000, 10000]
	max_features	['sqrt', 'log2']
Support Vector Machine	kernel	['poly', 'rbf', 'sigmoid']
	c	[50, 10, 1.0, 0.1, 0.01]
	gamma	['scale']

Subsequently, two hyperparameter search methods were used: GridSearchCV and RandomSearchCV, which will have the F1-Score metric defined as scoring and use a cross-validation of 10 folds and three repetitions. Once the models with the optimal hyperparameters were identified, we trained them using the complete training data set. Then, the validation data set was used, and the performance of these models was evaluated using metrics of the correlation matrix and prediction time.

4   Analysis and Interpretation of Results

The results of the training stage are shown in Table 3.

Better performance values were observed for the SVM and Random Forest models, with a slight difference but leaving behind the Naive Bayes model. Therefore, they were selected as the best models, and we proceeded with the hyperparameter adjustment and validation of the selected models, the results of which are detailed in Table 4.

Table 3. Comparative table of classification models in the training stage.

MODELS	Precisión		Recall		Accuracy		F1-Score	
	Smote	Augmentation	Smote	Augmentation	Smote	Augmentation	Smote	Augmentation
Random Forest	0.889	0.9103	0.924	0.934	0.904	0.921	0.906	0.922
Naive Bayes	0.873	0.932	0.924	0.898	0.895	0.917	0.898	0.915
SVM	0.911	0.916	0.9101	0.935	0.911	0.925	0.9108	0.925

Table 4. Comparative table of the best classification models with validation data.

MODELS	Precisión		Recall		Accuracy		F1-Score	
	Smote	Augmentation	Smote	Augmentation	Smote	Augmentation	Smote	Augmentation
Random Forest	0.659	0.652	0.879	0.934	0.827	0.827	0.753	0.759
SVM	0.667	0.674	0.909	0.935	0.836	0.845	0.769	0.785

When validating the models, a tendency towards false positive predictions is observed, indicated by the low precision values (SVM: SMOTE 0.6667 vs. Augmentation 0.6739; Random Forest: SMOTE 0.6591 vs. Augmentation 0.6522). However, an increase in F1-Score (SVM: SMOTE 0.7692 vs. Augmentation 0.7848; Random Forest: SMOTE 0.7532 vs. Augmentation 0.7595) stands out, indicating the models' ability to reach a balance in the accurate identification of positive cases. As for accuracy, a better performance is observed (SVM: SMOTE 0.8364 vs. Augmentation 0.8455; Random Forest: SMOTE 0.8273 vs. Augmentation 0.8273), indicating that both models predict correctly in more than 80% of the cases.

The application of Data Augmentation shows minimal but consistent improvements in model performance. Although SVM might be slightly superior in performance, the difference compared to Random Forest is minimal. Therefore, the prediction time was measured as shown in Table 5.

Table 5. Prediction time of the best classification models using the validation data.

MODELS	Prediction Time	
	Smote	Augmentation
Random Forest	0.061 s	0.606 s
SVM	0.004 s	0.001 s

It is observed that the SVM model has a considerably shorter prediction time than the Random Forest model, which can be essential in situations where fast and real-time anxiety detection is required.

5 Discussion

Table 4 shows a decrease in the metrics during model validation; this may be due to the limited amount of initial data (1100); although data augmentation was applied, it may have had a non-uniform impact on different instances of the dataset, affecting consistency in model generalization. On the other hand, despite attempts at mitigation with cross-validation and hyperparameter adjustment, possible overfitting could have occurred during training, resulting in less robust performance when confronted with new data during validation. These reasons could explain the observed reduction in model metrics in the validation phase.

For comparison, Table 6 shows the results obtained in other investigations.

Table 6. Comparative table of our results with other studies.

	This research	Priya et al. (2020)	Bhatnagar et al. (2023)	Haque et al. (2023)
Best algorithm	SVM	Random Forest	Random Forest	GaussianNB
Results	Accuracy: 0.845 Precision: 0.674 Recall: 0.935 F1 Score: 0.785	Accuracy:0,714 Precision: 0,431 Recall: 0,510 F1 Score: 0,470	Accuracy: 0,789	Accuracy: 0,79 Precision: 0.62 Recall: 0.46 F1 Score: 0.53

It can be seen that the SVM model implemented in this study significantly outperforms the results reported by others. This can be attributed, in part, to techniques such as data balancing, data augmentation, and hyperparameter fitting, which were not used in previous studies.

On the other hand, Table 5 shows an increase in SVM of 1.5% in F1-Score and 0.9% when using the GAN technique for data augmentation, compared to the study by Moreno et al. (2022), where a 4.08% improvement in accuracy and a 29% increase in F1-Score was achieved. Implementing the Smote technique to address the initial data imbalance may have attenuated the variation in results using data augmentation techniques, as observed by Moreno et al. (2022).

A relevant finding is the time efficiency of the SVM model, which achieves accurate predictions in a much shorter time than Random Forest, highlighting the importance of considering time efficiency when selecting classification models, especially in environments where speed and pre-accuracy are crucial.

These results provide a basis for future research in detecting and preventing anxiety in college students, emphasizing the importance of a balanced approach considering predictive accuracy and temporal efficiency.

6   Ethical Considerations

It is essential to mention that the data set with which the present research has been carried out has maintained the confidentiality of the patient's data since there was no access to these data during the data collection stage. For data collection, authorization was obtained from the Dean and the Head of the Psychology area of the Student Counselling and Orientation Unit (UNAYOE – Acronym in Spanish) of the Faculty of Systems and Computer Engineering through the Work Authorization Module (MAT – Acronym in Spanish) o guarantee the quality and reliability of the data. However, validations through expert judgment would be necessary to ensure the degree of generality obtained with the proposed model before its deployment as a decision support tool for mental health staff at the university.

Finally, the proposed model will be helpful for common cases, such as determining whether characteristics can lead us to conclude that the patient has or does not have anxiety, but not to help diagnose moderate or severe cases.

7   Conclusions and Future Works

The results highlight the SVM as the most efficient algorithm for the predictive model, reaching an accuracy of 0.845 and an F1-Score of 0.785; also noteworthy is its efficiency in prediction time, registering only 0.001 s to evaluate 110 cases. It is relevant to mention that, despite its outstanding performance, the model's accuracy was 0.673913, indicating the possibility of 33% false positive predictions. The effectiveness of data augmentation strategies in improving the performance of classification models on small data sets is also highlighted. Taken together, these results are promising for the detection of anxiety in college students.

For further research, it is recommended that false positive predictions be explored and addressed, in addition to including assessment data for multiple years, to examine the possibility of further improving model performance and accuracy in future research. Also, the study of class Anxiety can be extended by defining sub-levels such as Mild, Moderate, and Severe to provide a valuable tool to identify early and address anxiety-related challenges in college students.

References

1. Organización Mundial de la Salud. (8 de junio de 2022). Trastornos mentales. Recuperado de. https://www.who.int/es/news-room/fact-sheets/detail/mental-disorders
2. Pérez, G.: Salud mental de estudiantes en pandemia: investigación del consorcio de universidades permitirá tomar acciones de apoyo. Puntoedu (2021). https://puntoedu.pucp.edu.pe/comunidad-pucp/salud-mental-de-estudiantes-en-pandemia-investigacion-del-consorcio-de-universidades-permitira-tomar-acciones-de-apoyo/
3. Suaquita, M.: Estrés, ansiedad y depresión en estudiantes de Obstetricia de la Universidad Nacional Mayor de San Marcos durante la pandemia por COVID-19, año 2021 [Tesis de titulación, Universidad Nacional Mayor de San Marcos]. Repositorio CYBERTESIS (2022). https://hdl.handle.net/20.500.12672/17795

190   B. Vera-Leon et al.

4. Díaz, I., y De la Iglesia, G. (2019). Ansiedad: revisión y delimitación conceptual. Summa Psicológica UST, 16(1), 42–50. https://dialnet.unirioja.es/servlet/articulo?codigo=7009167
5. Riveros, Q.M., Hernández, V.H., Rivera B.J.: Niveles de depresión y ansiedad en estudiantes universitarios de Lima Metropolitana. Revista De Investigación En Psicología **10**(1), 91–102 (2007). https://doi.org/10.15381/rinvp.v10i1.3909
6. Castillo, C., Chacón, T. y Díaz-Véliz, G.: Ansiedad y fuentes de estrés académico en estudiantes de carreras de la salud. Investigación en educación médica **5**(20), 230–237 (2016). https://www.sciencedirect.com/science/article/pii/S2007505716000491
7. Martínez-Otero, V.: Ansiedad en estudiantes universitarios: estudio de una muestra de alumnos de la Facultad de Educación. Ensayos: Revista de la Facultad de Educación de Albacete **29**(2), 63–78 (2014). https://dialnet.unirioja.es/servlet/articulo?codigo=4911675
8. Vélez, J.I.: Machine Learning based Psychology: Advocating for A Data-Driven Approach. Int. J. Psychol. Res. **14**(1) (2021). https://doi.org/10.21500/20112084.5365
9. Urbanowicz, R., Meeker, M., La Cava, W., Olson, R. y Moore, J.: Relief based feature selection: introduction and review. J. Biomed. Inf. **85**, 189–203 (2018). https://doi.org/10.1016/j.jbi.2018.07.014
10. Cárdenas, J.: Clasificación de aceptación de campañas para una entidad financiera, usando random forest con datos balanceados y datos no balanceados. [Tesis de Maestría, Universidad Ricardo Palma]. Repositorio Institucional - URP (2019). https://hdl.handle.net/20.500.14138/2307
11. Villanueva, R.: Sistema inteligente basado en redes neuronales para la identificación de cáncer de piel de tipo melanoma en imágenes de lesiones cutáneas [Tesis de titulación, Universidad Nacional Mayor de San Marcos]. Repositorio CYBERTESIS (2021). https://hdl.handle.net/20.500.12672/17574
12. Sánchez, E., Hernández, Y., y Ortiz, J.: Técnicas de Optimización de Hiperparámetros en Modelos de Aprendizaje Automático para Predicción de Enfermedades Cardiovasculares [Paper presentation]. 9ª Jornada de Ciencia y Tecnología Aplicada, Cuernavaca, Morelos, Mexico, 2022, 16 al 18 de noviembre
13. Priya, A., Garg, S. Prerna, N.: Predicting anxiety, depression and stress in modern life using machine learning algorithms. Procedia Comput. Sci. **167**, 1258–1267 (2020). https://doi.org/10.1016/j.procs.2020.03.442
14. Bhatnagar, S., Agarwal, J., Sharma, O.: Detection and classification of anxiety in university students through the application of machine learning. Procedia Comput. Sci. **218**, 1524–1550 (2023). https://doi.org/10.1016/j.procs.2023.01.132
15. Verma, G., Verma, H.: Model for predicting academic stress among students of technical education in India. Int. J. Psychosoc. Rehabil. **24**(04), 2702–2714 (2020). https://doi.org/10.37200/ijpr/v24i4/pr201378
16. Javed, F., Gilani, S.O., Latif, S., Waris, A., Jamil, M., Waqas, A.: Predicting risk of antenatal depression and anxiety using multi-layer perceptrons and support vector machines. J. Personalized Med. **11**(3), 199 (2021)
17. Moreno-Barea, F.J., Franco, L., Elizondo, D., Grootveld, M.: Application of data augmentation techniques towards metabolomics. Comput. Biol. Med. **148**, 105916 (2022)
18. Haque, U.M., Kabir, E., Khanam, R.: Early detection of paediatric and adolescent obsessive–compulsive, separation anxiety and attention deficit hyperactivity disorder using machine learning algorithms. Health Inf. Sci. Syst. **11**(1), 31 (2023)
19. Gárate-Escamila, A.K., El Hassani, A.H., Emmanuel, A.: Classification models for heart disease prediction using feature selection and PCA. Inf. Med. Unlocked **19**, 100330 (2020). https://doi.org/10.1016/j.imu.2020.100330
20. Xiao, Y., Wu, J., Lin, Z.: Cancer diagnosis using generative adversarial networks based on deep learning from imbalanced data. Comput. Biol. Med. **135**, 104540 (2021). https://doi.org/10.1016/j.compbiomed.2021.104540

Synergizing First-Principles and Machine Learning: Predicting Steel Flatness in the Era of Digital Twins and Physics-Informed Intelligence

Nils Hallmanns[1]([✉]), Alexander Dunayvitser[1], Hagen Krambeer[1], Andreas Wolff[1], Roger Lathe[1], Colin Goffin[1], Monika Feldges[1], Pavel Adamyanets[2], and Christoph Evers[3]

[1] VDEh-Betriebsforschungsinstitut GmbH, Sohnstraße 69, 40237 Düsseldorf, Germany
Nils.Hallmanns@bfi.de
[2] Thyssenkrupp Steel AG, Duisburg, Germany
[3] Thyssenkrupp Hohenlimburg GmbH, Hagen, Germany

Abstract. Flatness in steel strips and sheets is a paramount quality metric in steel manufacturing, vital for ensuring product integrity and safety during production. Contemporary rolling processes adeptly maintain local flatness, but challenges arise particularly with thinner and high-strength steel grades. These complexities manifest as pronounced flatness deviations during subsequent processing and post-production phases. Merging traditional first-principle models with advanced machine learning algorithms, this new initiative achieves a comprehensive holistic grasp of the origins of flatness discrepancies. Taking into account the physical nature of the forming process. This synergy offers plausible prediction of cross-process flatness deviations while ensuring transparency and interpretability. Key advancements of this method are the development of predictive models, real-time flatness monitoring tools and a state-of-the-art digital twin structure connected to a network of software agents. Through sensitivity analysis, critical process parameters influencing flatness are identified. These outcomes not only elevate flatness control and product quality but also amplify plant efficiency. Ultimately, the research holds the potential for significant cost savings for steel producers, augmented product standards, and a dedicated focus on sustainable production with minimized environmental repercussions.

Keywords: Flatness in steel · Machine learning algorithms · Digital twin · Industry 4.0 · Big data · Predictive analytics · Smart manufacturing · Steel Industry

1 Introduction

The steel manufacturing sector plays a pivotal role in shaping the infrastructural backbone of modern societies. A core aspect of ensuring high-quality steel products lies in preserving the flatness of steel strips and sheets. Perfecting this flatness is crucial

© The Author(s), under exclusive license to Springer Nature Switzerland AG 2024
Á. Rocha et al. (Eds.): WorldCIST 2024, LNNS 989, pp. 191–197, 2024.
https://doi.org/10.1007/978-3-031-60227-6_17

for both product reliability and the safety of the manufacturing processes. However, as the industry ventures into producing thinner and more robust steel grades, maintaining consistent flatness becomes increasingly challenging.

In response to these challenges, the HATFLAT[1] research project, backed by the European Union, steps in. It focuses on proactive measures rather than just identifying flatness discrepancies. With the steel industry's shift towards Industry 4.0, leveraging accurate data and real-time analytics is essential. HATFLAT aims to pioneer predictive tools to anticipate flatness issues, assess risks, and guide operators towards making more informed production decisions. This approach blends the robustness of first-principle models with the adaptability of machine learning, offering advanced solutions for the evolving needs of steel flatness control.

2 Basics of Strip Flatness

In steel production, strip flatness is a key quality aspect, with its definition varying in different industrial contexts. For uniformity in this discussion, strip flatness encompasses variations from a perfectly flat shape, including defects like edge waves, center buckles, and camber, which can impact both the product's aesthetic and functional qualities [1].

Achieving desired flatness in hot-rolled steel strips involves managing critical factors like:

- **Roll Bending and Force**: Accurate application ensures balanced pressure distribution, essential for preventing flatness issues.
- **Roll Gap Control**: Consistent gaps between rolls are crucial for uniform strip thickness and flatness.
- **Thermal Effects on Rolls**: Changes in roll temperature during processing can affect strip flatness.
- **Material Properties and Stretching**: Inherent material differences and the effects of rolling pressures contribute to flatness variations.
- **Environmental and Operator Factors**: The rolling process is also influenced by external conditions and the decisions of the operators.

Given the complexity of these elements, precise control and ongoing adjustments are essential, particularly for products used in industries where precision is paramount. Maintaining these standards is vital for ensuring the quality of the final steel strips [2].

3 Digital Twin System

In order to gain a holistic understanding of possible flatness problems, a digital twin (DT) that allows tracking of products throughout the whole process is an integral part of the overall success of the project.

Digital Twins serve a dual role as digital representations, operating both as dedicated data repositories, a concept commonly termed as shadowing, and as interface endpoints for agents to monitor and engage with them upon detection of potential triggering events.

---

[1] **H**olistic **A**ssistance **T**ool for Cross-Process Analysis and Prediction of Strip and Plate **Flat**ness.

Consequently, Digital Twins can additionally function as instigators for model computations and enablers for the encapsulation of the tangible attributes of the object or physical asset being shadowed [3].

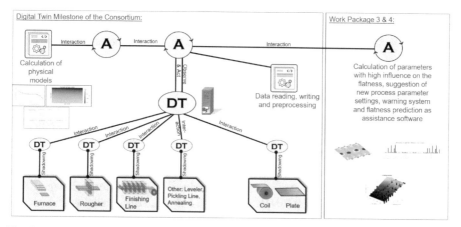

Fig. 1. Schematic representation of the digital twin system. The general concept depicted here shows the Digital Twin and its interaction with the software agents. For every product a cyber-physical representation is created and complemented along its process-chain. On the right side the agent responsible for the calculation of flatness prediction models is shown.

The schematic representation of the digital twin concept is delineated in Fig. 1. For each product traversing the process chain, a digital twin is instantiated. In its initial phase, this digital twin accumulates all pertinent data pertaining to the product and the corresponding physical asset it transits. Integration of this data amalgamates into a singular entity denoted as the Meta-Digital-Twin, centrally positioned within the diagram. This amalgamation bears resemblance to the commonly used simple product-oriented data management systems, which already provide a solid basis for efficient cross-process data analysis. Concurrently, as illustrated in Fig. 1, the primary stratum of digital twins exists as mere virtual reflections of their corresponding physical assets, while the meta digital twin maintains connectivity with a network of agents. These agents facilitate the continual updating of the cyber-physical representations and are endowed with both observation and action capabilities. Furthermore, each agent possesses a unique repertoire of algorithms tailored to its specific responsibilities.

4   Flatness Prediction Algorithm

The Flatness Prediction Algorithm serves as the nexus where data-driven insights and machine learning innovation coalesce to anticipate and rectify flatness discrepancies that can affect the end-use functionality of steel products.

The cornerstone of this algorithm is the sophisticated integration of diverse data sets that encapsulate the breadth of the production process. This integration is not a trivial task; it demands a concerted effort in pre-processing, where data is cleansed, transformed, and normalized to construct a coherent framework suitable for analysis [4]. The algorithm's robustness is further fortified by incorporating measurements from various points of the production line, including both real-time sensor data and historical records.

To elevate the predictive accuracy to the desired levels of precision, additional information provided by first principle models is included alongside measured data. This inclusion acts as a safeguard, addressing potential gaps in the measurements and ensuring robust predictions of steel flatness.

4.1 First Principle Models

In the pursuit of advancing flatness prediction capabilities for hot strip rolling, integrating established first principle models an essential approach. These models provide a foundational understanding of the rolling process, assimilating principles of physics and materials science to predict outcomes under various conditions. The integration strategy adopted herein diverges from the conventional Physics Informed Neural Network (PINN) approach. While PINNs typically incorporate differential equations into the learning process as a loss function [5], the here used methodology forges a different path. It utilizes the output from first principle physical models not as a constraint within the learning algorithm but rather as an additional input variable, enriching the data-driven predictive model. The specific first principle models employed in the initial instance are the Roll Gap model [6] and the thermal roll deformation model [7, 8]. Both are underpinned by robust numerical methods and mathematical modeling, and they operate under a set of assumed environmental conditions, taking process parameters as input variables.

4.2 Resulting Prediction Model

The flatness prediction model is shown as an schematic illustration in Fig. 2. In the context of the hot rolled strip use case, the BFI HR-Server [9] acts as a provider of both process data (u) from the rolling process and as an input for the first principle models. The encoder E2 within the autoencoder A2 plays a critical role in merging the data from the first principle models and the process data, effectively reducing the dimensionality and generating a compact representation known as the most important features of the process signals (z2).

These features z_2 are subsequently utilized by the regression network R_1 to establish a connection between specific flatness features z_1 extracted by the autoencoder A_1. By leveraging the decoder within A_1, these flatness features can be transformed into a complete flatness image P_v. By comparing this predicted flatness P_v with the real optically measured flatness P_m and minimizing the difference between them through tuning the weights and biases of R_1 accordingly, the performance of the flatness forecast model is optimized. This used model design is inspired by the "DeepGreen"-Model concept, utilizing dual autoencoders for simplifying complex systems [10].

After performing the training process on a small extraction of the industry data, the performance of the flatness prediction algorithm is tested (Fig. 3).

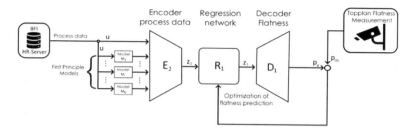

Fig. 2. Schematic of the concept the flatness prediction model is relying on. The incoming process data enables the physical models to be computed and together with these results everything is used as an input for the prediction model. The Encoder $E2$ reduces the data to a compact features space, $R1$ is transforming it to the reduced flatness feature space and finally, $D1$ is reconstructing the flatness image resulting in an analytical expression for the predicted flatness.

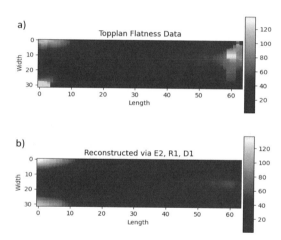

Fig. 3. Flatness comparison: a) Flatness measurement acquired by optical triangulation principal. b) Flatness prediction calculated through propagation of input signals through the model.

The evaluation is made by calculating the performance score with following equation (Eq. 1):

$$E_{rel} = \frac{1}{N} \sum_{i=1}^{N} \frac{v_{c,i} - v_{m,i}}{\max(v_m)} \qquad (1)$$

By averaging the relative error of all cuts for one specific strip, one obtains a quantitative value for the evaluation of the flatness prediction quality. In total this adds up to 75% for the shown example, which definitely also fits to the purely visual comparison of global flatness characteristics.

4.3 Sensitivity Analysis

In search of the underlying relationships affecting steel strip flatness in hot strip rolling, sensitivity analysis emerges as an indispensable tool to achieve explainable AI (XAI) [11]. It elucidates how variations in process parameters influence the final flatness of the steel strip. In the context of the hot rolled strip use case, the utilization of a physics informed artificial intelligence for flatness prediction offers an analytical representation of the measured flatness. Leveraging this framework shown in Fig. 1, one can compute partial derivatives with respect to each process parameter to determine corresponding weighting factors, like established feature importance metrics employed in state-of-the-art methodologies. To ensure reliable and meaningful derivation of these weighting factors, it is imperative that the flatness prediction model captures and incorporates the dynamic cross-process rolling transformation spanning from the oven to the final flat steel product. Furthermore, the various input factors comprising the multiple process parameters must be normalized to facilitate comparability across different quantities.

It is within this context that the formula used for calculating the weight value (Eq. 2) can be understood as a straightforward application of the chain rule similar to the concept elaborated in NeuralSens [12].

$$\frac{\partial F}{\partial G_k} = \frac{1}{C_{N,k}} \sum_{i,j,l} \sum_{m,n} \frac{\partial D1_{ij}}{\partial R1_n} \frac{\partial R1_n}{\partial E2_m} \frac{\partial E2_m}{\partial G_{kl}} \tag{2}$$

After calculating the sensitivity of all 200 input parameters across a data set of 20 strips the average sensitivity of the top 7 most sensitive parameters is depicted in Fig. 4. Basically all parameters influencing the flatness foreshadowed in the basic strip flatness chapter can be found here. Surprisingly Lateral Guidance Force seems to have more impact on flatness than Rolling Force.

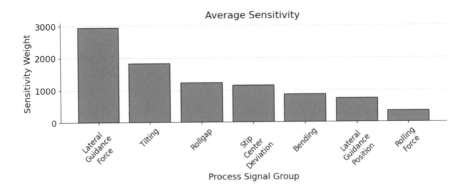

Fig. 4. Top 7 input parameter sensitivity weights averaged over a test data set of 20 strips.

5 Summary

This scientific paper introduces a cutting-edge approach for predicting and managing steel flatness in manufacturing by synergizing first-principle models with machine learning. The centerpiece is a Flatness Prediction Algorithm within a digital twin system, aimed at enhancing real-time monitoring and control in steel production. The paper underscores the role of sensitivity analysis in understanding the influence of process parameters on steel flatness. Looking ahead, the plan is to integrate the prediction model into the software infrastructure of industrial partners for test trials and in-depth sensitivity analysis, further refining flatness control. This methodological fusion promises significant improvements in product quality, operational efficiency, and cost-effectiveness in the steel industry.

References

1. Jelali, M., Müller, U., Wolff, A., Ungerer, W., Thiemann, G.: Advanced measurement and flatness control for hot strip mills. Rev. Met. Paris **99**(6), 517–522 (2002). https://doi.org/10.1051/metal:2002133
2. Molleda, J., Usamentiaga, R., García, D.F.: On-line flatness measurement in the steelmaking industry. Sensors **13**(8), 10245–10272 (2013). https://doi.org/10.3390/s130810245
3. Stefano, M., Marco, P., Alessandro, R.: About Digital Twins, agents, and multiagent. In: Autonomous Agents and Multiagent Systems, pp. 114–129. Springer International Publishing, Best and Visionary Papers (2022)
4. Alexandropoulos, S.-A. N., Kotsiantis, S. B., & Vrahatis, M. N. (2019). Data preprocessing in predictive data mining. The Knowledge Engineering Review, 34(e1). DOI: https://doi.org/10.1017/S026988891800036X
5. Raissi, M., Perdikaris, P., Kaniadakis, G.E.: Physics-informed neural networks: a deep learning framework for solving forward and inverse problems involving nonlinear partial differential equations. J. Comput. Phys. **378**, 686–707 (2019)
6. B. Berger, O. Pawelski and P. Funke, "Die elastische Verformung der Walzen von Vierwalzengerüste," Arch. Eisenhüttenwes., 1976, p. 351–356
7. M. P. Guerrero, C. R. Flores, A. Perez and R. Co-Las, "Modelling heat transfer in hot rolling work rolls," Journal of Materials Processing Technology, pp. 52–59, vol. 94 1999
8. X. Ye and I. V. Samarasekera, "The role of spray cooling on thermal behaviour and crown development in hot strip mills work rolls," Iron and Steelmaker, p. 4960, vol. 21 1994
9. J. Brandenburger, V. Colla, G. Nastasi, F. Ferro, C. Schirm and J. Melcher, "Big Data Solution for Quality Monitoring and Improvement on Flat Steel Production," IFAC-PapersOnLine, pp. 55–60 , 2016
10. Gin, C. R., Shea, D. E., Brunton, S. L., & Kutz, J. N. (2021). DeepGreen: Deep Learning of Green's Functions for Nonlinear Boundary Value Problems. Scientific Reports, 11, Article number: 21614
11. Adadi, A., Berrada, M.: Peeking Inside the Black-Box: A Survey on Explainable Artificial Intelligence (XAI). IEEE Access **6**, 52138–52160 (2018). https://doi.org/10.1109/ACCESS.2018.2870052
12. Pizarroso, J., Portela, J., Muñoz, A.: NeuralSens: Sensitivity Analysis of Neural Networks. J. Stat. Softw. **102**(7), 1–36 (2022). https://doi.org/10.18637/jss.v102.i07

Software Systems, Architectures, Applications and Tools

Solving the Relational Data Access Problem with Data Access Services

Ken Wang[✉] [iD]

BackLogic LLC, Chino Hills, CA 91709, USA
kywang123@gmail.com

Abstract. The object-relational impedance mismatch problem consists of two sub-issues: language impedance and data structure mismatch. Solution to the first issue is to abstract away the programming language from SQL; to the second issue is to abstract away the object-relational transformation from developer. This paper presents three standardized data access services, *query*, *command* and *repository*, collectively as a preferable solution to relational data access. These services are specified declaratively with SQL and JSON. The hard object-relational transformation is delegated to service engine. This paper also presents a reference implementation of the service engine for service execution. The complexity test proves that this reference service engine is capable of handling highly complex objects up to the *5x5* level on the complexity scale developed in this research.

Keywords: object · relational · database · data access · persistence

1 Introduction

Relational data access (RDA) is an old but not well-solved problem. ORM (object relational mapping) is intended to bridge the gap between the object and relational world, but has drawn as much criticism as praise over the years [1, 2]. Like raw SQL, it is incapable of gracefully handling even moderately complex objects. As a result, developers are forced to compromise on their object design, which takes a lot of flexibility and productivity away from them.

People have hoped that NoSQL database could provide a solution to the problem. However, as it stands now, relational database (RDBMS) is still the dominating database according to data from DB Engines Ranking [3] and will likely remain so solely because of the maturity and superiority of relational data model. This situation leaves us no choice but to find a better solution to relational data access.

In this research, we have reexamined the nature of object-relational impedance mismatch and realized that this impedance mismatch problem actually consists of two sub-issues: the *language impedance* between embedded SQL and the host language, and the *data structure mismatch* between the object model and the relational data set. We believe that the solution to the first issue is to abstract away the host language from SQL, or to separate SQL from the host language, and that the solution to the second issue is to

Á. Rocha et al. (Eds.): WorldCIST 2024, LNNS 989, pp. 201–212, 2024.
https://doi.org/10.1007/978-3-031-60227-6_18

abstract away the object-relational transformation from developer and delegate it to data access framework/library.

In this research, we have also reexamined the familiar DAO (data access object) and repository design patterns and realized that the whole RDA problem actually comprises only three basic data access patterns: *query*, *command* and *repository*.

These findings lead us to the idea of reducing the generic DAO and repository design patterns to three standardized data access services (DAS), namely *query*, *command* and *repository* services. These services will be specified declaratively with SQL and JSON, rather than being coded with an imperative programming language, and will be executed with a service engine residing on the data access tier, as opposed to a framework/library embedded in the data access layer. These DAS are conceptually simple and thus easy to learn and quick to develop. Most importantly, they delegate the hard object-relational transformation problem to service engine, while providing developers with the flexibility to design their object model according to application needs rather than framework constraints.

Due to space limitations, this paper provides only a brief description of the structure, composition, development process and test results of these DAS. The examples are available on GitHub [8] for public review.

To prove the concept of DAS, a reference service engine, along with a service builder tool, is implemented. Example services are developed with the service builder and tested with the reference service engine.

The success of the DAS concept is to a great extent determined by the capability of the service engine to handle complex objects. The complexity test shows that the reference engine is capable of handling objects at level *5 x 5* on the complexity scale developed in this research for measuring the complexity of an object structure. At this level, the object is a nested structure with 5 array structures in both the vertical and horizontal directions. It is unlikely that a real-world application would need data structures of this complexity, and thus this test virtually assures developers of the total flexibility in their object design.

2 Data Access Services

2.1 Query Service

Query service is for retrieving data from data source. It takes an input object of query parameters and returns an object or an array of objects as output.

Figure 1 illustrates the structure of the example query service *getCustomersByCity*. Table 1 lists the JSON and SQL files composing the query service. The *service.json* file is not a component of the service but contains the meta data of the service.

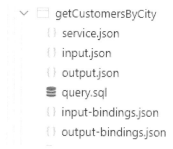

Fig. 1. Structure of query service

Table 1. List of query files

File	Description
input.json	json file for service input
output.json	json file for service output
query.sql	sql file for service query
input-bindings.json	json file for input bindings that map query parameters to input data fields
output-bindings.json	json file for output bindings that map output data fields to query columns

The input and output define what the query service is for; the SQL query implements the data access logic of the service. To specify the query service, the user first.

- Specifies the input and output for the service with JSON in the input and output files; then
- Composes the SQL query to retrieve data in the query file; and lastly
- Generates the input and output bindings from the service input, output and query.

The last step is to be assisted by service builder, a development tool for DAS, and be done with a click of a button. The output, query and output-bindings code for the *getCustomersByCity* service are listed below for illustration.

Output:

```
[{
    "customerNumber": 1,
    "customerName": "Land of Toys Inc.",
    "address": {
        "addr": "NYC",
        "city": "NYC"
    }
}]
```

Query:

```
select customerNumber, customerName, address, city
   from customers where city = :city
```

Output-Bindings:

```
[
    {"field": "..customerNumber", "column": "customerNumber"},
    {"field": "..customerName", "column": "customerName"},
    {"field": "..address.addr", "column": "address"},
    {"field": "..address.city", "column": "city"}
]
```

The query service provides developer with the flexibility to retrieve any ad hoc data structure per application needs.

2.2 Command Service

The command service is for writing and changing data in data source. The basic command service takes an input object of SQL parameters and returns nothing. The advanced version appends a query to the command. Figure 2 illustrates the structure of the basic command service *cloneProductLine*.

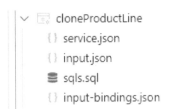

Fig. 2. Structure of command service

Table 2 lists the JSON and SQL files composing the command service.
To specify the command service, the user first.

- Specifies the input object for the service with JSON in the input file; then
- Composes a sequence of DML statements to be executed for the command service in the *sqls* file; and lastly
- Generates the input bindings from the input and SQL statements.

Table 2. List of command files

File	Description
input.json	json file for service input
sqls.json	sql file for service DML statements
input-bindings.json	json file for input bindings that map sql parameters to input data fields

The JSON and SQL code for the *cloneProductLine* service are not included in this paper but can be found on GitHub [8], along with many other command examples.

The command service provides developer with the flexibility to perform any ad hoc data loading and any ad hoc data changes per the application needs.

2.3 Repository Service

Repository service is for CRUD operations of aggregate root object. It includes both read operations for querying the object and write operations for creating, updating and deleting the object. Figure 3 illustrates the structure of the repository service *Order* .

Fig. 3. Structure of repository service

Table 3 lists the JSON and SQL files composing the repository service.

The object component defines what the repository service is for; the read component specifies a query service for the object; the write component comprises a set of table and column bindings (aka mappings) for dynamic generation of insert, update and delete statements. To specify the repository service, the user first.

Table 3. List of repository files

Component	File	Description
object	object.json	json file for service object
read	input.json	json file for read input
	query.sql	sql file for read query
	input-bindings.json	json file for input bindings that map query parameters to input data fields
	output-bindings.json	json file for output bindings that map object data fields to query columns
write	tables.json	json file for table bindings that map object structures to tables
	orders.columns.json	json file for column bindings that maps object data fields to table columns
	orderdetails.columns.json	Same as above. One file per table

- Specifies the object for the service with JSON in the object file; then
- Specifies the query service for the read component (the output of the query service is the object and needs not to be specified again); and lastly
- Generates the table and column bindings from the object, query and DB meta.

The last step is to be assisted by service builder and be done with a click of a button. The JSON and SQL code for the *Order* service are not included in this paper but can be found on GitHub [8], along with many other repository examples.

The read component of the repository service is designed to be a dynamic query, implying that the *WHERE* clause of the query will be dynamically modified at execution, based on the input parameters coming from the service call, to accommodate the different query needs for the object, such as *getOrderById*, *getOrdersByCustomer*, *getOrdersByStartAndEndDates*, etc.

The write component of the repository service is to be derived by design from the read component. The tables to write are derived from the query; the column data are both derived from the query and retrieved from database meta data; and the table and column bindings are helped by the existing output bindings. For this purpose, a set of rules is imposed on the read query. For example, the table join must be in ANSI format; the tables must be aliased; the columns must be prefixed with table alias; the alias for reference table (aka read-only table) must start with "_", so that it can be excluded from the table bindings, etc. However, most of these rules are also best practices.

The repository service is for persistence of domain objects. Because of its capability of handling complex aggregate root object, it is expected to be a good fit for domain-driven design (DDD) that is driving microservices development these days.

3   Reference Engine

Just like SQL is executed by SQL engine, DAS is executed by service engine. The concept of DAS hinges on the successful implementation of the service engine. Additionally, the concept of DAS is bundled with a builder tool for automatic generation of input, output, table and column bindings. Therefore, to prove the concept of DAS, we have implemented a reference service engine, along with a service builder.

3.1   Service Builder

The service builder is implemented as a VS Code extension [7], so that the SQL and JSON can be developed in specialized SQL and JSON editors, respectively. The SQL editor is to be connected to the data source, so that the SQL development can be done in a database-centric environment.

An important feature of this service builder is its capability of generating input and output bindings for query and command services and its capability of generating table and column bindings for repository service. For this purpose, a SQL parser and a fuzzy data field matcher have been installed with the service builder.

A slew of example query, command and repository services [8] have been developed with this service builder in a streamlined process, as described in the previous section.

3.2   Service Engine

The service engine is implemented as a library component and embedded in the service builder [7], so that the service can be tested as it is developed. In this research, the service engine is tested with the various example query, command and repository services developed using the service builder, and it has worked as designed for all.

However, the real success of the service engine is measured by its capability of handling complex objects, which is what distinguishes DAS from other approaches. Hence, we have developed a complexity scale for measuring object complexity and carried out a complexity test on the service engine, as discussed in the following sections.

4   Complexity Scale

The difficulty of object-relational transformation increases with the complexity of the target object structure. To gauge this difficulty, a five-level complexity scale is developed in this research for measuring the complexity of the object structure, as described in the following.

At level 0×0, the object is the same as the relational data set.

At level 1×1, the object is different from the relational data set but does not contain any array structure. The object may include nested object structure vertically and horizontally. Nonetheless, a skilled developer can easily handle this type of objects.

At level $1 \times n$, the object includes nested array structure only vertically, as illustrated below:

```
{
    "id1": 1,
    "a1": [{
        "id2": 2,
        "a2": [{
            "id3": 3
            ...
        }]
    }]
}
```

A skilled developer can comfortably handle objects at level *1 x 2*. Beyond this level, it would be a hard stretch and may require the infamous N + 1 technique.

At level **n x 1**, the object includes nested array structure only horizontally, as shown in the following:

```
{
  "id1": 1,
  "a2": [{"id2": 1}],
  "a3": [{"id3": 1}]
   ...
}
```

Horizontally nested array structure is more difficult to handle than vertically nested array. Even a *1 x 2* structure is quite hard to a skilled developer.

At level **n x n**, the object includes both vertically and horizontally nested array structures. This is the hardest level to deal with.

This 5-level complexity scale could be used as a standard for measuring the complexity of an object and the capacity of an RDA solution. Note that, at all levels, the difficulty of object-relational transformation is measured by transforming a relational data set into an *array* of the objects.

5   Complexity Test

5.1   Test Service

The capability of service engine to handle complex objects is measured by its capability to *query* complex objects. Therefore, a set of specially designed query services are used for the test. The test service does not take any input but returns an array of objects of different complexity per the test level.

For example, the test output for a *3 x 3* test would look like:

```
[{
  "id11": 1,
  "a2": [{
    "id22": 1,
    "b2": [{
      "id23": 1
    }]
  }],
  "a3": [{
    "id32": 1,
    "b2": [{
      "id33": 1
    }]
  }]
}]
```

It contains 3 array structures in both the vertical and horizontal directions. To provide data for the test output, a test query is prepared for the test service to produce a data set of 5 data fields and 8 rows for the *3 x 3* test, as shown in Table 4.

5.2 Test Result

Once executed, the *3 x 3* test service is expected to return a data structure the same as the test output defined for the test service, but with two records in each of the 5 array structures. The test result has come out exactly as expected.

Table 4. Data records from test query.

Id11	Id22	Id23	Id32	Id33
1	1	1	1	1
1	1	2	1	2
1	2	3	2	3
1	2	4	2	4
2	3	5	3	5
2	3	6	3	6
2	4	7	4	7
2	4	8	4	8

The *3 x 3* test is a scaled down version for illustration. For the real test, we have tested the service engine with objects at various levels of complexity, including:

- 1 x 5
- 5 x 1, and
- 5 x 5

The results all come out as expected and are saved in GitHub [9] for public access. Therefore, we may conclude that the reference implementation of service engine is capable of handling complex objects up to the 5 x 5 level, and that the concept of DAS is feasible for handling complex objects.

6 Advantages

Table 5 compares DAS with Spring JDBC and JPA, which represent the two mainstream approaches at the moment: raw SQL and ORM. Compared to JDBC and JPA, DAS offers some unique advantages:

- DAS is scalable, meaning that as the query and, especially, the object gets complex, the level of effort required do not increase dramatically.
- DAS is platform independent. It works not only with Java but also with other applications.
- DAS requires only SQL and JSON skills and thus presents an opportunity to delegate the data access development to a SQL developer, just like the presentation layer is delegated to front-end developers.
- DAS development is highly efficient. The SQL is developed in a database-centric environment; the data mapping is largely automated; and the developer needs not to concern about database session and transaction, etc.
- DAS is safe from SQL injection. The SQLs are fully parameterized and, by default, are never altered through string appending or substitution.

Architecturally, DAS separates the data access layer from the application and makes data access a backing service to the application. To serverless or microservices applications, DAS may be deployed to provide not only data access but also connection polling services.

Table 5. Comparison of DAS with JPA and Spring JDBC.

	Spring JDBC	JPA	DAS
Architecture			
Application architecture	3-tier	3-tier	4-tier
object-data model coupling	loose	tight	loose
platform dependency	Java	Java	None
Scalability			
with SQL complexity	Fully scalable	Semi-scalable	Fully scalable
with object complexity	Not scalable	Not scalable	Highly scalable
Development			
paradigm/environment	Java	Java	SQL
skill requirements	Java, SQL	Java, QL, SQL	SQL, JSON
coding style	Imperative	Imperative	Declarative
mapping automation	By convention	By convention	Fussy matching
Security			
SQL injection	Up to developer	Generally safe	Safe

7   Summary

A small set of standardized data access services, namely query, command and repository services are presented as a preferable solution for the hard relational data access problem. Different from all traditional approaches, these services are specified declaratively with SQL and JSON, and are capable of handling highly complex object structures, as the hard object-relational transformation problem is abstracted away from developer and delegated to service engine. A reference service engine, along with a service builder tool, is implemented and tested with a slew of example query, command and repository services. A complexity scale is developed for measuring the complexity of object structure, as well as the capability of RDA solutions. The reference service engine is capable of handling objects at level *5x5* on above complexity scale, potentially covering all realistic application needs.

References

1. Fowler, M.: ORM Hate. https://www.martinfowler.com/bliki/OrmHate.html. Accessed 09 Nov 2023
2. Neward, T.: The Vietnam of computer science. https://www.semanticscholar.org/paper/The-Vietnam-of-Computer-Science-Neward/331e490c55ee72d6011bbceb323c03f0572a5235 (2006). Accessed 09 Nov 2023
3. DB-Engines DB-Engines Ranking, https://db-engines.com/en/ranking. November 2023, Accessed 09 Nov 2023

4. Bauer, C., King G.: Java Persistence with Hibernate. Manning Publications Co., 209 Bruce Park Ave. Greenwich, CT 06830 (2007)
5. Ireland, J.: Object-relational impedance mismatch: a framework based approach, PhD thesis The Open University (2011)
6. Oracle Data Access Object. https://www.oracle.com/java/technologies/data-access-object.html. Accessed 09 Nov 2023
7. GitHub Service Builder. https://github.com/bklogic/ServiceBuilder. Accessed 09 Nov 2023
8. GitHub data access service example. https://github.com/bklogic/data-access-service-example. Accessed 09 Nov 2023
9. GitHub Complexity Test. https://github.com/bklogic/complexity-test. Accessed 13 Nov 2023

Baseline Proposal of User Experience Practices in the Software Engineering Process: Global MVM Case

Patricia Elena Gómez Muñoz[1]([✉]), María Clara Gómez Álvarez[2], and Ricardo Alonso Gallego Burgos[1]

[1] Digital Experience Unit, Global MVM, Medellín, Colombia
{patricia.gomez,ricardo.gallego}@globalmvm.com
[2] Engineering Faculty, Universidad Nacional de Colombia, Medellín, Colombia
mcgomez@unal.edu.co

Abstract. Software development companies must consider the satisfaction of interested parties, especially users, and within quality attributes including user experience (UX). This factor is relevant since the success or failure of the software will depend on the ease with which users can handle it and interact with it in a way that generates a positive perception, achieves acceptance and allows more significant use. Additionally, the joint participation scenarios that software companies must face due to opening processes through co-creation with clients, users, companies and other agents, increase the complexity of their environment. Thus, under this scenario, research is carried out to incorporate UX practices into the software development life cycle of the Global MVM company in a way that allows it to face current challenges, optimize and improve its portfolio of products and services and a better approach to the needs of its clients and users is achieved.

Keywords: User Experience · Software Development · Open innovation · Competitiveness

1 Introduction

Software development companies currently face various challenges, among them are the industry's competitive dynamics and the high demand for services and products with a highly innovative technological component in technology, among other aspects [1]. All these scenarios are decisive for any firm that operates in the software industry. Added to this, the joint participation scenarios that these companies must face through opening processes through co -creation with clients, users, companies and other agents, increases the complexity of the environment in this industry [2].

Nowadays, it has been identified that the software development process focuses a great effort on its product quality assurance, becoming an essential factor in the industry's evolution [3]. Increasingly, software systems must consider stakeholders satisfaction, especially users, and user experience (UX) is included in the quality attributes. UX refers to how users interact with a product or service. It accounts for user perceptions before, during and after use. A good user experience goes beyond what users express they want from a product (or service) and focuses on what they need instead.

User experience design in the software development industry aims to improve customer satisfaction and loyalty through utility, ease of use, and the pleasure of interacting with a product [4]. This has led companies to involve more UX practices in the software development lifecycle, understanding that users' perception and experience with the products and services they deliver affect the impact on the market [5, 6].

Global MVM is no stranger to these challenges and understands the challenges it must face to optimize and improve its portfolio of products and services, identifying the need to strengthen the software development life cycle to achieve a closer approach to the needs of its customers and users. In this sense, this research has employed the case study method [7], It is based on the research question: ¿How can user experience (UX) practices enable software engineering processes in open innovation scenarios to develop high-value solutions for their stakeholders? The research hypothesis is: a baseline set of UX practices integrated to the software engineering process in open innovation scenarios enable the construction of solutions that add value to customer and user business processes.

The unit of analysis is composed of the Colombian software development company Global MVM with its processes, for which UX practices are to be incorporated. Some of the benefits obtained for this enterprise from the incorporation of the proposed baseline are: (1) having an approved and disseminated process to improve the adherence and coverage of UX practices at the organizational level and (2) the incorporation of instruments associated with the process to leverage the process activities.

The article is structured as follows: Sect. 2 presents the methodology followed; Sect. 3 describes the background or previous work; Sect. 4 presents the proposed baseline; Sect. 5 details the results obtained; and finally the conclusions and future work are presented in Sect. 6.

2 Methodology

The unit of analysis of this research is composed of the Colombian company Global MVM with its software development processes, for which it is required to incorporate UX practices to deliver digital products that add value to its customers and end users of the electricity sector in Colombia and the region. This research used the case study method [7] to analyze the problem in the specific unit and contrast it with what was found in the systematic literature review (SLR) on practices, models, principles and characteristics of UX in software development processes in open innovation scenarios, and within the case study, validation with experts was used as an instrument.

The development of this project was carried out through the phases shown in Fig. 1:

A brief description of each of the phases is given below:

Phase 1. Execution of a systematic literature review (SLR) to establish practices, models, principles, and characteristics of the UX that can be applied to the software development process in open innovation scenarios, allowing to elaborate the state of the art and to have an appropriation of fundamentals related to the object of study.

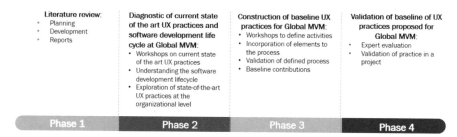

Literature review:	Diagnostic of current state of the art UX practices and software development life cycle at Global MVM:	Construction of baseline UX practices for Global MVM:	Validation of baseline of UX practices proposed for Global MVM:
• Planning • Development • Reports	• Workshops on current state of the art UX practices • Understanding the software development lifecycle • Exploration of state-of-the-art UX practices at the organizational level	• Workshops to define activities • Incorporation of elements to the process • Validation of defined process • Baseline contributions	• Expert evaluation • Validation of practice in a project
Phase 1	Phase 2	Phase 3	Phase 4

Fig. 1. Proposed methodology to develop the baseline.

Phase 2. Current diagnosis of software development lifecycle practices at Global MVM by conducting information gathering questionnaires that allowed the identification of methods, artifacts, techniques, and UX models used at Global MVM, as well as the activities carried out.

Phase 3. Construction of the UX baseline by identifying the process cycle, activities and artifacts, thus, allowing the construction of the baseline of UX practices and incorporating them into Global MVM's software engineering processes.

Phase 4. Validation of the defined baseline using the Delphi method to determine that it applies to the contexts and meets the needs of the Global MVM company. Additionally, the practices proposed in the UX baseline are evaluated in a pilot project where they were applied.

3   Background

As part of the research, a systematic literature review (SLR) was carried out, taking 26 articles that were used as theoretical and conceptual support to appropriate foundations. It focused on the search for relevant studies related to practices, models, principles, and characteristics of UX in software development processes in open innovation scenarios, evaluating and synthesizing the contributions. In the literature, it was found that the software development process that best integrates with UX is the one that incorporates agile frameworks [8]. Also, the collaborative relationship between the roles participating in software development is essential to have a common understanding of the problem being addressed [9]. Additionally, several challenges must be faced when integrating a software development process with UX, such as having clarity of the concept of UX so that it is not confused with usability [9], enabling collaboration between development and UX teams on the day-to-day [10], clearly differentiate the role of client and user [11, 12], prioritize UX activities [13], among others. It is important to mention that all these findings were used for the baseline proposal by the company Global MVM.

Throughout its history, Global MVM has considered using open innovation practices as a critical element to increase its competitiveness. Users have become an essential part of the innovation process in this environment. The UX, as such, has been established as a new driver since now; instead of being focused on the features of a product or service, it is necessary to ensure that users are involved in the conceptualization, design, and implementation activities [14].

Over the last decade, open innovation has become one of the most researched topics in innovation management [15–17]. Its main objective is to incorporate external knowledge (customers, suppliers, technology partners, universities, research centers) within the organization, to generate learning that allows the generation of value. This process also includes exploring ideas and intellectual property to reach the market faster than competitors [18].

In general, some UX is provided in software products, whether or not it has been explicitly worked out during the development life cycle, and although some user experience of software functionality cannot be guaranteed, it is important to consider principles, practices, models so that the criteria on which a well-qualified UX is based are more likely to be met [19]. In fact, simply applying these UX associated features in isolation is not sufficient [20]. As well as the methods and practices used to support other software quality characteristics must be integrated into the development processes and considered transversally in the projects. UX and software development have different approaches that represent challenges for their integration: initially, the clarity of the UX concept so that it is not confused with usability [9], the collaborative approach of the work between teams that allows understanding of the why and what of what is being done [21], among others.

Consistent with this context, Global MVM identified the need to integrate UX practices in its software development cycle to provide better quality products, provide a service that adds value to its customers; and obtain better acceptance of these products by end users. Likewise, it sought to give more importance to users by including them as an important part of the software engineering process, thus achieving that by incorporating these UX practices, solutions that add value to their business processes are delivered.

The term baseline has several definitions, focusing on the definition found in Sebok, which indicates it is "A specification or product that has been formally reviewed and agreed upon, which then serves as the basis for further development and can only be changed through formal change control procedures" [22].

Considering this existing problem around UX and business needs, this research is realized for the definition of the baseline of UX practices that should be used during the software development life cycle for its incorporation into the processes of the Global MVM company. The advantages of this integration are that the roles execute a single process, allowing a collaborative approach with agents external and internal, in addition to the adaptation to the business culture.

4   Solution

Based on the phases established in the methodology, after the systematic literature review, a diagnosis of the current state of Global MVM's UX practices was carried out, as well as an understanding of its software development process. Based on what was identified, the baseline of UX practices to be incorporated was proposed.

The following is a brief description of the products obtained in the different phases of the methodology.

4.1 Diagnostic

The diagnosis was carried out by conducting workshops to determine the state of UX practices and an understanding of the software development process at Global MVM, and a survey to explore the state of UX practices at the organizational level (See Table 1). It showed a defined software development life cycle based on best practices (Agile, Devsecops, test automation, among others) and methodologies (ISO9001:2015, CMMI DEV, Design Thinking, among others) where the best fit to the company's needs has been taken. The process is called *Development service delivery model*. However, for UX, it was found that there was no formal or approved process, that the activities were not disclosed, and there was no adherence to the practices, and that there was a lack of knowledge of models, techniques, artifacts, and methods. This made it possible to corroborate the relevance of this research.

Table 1. Technical data of the study of the current state of UX practices at Global MVM.

Sample size	35 Persons
Technique used to collect information	Online questionnaire
Information collection date	November-December de 2022
Target audiences	Group 1 Digital Experience Unit Team Group 2 Architects and technical leaders, Group 3: Developers Group 4: Unit and project leaders, agile facilitators, and business specialists
Number of survey questions	Group 1: 40 questions, Group 2: 26 questions Group 3: 21 questions, Group 4: 44 questions

4.2 Baseline Construction

Based on what was identified in the background and the results of the diagnosis, a set of good UX practices was taken, to solve the missing elements and to adapt them to the software development process, thus defining the baseline of UX practices for Global MVM. The definition of the baseline involved the realization of several activities that included workshops, the incorporation of UX practices into to the processes, the validation of the defined process and the consolidation of the contributions made by the proposed baseline. The Digital Experience Unit of Global MVM is in charge of carrying out the activities related to UX, which is why, initially, workshops were held with its members: unit leader, specialists, and UX designers. This made it possible to discover the objective of the process and its scope. The process *Experience design and innovation* was defined based on Design thinking, with some stages where each has associated activities, responsibilities, and artifacts to carry them out (See Fig. 2).

The defined scope comprised three pillars: User Experience (UX)-User Interface (UI), Experience Design and Service Experience, in which the main objective is to generate the best customer experience throughout the entire process.

Fig. 2. Stages of the Experience design and innovation process proposed for Global MVM.

The main objective of the process is to provide comprehensive discovery capabilities, based on customer and user-centric thinking to understand their needs, problems, and challenges, and to enable the design of feasible, scalable, innovative, and cost-effective digital solutions with high impact for Global MVM's business and its stakeholders, delivering valuable experiences.

After the workshops, it was analyzed and determined how to introduce the Experience Design and Innovation process in Global MVM's software development process called Development Service Delivery Model, achieving its incorporation, thus forming the baseline of UX practices for Global MVM. Finally, once the baseline is defined, we proceed with its publication and dissemination for the knowledge of all Global MVM stakeholders.

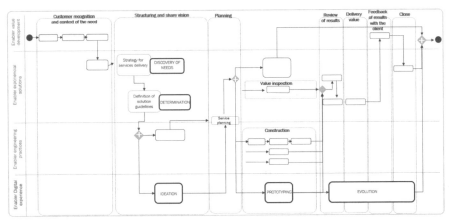

Fig. 3. Diagram of the software development process for Global MVM, incorporating the defined stages of the UX process.

Figure 3 shows the software development process with integrated UX practices (baseline). The stages of the process proposed *Experience design and innovation* are highlighted in dark purple, showing where they were included: the *Discovery of needs* is part of the *Strategy of services delivery*, the *Determination* is integrated within the *Definition of solution guidelines*, later are the *Ideation, Prototyping,* and *Evolution* which are not included within another sub-process.

The diagnosis identified different UX gaps related to the process, its homologation, dissemination, and adherence, and lack of knowledge of techniques, artifacts, and methods. The elements proposed in the baseline contributed to covering several of these gaps, such as having a standardized, disclosed and used process to improve the adherence and coverage of UX practices at the organizational level and the incorporation of several instruments associated with the process to leverage UX activities in the different stages of Global MVM's software development process (see Table 2). For example, techniques included heuristic analysis, navigation tree, card sorting, eye tracking, web analytics, and interview questionnaire; artifacts included customer journey map, stakeholders and interested parties map; methods included user research questionnaire, A/B test, think aloud test, and others.

Table 2. UX gaps identified in Global MVM.

Gap	Baseline contribution
UX practices are not integrated into processes, particularly the software development lifecycle defined	Process of Experience design and Innovation defined and integrated to the software development lifecycle
Lack of knowledge of UX artifacts, techniques and models	Definition of instruments associated with the Experience and Innovation Design process involving UX artifacts, techniques and models
Organizational level lack of knowledge about the use of UX practices	Dissemination of the Experience Design and Innovation process throughout the organization

5   Results

The validation of the baseline was carried out through two work axes: (1) expert assessment, where the proposed baseline of UX practices was presented and how it was incorporated into Global MVM's software engineering processes and (2) a pilot project where the practices were incorporated and it was reviewed how the defined practices are applied and what evaluation internal and external users of Global MVM make of the general application of the UX practices baseline.

5.1   Expert Evaluation

The expert evaluation was carried out using the Delphi method, which included the participation of three external experts and one internal to Global MVM. The group of experts consisted of professors with PhDs and knowledge in UX and usability, and the internal expert with knowledge in UX and usability and the software engineering process carried out particularly in Global MVM (See Table 3).

The results of the experts evaluation of the UX practices baseline for Global MVM incorporated into the software development process showed that it is generally rated as

Table 3. Expert evaluation technical data.

External experts	Three persons
External experts profile	Professors with PhDs and knowledge in UX and usability
Internal experts	One person
Internal experts profile	Knowledge in UX and usability and the software engineering process in Global MVM
Target audiences	Experts who participated in the evaluation (4)
Number of questions in the questionnaire	Eight questions

High or Medium (See Table 4). The experts did not cast votes on the Low scale. Within the results, observations emerged indicating that it would be important to identify the UX concept on which the baseline definition is made and to determine the approach, i.e. how the user is viewed within the baseline, whether as the center of the process, as a customer, or another user approach.

They also mentioned the importance of highlighting UX practices within the process outline to make it easier to institutionalize them in the organization.

Table 4. Results of the baseline assessment of UX practices.

Concept	Qualification
Baseline of UX practices	50% High and 50% Medium
Discovery stage activities	75% High
Determination stage activities	75% High
Ideation stage activities	75% High
Prototyping stage activities	100% High
Evolution stage activities	50% High and 50% Medium
Instruments	75% High

5.2 Validation of UX Practices

The validation of the practices proposed in the UX baseline for Global MVM was carried out by selecting a pilot project where the practices were applied and specifically evaluating one of the phases proposed in the process and, on the other hand, a survey was used to determine the evaluation made by internal and external users of Global MVM of the general application of the UX baseline.

The application of the baseline practices proposed in a Global MVM project was evaluated by executing the activities proposed in the *Prototyping* phase. (see Fig. 4).

The project where the evaluation was performed consisted of the development of a web application and a mobile application for the commercial management of non-regulated clients in the energy sector. No observations or improvements on the activities and artifacts defined for the *Prototyping* phase of the proposed baseline emerged from this validation.

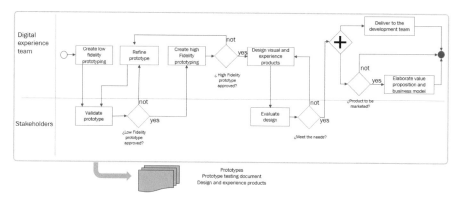

Fig. 4. Baseline of UX practices evaluated in a project - Prototyping Subprocess.

As for the survey applied, its focus was to validate the adherence and impact of UX practices in the dimensions of usability, experience, satisfaction, and value contribution (see Table 5). The survey was applied in a general way once the dissemination of the process and the beginning of the appropriation of the practices defined in the baseline were completed, where a total of 66 responses were received, of which 41 correspond to internal clients or users of Global MVM and 25 to external clients. Of this sample, 5 responses were related to the pilot project where the prototyping phase was evaluated (4 internal, 1 external).

Table 5. Survey technical data applied to evaluate the application of the defined baseline.

Sample size	66 persons
Technique used to collect information	Online questionnaire
Target audience	Users, internal and external clients of Global MVM
Number of survey questions	16 questions

Within the analysis of the survey results, one indicator that is evaluated is the adherence to UX practice in Global MVM, the NPS. The results show that 47 of the 66 users who participated in the assessment of adherence to UX practice are promoters. This means that respondents believe that UX products and artifacts add value to the processes, capabilities, and strategy in the context of each product or service provided. This aspect is fundamental for Global MVM's purpose of strengthening these practices, thus making UX capabilities recognized by the market, including internal customers. In second place are the passive users, since the indicator showed that 19 of the 66 users consider it an evolving practice, and that, over time, Global MVM will consolidate this UX capability. There are no detractor users, indicating that users are satisfied with the design, UX products, and artifacts produced at Global MVM (See Fig. 5).

The pilot project shows that 40% are promoters and 60% are passive, there are no detractors, which indicates that the pilot project still has improvements to make to the process and that the practices are not fully consolidated.

Promoters	47
Passives	19
Detractors	0

0

71

-100   +100

NPS®

Fig. 5. Results of the NPS Indicator associated with the adherence to the UX practice.

The results of the expert evaluation and the validation of the UX practices defined in the baseline show, on the one hand, that the software development process is considered to be strengthened by incorporating of this baseline and that the stages described in the process make an important contribution to it. Likewise, in applying of the activities of one of the phases defined in the baseline to a specific project, they indicate that they are satisfied with the UX services offered and that they are useful.

Although the baseline of UX practices was applied to the Global MVM company, the proposal is generic. Other software companies can adopt it because it includes best practices related to user experience in software products.

6   Conclusions

In Phase 1 of this research, a literature review was conducted to clarify relevant concepts such as open innovation, UX, baselines, and other topics such as methodologies, practices, and methods, and thus have information that would allow its incorporation into the context of the Global MVM company and adapt them to its software engineering cycle. The literature highlights that there must be an understanding of the Engineering and UX processes to make the integration and that, in the current agile trend, the application of UX practices must not affect what is sought, which is to deliver in a short time and with value.

After the SLR, we proceeded with understanding of the activities associated with UX, the software development process and the software engineering practices used in Global MVM by conducting workshops, thus identifying tasks, artifacts and methods used. Additionally, the current state of UX practices was determined. This allowed us to assess the maturity status of the process, which indicated several aspects to strengthen and standardize in methods, models, techniques, artifacts, among others.

Based on the results of the activities carried out in the diagnosis, we proceeded with the construction of the baseline of UX practices for Global MVM. As part of its definition, the scope, the phases or stages that would be part of it, its activities and the support tools to be used were established. The sub-processes identified were *Discovery, Determination, Ideation, Prototyping, Evaluation.* The activities were also diagrammed and incorporated into the Global MVM software development process. Finalizing with the dissemination of the process and the beginning of the application of the practices.

Next, we proceeded validating of the baseline of UX practices defined for Global MVM, for which experts are convened with whom the validation is performed. The results of this validation showed that in general there is a High or Medium rating on the proposal presented, indicating that it is a contribution that strengthens the software development process, which also allows incorporating and adjusting some elements to improve the baseline presented, among which are the form of presentation of the sub-processes incorporated, the scope of the process, among others.

Additionally, one of the practices defined in a pilot project was evaluated, selecting the Prototyping sub-process, where the proposed flow was followed, allowing the application of the defined activities and their artifacts. The adherence to the practices defined in the baseline was also reviewed through a survey with internal and external users whose main findings are that an important contribution of the practices is identified but that a consolidation of the process is still lacking.

Finally, based on the results of the validation by experts and the survey with internal and external users, it can be concluded that the baseline of UX practices incorporated into the software engineering process of Global MVM allows this process to be strengthened, achieving an increase in the quality of products and services delivered by Global MVM, improvement in the perception of users because they are more actively involved in the process, among others.

This research work allowed the definition of a baseline of UX practices incorporated into Global MVM's software development process. Based on this product, the following future work is planned: (1) Design a training session on the new process based on gamification as one of the standardization mechanisms of this process within the company, (2) Perform another validation of the process adherence and collect improvement opportunities to refine the activities within the proposed phases, as well as the instruments, (3) Include metrics to identify the behavior of the process, work on the service experience approach going beyond the product and develop an evaluation with the clients to assess the contribution of all the UX phases incorporated and determine if they add value to the products that are being delivered, and (4) Invite other companies to use the baseline, evaluate the results of their application and compare company processes results.

References

1. Cuéllar, M.C.: Los desafíos de las empresas de software frente a la globalización. Revista de Ingeniería, 86–90 (2013). https://doi.org/10.16924/REVINGE.38.13
2. Morales Peña, G.A., Freire Morán, J.F., Morales Peña, G.A., Freire Morán, J.F.: La innovación tecnológica: creando competitividad en las empresas desarrolladoras de software. Podium, 139–154 (2021). https://doi.org/10.31095/PODIUM.2021.39.9

3. Soraluz, A.E., Vallez Coral, M.A., Levano Rodriguez, D.: Desarrollo guiado por comportamiento: buenas prácticas para la calidad de software, Barranquilla (2021)
4. Kujala, S., Roto, V., VäänänenVainioMattila, K., Karapanos, E., Sinnelä, A.: UX Curve: a method for evaluating long-term user experience. Interact. Comput. **23**, 473–483 (2011). https://doi.org/10.1016/J.INTCOM.2011.06.005
5. Jurca, G., Hellmann, T.D., Maurer, F.: Integrating agile and user-centered design: a systematic mapping and review of evaluation and validation studies of agile-UX. In: Proceedings - 2014 Agile Conference, AGILE 2014, pp. 24–32 (2014). https://doi.org/10.1109/AGILE.2014.17
6. Acuña, C., Pinto, N., Tomaselli, G.P., Tortosa, N.: Evaluación del impacto de las emociones en la calidad de software desde el punto de vista del usuario. http://sedici.unlp.edu.ar/handle/10915/103909. Accessed 12 Mar 2022
7. Fong Reynoso, C.: El estudio de casos en la preparación de tesis de posgrado en el ámbito de la PYME. In: Conference: Estableciendo puentes en una economía global, vol. 34 (2008)
8. Alhammad, M.M., Moreno, A.M.: A Gamified framework to integrate user experience into agile software development process. Universidad Politécnica De Madrid Escuela Técnica Superior De Ingenieros Informáticos (2020)
9. Kashfi, P., Feldt, R., Nilsson, A.: Integrating UX principles and practices into software development organizations: a case study of influencing events. J. Syst. Softw. **154**, 37–58 (2019). https://doi.org/10.1016/J.JSS.2019.03.066
10. Da Silva, T.S., Silveira, M.S., Maurer, F., Silveira, F.F.: The evolution of agile UXD. Inf. Softw. Technol. **102**, 1–5 (2018). https://doi.org/10.1016/J.INFSOF.2018.04.008
11. Pillay, N., Wing, J.: Agile UX: integrating good UX development practices in Agile. In: 2019 Conference on Information Communications Technology and Society, ICTAS 2019 (2019). https://doi.org/10.1109/ICTAS.2019.8703607
12. Alzayed, A., Khalfan, A.: Analyzing user involvement practice: a case study. Int. J. Adv. Comput. Sci. Appl. **12** (2021)
13. Rukonić, L., Kervyn de Meerendré, V., Kieffer, S.: Measuring UX capability and maturity in organizations. In: Marcus, A., Wang, W. (eds.) HCII 2019. LNCS, vol. 11586, pp. 346–365. Springer, Cham (2019). https://doi.org/10.1007/978-3-030-23535-2_26
14. Curley, M., Salmelin, B.: Open Innovation 2.O: A New Paradigm. OISPG White paper (2013)
15. Christensen, J.F., Olesen, M.H., Kjær, J.S.: The industrial dynamics of Open Innovation— Evidence from the transformation of consumer electronics. Res. Policy **34**, 1533–1549 (2005). https://doi.org/10.1016/J.RESPOL.2005.07.002
16. Gassmann, O.: Editorial: opening up the innovation process: towards an agenda. R D Manage. **36**, 223–228 (2006). https://doi.org/10.1111/J.1467-9310.2006.00437.X
17. Chiaroni, D., Chiesa, V., Frattini, F.: Unravelling the process from Closed to Open Innovation: evidence from mature, asset-intensive industries. R&D Manage. **40**, 222–245 (2010). https://doi.org/10.1111/J.1467-9310.2010.00589.X
18. Gassmann, O., Enkel, E.: Towards a Theory of Open Innovation: Three Core Process Archetypes. https://www.alexandria.unisg.ch/274/. Accessed 07 Mar 2022
19. Hassenzahl, M.: User experience (UX): towards an experiential perspective on product quality. In: ACM International Conference Proceeding Series, pp. 11–15 (2008). https://doi.org/10.1145/1512714.1512717
20. Abrahão, S., Juristo, N., Law, E.L.C., Stage, J.: Interplay between usability and software development. J. Syst. Softw. **83**, 2015–2018 (2010). https://doi.org/10.1016/J.JSS.2010.05.080
21. Zaina, L.A.M., Sharp, H., Barroca, L.: UX information in the daily work of an agile team: a distributed cognition analysis. Int. J. Hum. Comput. Stud. **147**, 102574 (2021). https://doi.org/10.1016/J.IJHCS.2020.102574
22. SEBoK- Guide to the systems engineering body of knowledge. https://sebokwiki.org/wiki/Baseline_(glossary. Accessed 22 Mar 2023

Exploring the Convolutional Neural Networks Architectures for Quadcopter Crop Monitoring

Oliviu Gamulescu, Monica Leba[(✉)] [iD], and Andreea Ionica [iD]

University of Petrosani, Petroșani, Romania
monicaleba@upet.ro

Abstract. Utilizing Unmanned Aerial Vehicles (UAVs) equipped with Convolutional Neural Networks (CNNs) has become pivotal in revolutionizing crop monitoring. This synergy offers unparalleled advantages in precision agriculture by enabling real-time, high-resolution data collection and analysis. UAVs facilitate rapid and cost-effective aerial surveillance, capturing intricate details of crop health, pest infestations, and growth patterns. CNNs, a subset of deep learning, enhance image processing, allowing for automated and accurate identification of anomalies. This integration optimizes resource management, reduces environmental impact, and enhances overall crop yield predictions. The combination of UAVs and CNNs is a transformative approach, empowering farmers with actionable insights for informed decision-making in modern agriculture. The paper is part of the current trend of UAVs use in agriculture, with the main purpose of evaluating the results obtained for crop monitoring by training three different architectures of CNNs, namely the simple sequence, SqueezeNet and GoogleNet. The research was carried out on a number of four identified crops prevalent in the studied area and allows, based on the results obtained, to choose the most suitable variant of CNN taking into account training precision, classification accuracy, initial costs and in-site use.

Keywords: drone · CNN · UAV · multi-criteria choice · agricultural

1 Introduction

The practical importance of this research is related to the recognition of the crop type for real time monitoring of their status providing a clear crop record. The use of drones is increasing in various fields, including agriculture.

There are several countries that have made a significant contribution to the research and development of the use of drones in agriculture. These countries have invested in technology and developed innovative solutions to support modern agriculture. Some of these include:

- The US is a leader in the development and implementation of drone technology in agriculture. Many companies in Silicon Valley and elsewhere in the country have invested in advanced solutions for mapping and monitoring agricultural fields.

- Canada is known for developing effective agricultural drones and extensive research in the field. Canadian universities and research institutions are involved in studies on the use of drones for crop and soil monitoring.
- Due to the large size of agricultural holdings in Australia, drones have become essential for effective surveillance and management of agricultural land. Studies have been conducted into the use of drones in pest detection and animal herding.
- The Netherlands has a long tradition in precision agriculture and is a pioneer in the use of drones for field monitoring and resource management. Here, drones are used to map water stress and support sustainable agriculture.
- Given the difficult climatic conditions, Israel has developed advanced technologies for efficient irrigation and crop management. Drones are used to monitor soil moisture and precisely apply water and fertilizers.
- France has been involved in the use of drones for precision agriculture and is a research center for farmland and crop mapping technologies.

These are just a few examples of countries that have made a significant contribution to the research and development of drones in agriculture. However, technology is still evolving, and many countries are engaging in research and development to manage and improve agricultural productivity.

The research objectives assumed in this paper are: Analysis of crop type identification possibilities (using different convolutional neural network architectures) and Analysis of acting alternatives (using multicriteria analysis). In order to address these research objectives, the paper was structured as follows: Literature review, Materials and methods, Results and findings and Conclusion.

2   Literature Review

The literature presents a comprehensive analysis of how farming systems related technologies are developed. As precision agriculture evolves and grows, so do the potential and opportunities. The new field of unmanned aerial vehicles (UAV) demonstrates this. There are several sectors where the UAV is welcomed, but in the agricultural sector it has proven to be of great value for crop yield and biomass estimation [1].

[2] discusses a study focused on the application of high-resolution UAV images for image classification in agricultural environments. To solve this problem, researchers conducted a study on the application of deep learning algorithms that can be used to analyze agricultural land where there are UAV datasets and small-scale composite coverage in Korea.

Also in Korea, [3] presents the impact of incorporating texture information on crop classification using machine learning techniques and UAV images.

Excessive use of agrochemicals to control weed infestation has serious associated agronomic and environmental repercussions [4]. An adequate amount of pesticides/chemicals is essential to achieve the desired smart agriculture and precision agriculture. In this regard, targeted weed control will be a critical component, significantly helping to achieve the objective. A prerequisite for such control is a robust classification system that could accurately identify weed crops in a field. In this regard, UAVs can acquire high-resolution images that provide detailed information for weed distribution and provide a cost-effective solution. Most established classification systems implementing UAV imagery are supervised, relying on image labels.

Reference [5] is focused on the use of UAV images for the classification of different types of crops using a combination of deep learning techniques and spatial - spectral features. The CNN model used is a simplified version of the AlexNet architecture.

[6] aims to differentiate between different varieties of pistachio and separate weeds from trees using a combination of Landsat 8 satellite imagery and UAV imagery.

[7] focuses on the development of a solution for monitoring agave crops, taking advantage of the opportunity to obtain a high spatial resolution, which is provided by low-cost UAVs. For classification, there is used an unsupervised approach: K-means. The proposed approach allows for an inspection of agave crops.

3 Materials and Methods

For the recognition of crops, the use of a Parrot Bebop2 drone was chosen, which has both a high-performance video camera and software for planning routes to explore autonomously based on GPS landmarks. [8] Another advantage is the fact that it can be easily integrated into MatLab, as there is a dedicated tools package. For the identification of crops, 4 categories specific to the analyzed area, a mountain type are in the south of Romania, were chosen, namely sunflower, wheat, corn and alfalfa. These cultures predominate in this area, the rest being covered by the alpine gap. In order to develop the recognition system based on a convolutional network, several flights were made in the area, after which approximately 500 images were obtained. Next, a smaller number (10%) of these images were chosen to train the network, given that the training was done on a CPU-only computing system without a GPU and the purpose was to prove the concept.

In the development of the automatic crop recognition system, three CNN architectures were chosen. The first architecture is a simple sequential architecture that consists of 2 levels of convolution and a final classifier in the 4 categories. This network was trained from scratch using the previously collected image set directly. The second is a pretrained SqueezeNet architecture where the final convolution and classification layers have been modified to retrain the network on the crop image set. The third is a pretrained GoogleNet architecture where the final convolution and classification layers have been also modified to retrain the network on the crop image set. The 3 types of architectures were chosen because each has its own advantages, the first allows customization and adaptation to the type of processor used for implementation, and the second and third are more complex ensuring greater precision and a pre-optimization of the implementation. The results of the use of the three types of architectures are analyzed according to: the accuracy obtained during training and validation, the confusion matrix and the accuracy of the recognition of completely new images.

The analysis of the 3 architectures for the considered data set was carried out using Deep Network Designer from MatLab in which each of the CNNs was visually designed. Then the basic code of the network was generated, the training with the previously captured and labeled image dataset was added, together with the code of generating the confusion matrix and for testing the network on new images recognition.

For choosing the optimal alternative a multi-criteria model was developed and applied taking into account the following criteria: training accuracy, classification errors, investment costs and in-site/remote use.

The mathematical model for multi-criteria analysis considers the following:

Matrix $E[i,j]_{n \times m}$ containing the values resulting from the analysis of each type i of CNN based on the criterion j, where n = number of CNNs considered and m = number of criteria.

$$\text{Matrix} \quad T[j]_m = \begin{cases} -1, \textit{for minimum criterion} \\ +1, \textit{for maxmum criterion} \end{cases} \tag{1}$$

Matrix $W[j]_m$, containing the importance coefficients related to each criterion. There is computed a new matrix U as follows:

$$U[i,j]_{n \times m} = \frac{E[i,j] \cdot T[j] + \max_i E[i,j] \cdot \frac{1-T[j]}{2} - \min_i E[i,j] \cdot \frac{T[j]+1}{2}}{\max_i E[i,j] - \min_i E[i,j]} \tag{2}$$

and based on it the usefulness of each CNN is determined based on the considered criteria, by:

$$V[i] = \sum_{j=1}^{m} U[i,j] \cdot W[j] \tag{3}$$

Finally, the network for which the V value is maximum is chosen.

4   Results and Findings

From the set of images collected in the monitoring flights, different images were chosen for each crop type, resized to (227 × 227) for the SqueezeNet and to (224 × 224) for the GoogleNet, and only the 3 standard RGB color channels were considered. Each image has been labeled with the name of the culture it contains. In Fig. 1 are shown images distribution and some examples.

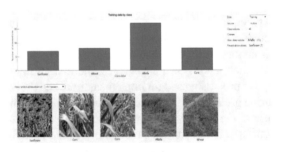

Fig. 1. Crop dataset

The first CNN architecture is a sequential one with 2 convolution layers and one classification layer. The network was initially trained with 20 epochs, and the accuracy obtained was 90% (see Fig. 2).

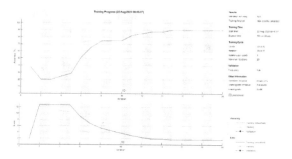

Fig. 2. Training precision

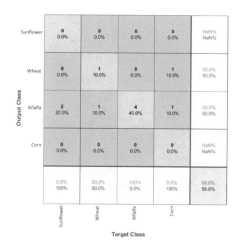

Fig. 3. Confusion matrix

From the set of 20% validation data, the confusion matrix was obtained as in Fig. 3. It can be seen that although the training accuracy was 90%, the validation only has 50% accuracy. In addition, when testing on completely new images, the accuracy of 50% was preserved. In Fig. 4, an incorrect classification is presented for the case Corn.

Fig. 4. Wrong classification of corn

Given that the accuracy was only 90% in training for the 20 epochs case, respectively 50% in validation, a new training was carried out for the same network architecture, increasing the number of epochs from 20 to 30.

The accuracy in this case for training reached 95% (see Fig. 5) and for validation 80% according to the Confusion matrix (see Fig. 6).

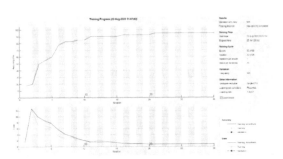

Fig. 5. Training precision

Confusion Matrix

Output Class	Sunflower	2 20.0%	0 0.0%	0 0.0%	0 0.0%	100% 0.0%
	Wheat	0 0.0%	1 10.0%	0 0.0%	0 0.0%	100% 0.0%
	Alfalfa	0 0.0%	0 0.0%	4 40.0%	1 10.0%	80.0% 20.0%
	Corn	0 0.0%	1 10.0%	0 0.0%	1 10.0%	50.0% 50.0%
		100% 0.0%	50.0% 50.0%	100% 0.0%	50.0% 50.0%	80.0% 20.0%
		Sunflower	Wheat	Alfalfa	Corn	

Target Class

Fig. 6. Confusion matrix

When testing the 30 epochs CNN using completely new images from a set of 5 images, 4 were correctly classified, maintaining the precision from the confusion matrix.

To further increase the accuracy, a new pre-trained CNN architecture, namely SqueezeNet, was chosen. This network works with image inputs of 227×227 pixels and RGB channels. We chose the SqueezeNet variant because it is a convolutional neural network that performs better than AlexNet, but with 50 times fewer parameters. The network is pre-trained with the ILSVRC-12 challenge ImageNet Dataset which is a different image set than crop images which means the model needs to be retrained to correctly classify crops. For this, the 9th convolution layer was replaced with a new one and also the final classifier in such a way as to achieve the classification in the 4 researched classes. Considering these 2 changes and the use of the own Dataset, the network was retrained. The retraining time obtained was approximately 20 times shorter than for the previous network.

The accuracy obtained after training with 10 epochs and 3 iterations per epoch is 100%, and the validation accuracy is 80% (see Fig. 7).

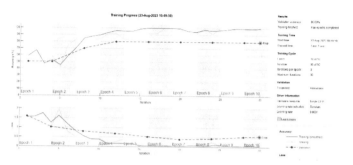

Fig. 7. Training precision

However, from the Confusion matrix for the 20% validation data, a precision of 100% is observed (see Fig. 8).

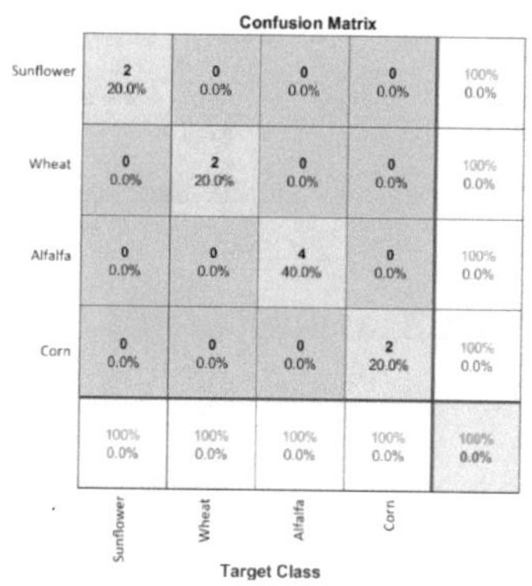

Fig. 8. Confusion matrix

Testing on completely new images we got 7 out of 8 correct classifications (see Fig. 9).

The third variant was also a pre-trained CNN, namely GoogleNet. This network works with image inputs of 224×224 pixels and RGB channels. The architecture of GoogleNet has 22 layers connected in such a way that we have a good efficiency of calculations. The architecture of GoogleNet is made so that it can also run with low computing resources. The network is pre-trained with over 1 million images and can perform classification into 1000 categories. However, these images are different from the crop images, which means that the GoogleNet model also needs to be retrained to correctly classify the crops. For this, the fully connected layer was extracted and both it

Fig. 9. Correct classification

and the classification layer were replaced with new layers in such a way as to achieve the classification in the 4 researched classes.

Considering these changes and using the own Dataset, the network was retrained. The retraining time obtained was approximately 2 times shorter than for the previous network (see Fig. 10).

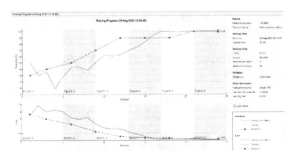

Fig. 10. Training precision

For the 20% validation data in the confusion matrix, an accuracy of 80% was obtained (see Fig. 11).

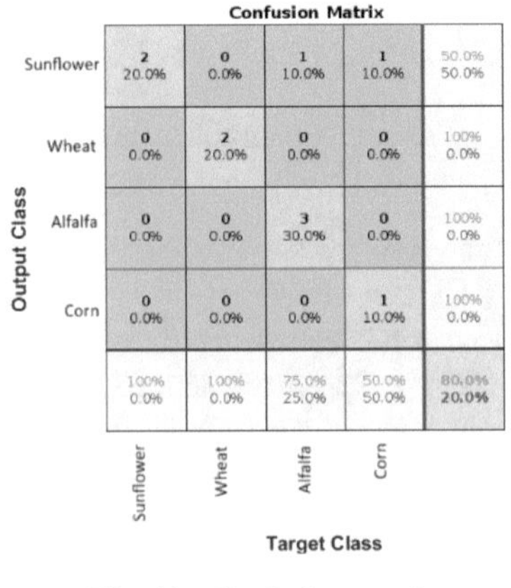

Confusion Matrix

	Sunflower	Wheat	Alfalfa	Corn	
Sunflower	2 20.0%	0 0.0%	1 10.0%	1 10.0%	50.0% 50.0%
Wheat	0 0.0%	2 20.0%	0 0.0%	0 0.0%	100% 0.0%
Alfalfa	0 0.0%	0 0.0%	3 30.0%	0 0.0%	100% 0.0%
Corn	0 0.0%	0 0.0%	0 0.0%	1 10.0%	100% 0.0%
	100% 0.0%	100% 0.0%	75.0% 25.0%	50.0% 50.0%	80.0% 20.0%

Output Class / Target Class

Fig. 11. Confusion matrix

Testing with new images instead provided 8 out of 8 correct classifications (see Fig. 12).

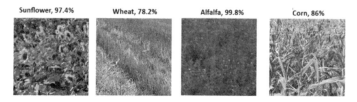

Fig. 12. Classification results

Since only the training accuracy and the classification errors are not enough to choose the optimal solution, it was necessary to introduce new criteria such as investment costs, and in situ/remote use.

Having 4 choice criteria, a multicriteria decision model was applied based on the mathematical support presented in Materials and methods section.

Thus, we have n = 3, representing the 3 considered CNN architecture versions and m = 4, representing the 4 evaluation criteria: Training precision, Classification precision, Initial costs and In-site use.

$$E = \begin{bmatrix} 95\% & 80\% & 1K & 0.9 \\ 100\% & 80\% & 3K & 0.7 \\ 100\% & 100\% & 2K & 0.8 \end{bmatrix} \tag{4}$$

$$T = \begin{bmatrix} +1 & +1 & -1 & +1 \end{bmatrix} \tag{5}$$

$$W = \begin{bmatrix} 0.2 & 0.4 & 0.3 & 0.1 \end{bmatrix} \tag{6}$$

Using the above values, we compute the matrix:

$$U = \begin{bmatrix} 0 & 0 & 1 & 1 \\ 1 & 0 & 0 & 0 \\ 1 & 1 & 0.5 & 0.5 \end{bmatrix} \tag{7}$$

And the final CNN architecture evaluation based on the 4 criteria:

$$V = \begin{bmatrix} 0.4 \\ 0.2 \\ 0.8 \end{bmatrix} \tag{8}$$

So, the GoogleNet architecture is the best suited for the considered purpose.

5 Conclusion

The research presents the analysis of the use of three CNN networks in the identification of crop types based on images collected by the Parrot Bebop 2 drone. The research has practical utility being reproducible for any type of crop analysis, providing a useful tool in practical decisions according to the criteria mentioned with the possibility of modifying them according to the particularities of the area, respectively the studied problem. It can be used to analyze the stage of crops, the evolution from sowing to harvesting or in case of unforeseen situations, such as floods, pests or other.

There were identified two limitations, namely: dataset enhancement for CNN in crop identification and multi-criteria analysis and multi-criteria decision making (MCDM) methods. While the Convolutional Neural Network (CNN) used for crop identification has shown promising results, there is room for improvement in the dataset. The current dataset might be limited in its diversity, potentially affecting the CNN's ability to accurately identify a broader range of crop types. To enhance the model's performance, acquiring a more extensive and diverse dataset could be beneficial. This could include incorporating samples from different geographical regions, various growth stages, and under different environmental conditions. By enriching the dataset, the CNN could be trained on a more representative set of examples, leading to improved accuracy and generalization across a wider variety of crops. The multi-criteria analysis employed in the paper relies on specific decision-making methods. However, it's acknowledged that there exist more advanced and rigorous Multi-Criteria Decision Making (MCDM) methods that leverage higher complexity algorithms. These advanced MCDM techniques may offer a more sophisticated approach to decision-making processes, considering a broader array of criteria and their interdependencies. This could involve incorporating algorithms that better capture the nuances and relationships among diverse criteria, contributing to a more robust decision-making process. However, it's essential to weigh the benefits of increased complexity against computational resources and practical applicability in specific contexts.

References

1. Schueller, J.K.: Intensive farming systems: efficiency and innovation for sustainability— the case of North America. In: 25th Annual Meeting of the Club of Bologna EIMA International (2014)
2. Choi, S.-K., Lee, S.-K., Kang, Y.-B., Seong, S.-K., Choi, D.-Y., Kim, G.-H.: Applicability of image classification using deep learning in small area: case of agricultural lands using UAV image. JKSGPC 38(1), 23–33 (2020)
3. Kwak, G.-H., Park, N.-W.: Impact of texture information on crop classification with machine learning and UAV images. Appl. Sci. 9(4), 643 (2019)
4. Khan, S., Tufail, M., Khan, M.T., Khan, Z.A., Iqbal, J., Alam, M.: A novel semi-supervised framework for UAV based crop/weed classification. PLoS One 16(5), e0251008 (2021)
5. Fan, C., Lu, R.: UAV image crop classification based on deep learning with spatial and spectral features. In: IOP Conference Series: Earth and Environmental Science, vol. 783, 2nd ICGMRS (2021)
6. Reza, H., Malamiri, G., Aliabad, F.A., Shojaei, S., Morad, M., Band, S.S.: A study on the use of UAV images to improve the separation accuracy of agricultural land areas. Comput. Electron. Agric. 184, 106079 (2021)
7. Calvario, G., Sierra, B., Alarcón, T.E., Hernandez, C., Dalmau, O.: A multi-disciplinary approach to remote sensing through low-cost UAVs. Sensors 17(6), 1411 (2017)
8. Gamulescu, O.-M., Rosca, S.-D., Leba, M., Ionica, A.: Agricultural land management using drones. In: 17th edition International Technical-Scientific Conference, Modern Technologies for the 3rd Millennium, Oradea (Romania), p. 39. EdLearning (2018)

The State of Smalltalk to Java Transformation: Approaches Review

Marek Bělohoubek$^{(\boxtimes)}$ and Robert Pergl

Faculty of Information Technology, Czech Technical University in Prague,
Prague, Czech Republic
{marek.belohoubek,robert.pergl}@fit.cvut.cz
http://ccmi.fit.cvut.cz

Abstract. The world of software engineering is an ever-changing landscape of tools, languages, and technologies. The most difficult kind of technological transition is the transfer of a system from one programming language to another, which might be necessary for increasing number of legacy systems, as their original architecture cannot cope with modern requirements. In this paper, we aim to provide readers with an overview of the current state of research on transformation from Smalltalk to other object-oriented languages, with the main focus on transformation to Java. We first compare the two languages and then continue with section containing analysis of existing works in three subsections. All of the projects were then compared in a separate section, with a special focus on the level of automation, availability of the tools, compatibility with current technologies, and expandability of the translated application.

Keywords: Smalltalk · Java · code translation · Smalltalk to Java transformation

1 Introduction

The field of software engineering is a dynamic environment that constantly evolves with the introduction of new tools, languages, and technologies.

Initially, systems are developed using a specific set of technologies. However, due to various factors such as performance enhancements, security updates, and changes in functional requirements, the system is often compelled to incorporate new libraries, adopt the latest technologies, or even undergo a complete overhaul of its underlying architecture.

As the frequency of alterations increases [1], it is necessary to create new systems with these changes in mind, preparing them for future technological improvements and therefore ensuring that it will be possible to make such changes with minimal effort [2].

The particularly difficult kind of technological transition is the transfer of a system from one programming language to another, which might be necessary for increasing number legacy systems, as their original architecture cannot cope with modern requirements. These changes are not common due to their complexity

Á. Rocha et al. (Eds.): WorldCIST 2024, LNNS 989, pp. 235–241, 2024.
https://doi.org/10.1007/978-3-031-60227-6_21

and the amount of labour required, particularly when there are no automated tools to help with the conversion of source code between languages.

Modern established programming languages are well supported with tools (such as [3]) that can do partial or in some cases even complete transformation automatically, but sadly Smalltalk and its dialects are severely lacking in this department.

In this paper, we aim to provide readers with an overview of the current state of research on transformation from Smalltalk to other object-oriented languages, with the main focus on transformation to Java.

2 About Languages

Contents of this section cover both Smalltalk and Java, namely their basic principles, similarities, and differences.

2.1 Smalltalk

Smalltalk is a dynamically typed, purely object-orientated programming language that was conceived in the 1970 s as Smalltalk-72, and the first release to the general public came in the form of Smalltalk-80 in the 1980 s [4].

Distributed as an image, Smalltalk comes with its own development environment and virtual machine, which means that any application written in Smalltalk requires the image to be run.

There are multiple dialects of Smalltalk with their own implementations of the Smalltalk virtual machine, such as the commercial Visual Works [5] systems or the open-source projects Squeak [6] and Pharo [7].

2.2 Java

Java is a statically typed, object-oriented, high-level programming language that was developed in the 1990 s and first released in 1996 as Java 1.0 [8].

Java Virtual Machine (JVM) is a key feature of Java that enables applications written in Java to be executed on any computer architecture, regardless of the platform. This allows the programmer to be shielded from the platform-specific details.

2.3 Language Differences

There are three major differences between Smalltalk and Java, each of which complicates the translation to such a degree that we talk about transformation instead.

First, they differ in their approach to typing and types. Java is statically typed, this means that all types of variables, method arguments, and return values have to be specified before compilation. Smalltalk, on the other hand, is dynamically typed and therefore checks for type information during run-time only.

The second distinction is found in the discrepancy between the Smalltalk meta-classes and the class features in Java. In Smalltalk, every object has a reference to its metaclass, which contains the methods and variables specific to the class.

In contrast, Java uses the static keyword to declare class features, which does not involve an underlying object, and as a result such features cannot be accessed via reference.

Lastly, while all types in Smalltalk are subtypes of Object, Java also contains eight primitive types. These types contain only values and are absent of any Object-like behaviour (such as methods, initialisation, etc.).

Resulting problems force any transformation tool to either inferring all types used in the code to be used in the statically typed Java, and/or provide an additional framework to simulate Smalltalk-like dynamic behaviour.

3   Previous Work

This section consists of three subsections: Foundations, Existing works, and Alternative solutions.

3.1   Foundations

In this subsection, we take a look at the most important projects and publications that laid the foundations for implementation of Smalltalk to Java translation.

The foundations of Smalltalk to Java transformation are actually older than Java itself, because most of the earliest projects that focused on transforming Smalltalk source code to Java draw knowledge from previous research on translating source code from Smalltalk-80 to Objective-C and later to C/C++.

Smalltalk-80 version 2 was released to the general public in 1983 along with documentation by Goldberg and Robson [9], and just four years after its release, came the first tool for translating the Smalltalk code to Objective-C, which was named Producer [10]. It was able to translate the source code, but there were several major functional limitations (missing support for blocks and garbage collection) stemming from the absence of those functionalities in Objective-C.

The next major project was the SPiCE system by Kazuki Yasumatsu and Norihisa Doi [11]. SPiCE was the first tool that allowed the transformation from Smalltalk to C through translation of the source code. It supported both block and garbage collection, and from its inception it was capable of executing translated applications in C at similar speeds to Smalltalk.

Java was introduced to the general public in 1995, and it became the focus of multiple research projects as a possible target language for Smalltalk translation. One of the first works dealing with this topic came out in 1998 [12] and focused on translation issues. It proposed solutions for simulating Smalltalk class hierarchy with Java classes and for mapping of instance and class methods between the languages.

This research was followed by another article published in 2003 called Translating Smalltalk blocks to Java [13], which provides an improved method for the translation of blocks. They are now translated using anonymous inner classes with a "Smalltalk-like" interface. Therefore, the transformed code has exactly the same functionality as the original.

It should be noted that this article was written before lambda expressions were added to Java (Java 8 was released in 2014). While lambda expressions cannot fully simulate Smalltalk blocks (for example, assignments to outside variables are forbidden), they can be used as a Java friendly substitution for majority of most common use cases.

3.2   Transformation via Translation

This subsection introduces the reader to projects tools and publications that solve the Smalltalk to Java Transformation through translation of the source code.

The most comprehensive work on Smalltalk to Java transformation using the aforementioned method is dissertation of R. L. Engelbert [14]. Building on the authors previous work [12] it goes in full detail on the principles and methods used for the translation of the source code.

Notable sections focus on mapping of classes, methods, Smalltalk blocks and most importantly on bridging implementation the gap, caused by mismatch of typing styles (dynamic typing in Smalltalk vs. static typing in Java). Three possible solutions are suggested:

– *usage of type inference* - considered too unreliable to be used in automated translation,
– *reflection-based runtime method binding* - capable of producing fully working code at a significant detriment to performance,
– *superclass-based runtime method binding* - reliable solution with the speed of execution effectively equivalent to the original Smalltalk code.

Unsurprisingly considered the best option, superclass-based runtime method binding is implemented by using a single common Java superclass of all translated Smalltalk classes. This superclass contains a generic implementation of every single method used by its subclasses, which are then overridden by concrete implementations in the subclasses.

This results in performance of the translated code being effectively equal to the original code, but it greatly complicates future expansions of the translated code, due to creation of massive "SmalltalkObject" superclass and its ubiquitous usage in the translated code.

The next major project that used the translation of the source code was the Master's thesis of M. Skarsaune [15]. It consisted of implementing tools for Smalltalk to Java translation following the methods and principles defined in

R. L. Engelbert's work and using them to successfully translate existing Smalltalk system SICS.

These tools were implemented for IBM VisualAge Smalltalk and thus contain some modifications of the original design, due to the differences between the Smalltalk dialects. Sadly due to the close relation between the work and real business system, some of the implementation details had to be hidden and in some instances there are several proposed solutions for particular problems, but there is little to no indication which solution was used.

3.3   Alternative Solutions

This subsection focuses on tools and projects that don't translate Smalltalk source code directly to Java source code, but either use some special "translation dialect" of Smalltalk (usually extended by introduction static typing) or generate special code that is then run on Java Virtual Machine.

Bistro Smalltalk [16] is a programming language that was created as a combination of Smalltalk and Java. It's syntax is similar to the standard Smalltalk, with several new constructs added from Java. These constructs include: optional type specification for methods and variables, support of variable initialisation in declarations, and ability to declare local variables outside of special sections at the start of a method/block.

Bistro runs on Java Virtual Machine and uses Java runtime reflection to simulate Smalltalk's dynamic methods, which preserves all of their functionality but causes non-negligible performance reduction.

The last advantage of using Bistro is its ability to be used as a migration tool, by first using Bistro utilities to translate source code from standard Smalltalk to Bistro and then from Bistro to Java.

Athena Smalltalk [17] is a system that allows the embedding of the Smalltalk image into the Java application and running it on a Java Virtual Machine. It was created as a successor to the Java-based Smalltalk interpreter called SmallWord [18]. Athena is based on an open-source dialect of Smalltalk - Squeak and, while it represents an important example of Smalltalk and Java integration, it is now severely outdated (the last available version was written for Java 1.6).

Lastly there is a commercial solution by Synchrony Systems, which allows generation of fully functional Java code and even major architectural changes during the transformation, but relies on static type inference requiring a higher amount of manual input compared to the previous methods.

4   Comparisons

In this section, we are comparing all previously mentioned solutions and discussing their pros and cons.

There are four main categories for the comparison: level of automation, availability of the tools, compatibility with the current technologies, and most importantly expandability of the translated application.

The system proposed by R. L. Engelbert is very robust with the possibility of a very high level of automation. Sadly it does not come with the tools themselves and although code translated by it provides excellent simulation of Smalltalk functionality in Java, it also creates environment that is not particularly easy to work with during further development of the translated system.

The implementation by M. Skarsaune has many of the same qualities as R. L. Engelbert's system (high level of automation and excellent simulation capabilities), but it also shares most of it's downsides, mainly the complications for further development and inability to access the tools (although in this case it is caused by them being commercial secret).

Bistro Smalltalk on the other hand provides tools that are easy to access and that produce slightly more development-friendly code, but this is offset by lower level of automation and the fact that the tools haven't been maintained for quite a while.

Athena Smalltalk can be used to run Smalltalk code on Java Virtual Machine and provides easy-to-access tools, but they are greatly outdated and continuous development is therefore nigh impossible.

Synchrony Systems transformation solution promises great compatibility with the current technologies, excellent expandability of the translated applications, and good availability of the tool (for a price); however, it also requires more significant manual input as a result of static type inference.

5   Conclusion

In this article, we have analysed the current state of Smalltalk to Java transformation, by comparing the two languages and analysing existing publications and tools.

Although there are several works focused on this topic, there seems to be a lack of usable tools, with most of them being either proprietary, outdated, or not implemented at all.

Based on this analysis, we believe that there is a space for further research and development that would address all features of modern versions of both languages as well as the latest results of research aimed at Smalltalk type inferring.

Acknowledgement. This research was supported by the grant of Czech Technical University in Prague No. SGS23/206/OHK3/3T/18.

References

1. Kurzweil, R.: The law of accelerating returns. In: Teuscher, C. (eds.) Alan Turing: Life and Legacy of a Great Thinker. Springer, Berlin, Heidelberg (2004). https://doi.org/10.1007/978-3-662-05642-4_16
2. Dvořák, O., Pergl, R.: Tackling rapid technology changes by applying enterprise engineering theories. Sci. Comput. Program. **215**, 102747 (2022)
3. CodeConvert AI - Convert code with a click of a button. [N.d.]. https://www.codeconvert.ai/

4. Kay, A.C.: The early history of smalltalk. In: History of Programming Languages—II. New York, NY, USA: Association for Computing Machinery, pp. 511–598 (1996). ISBN 0201895021. https://doi.org/10.1145/234286.1057828

5. Custom Software Application Development Services - Cincom VisualWorks®—Cincom Smalltalk® (2023). https://www.cincomsmalltalk.com/main/products/visualworks/

6. SQUEAK.ORG. Squeak/Smalltalk. [N.d.]. https://squeak.org/

7. Pharo - Welcome to Pharo! [N.d.]. https://pharo.org/

8. Cosmina, I.: An Introduction to Java and Its History. In: Java 17 for Absolute Beginners: Learn the Fundamentals of Java Programming. Berkeley, CA: Apress, pp. 1–31 (2022) ISBN 978-1-4842-7080-6. https://doi.org/10.1007/978-1-4842-7080-6_1

9. Goldberg, A., Robson, D.: Smalltalk-80: the language and its implementation, Addison-Wesley Longman Publishing Co., Inc., (1983)

10. Cox, B.J., Schmucker, K.J.: Producer: a tool for translating Smalltalk-80 to objective-C. In: Conference Proceedings on Object-Oriented Programming Systems, Languages and Applications, pp. 423–429 (1987)

11. Yasumatsu, K., Doi, N.: SPiCE: a system for translating Smalltalk programs into a C environment. IEEE Trans. Softw. Eng. **21**(11), 902–912 (1995). https://doi.org/10.1109/32.473219.

12. Engelbrecht, R.L., Kourie, D.G.: Issues in translating Smalltalk to Java. In: Koskimies, K. (ed.) CC 1998. LNCS, vol. 1383, pp. 249–263. Springer, Heidelberg (1998). https://doi.org/10.1007/BFb0026436

13. Engelbrecht, R.L., Kourie, D.G.: Translating Smalltalk blocks to Java. IEE Proc. Softw. **150**(3), 203–211 (2003)

14. Engelbrecht, R.L., et al.: Implementing a Smalltalk to Java translator. 2006. PhD thesis. University of Pretoria

15. Skarsaune, M.: The SICS Java Port Project: automatic translation of a large system from Smalltalk to Java. 2008. MA thesis

16. BOYD, N.S.: Bistro overview (2010) https://bistro.sourceforge.net/overview.htm

17. The Athena Smalltalk (2008). http://www.bergel.eu/athena/

18. Budd, T.: Small World: A tiny Smalltalk-80 like interpreter. [N.d.]. https://web.engr.oregonstate.edu/~budd/SmallWorld/ReadMe.html

A Self-adaptive HPL-Based Benchmark with Dynamic Task Parallelism for Multicore Systems

Cassiano Rista[1](✉), Marcelo Teixeira[2], and Mauro Fonseca[1]

[1] Federal University of Technology - PR (CPGEI), Curitiba, Brazil
{rista,maurofonseca}@utfpr.edu.br
[2] Federal University of Technology - PR (PPGEEC), Pato Branco, Brazil
mtex@utfpr.edu.br

Abstract. Benchmarks play a crucial role in enhancing systems, applications, and technologies across multiple domains. In scientific computing, traditional benchmarks such as HPL (High-Performance Linpack) require parameters such as the number of processes, input data size, and other execution details to reflect the characteristics of the system under test, which leads to the adoption of fine-tuning strategies. In this sense, this paper presents a novel implementation of the HPL, originally used for evaluating distributed-memory parallel architectures. The proposed benchmark, named SA-HPL, enhances HPL with self-adaptive features to facilitate the evaluation of shared-memory parallel architectures. The inclusion of novel schedulers for dynamic task parallelism, combined with simplified configuration settings, makes the SA-HPL well-suited to provide enhanced performance and efficiency without requiring fine-tuning configurations. Results show that our approach improves the throughput to 19534.50 MFLOPS and maintains an efficiency level of 74.70% compared to the baseline (with unbalanced workload) in a workstation setup, which recorded 16267.60 MFLOPS and 61.23%, respectively.

Keywords: Self-Adaptive · Benchmark · Dynamic Task Parallelism

1 Introduction

Recent advances in microprocessor technologies and high-performance multicore computing have made parallel processing more accessible, either in industry or academia. By harnessing the parallelism of multicore systems, it becomes possible to meet high-performance computing requirements, analyze complex problems, simulate large-scale phenomena, and derive accurate and timely results.

However, the utility of multicore systems depends on the development of applications that can efficiently take advantage of such resources. Using these systems with traditional benchmarks for scientific computing [1,3,8], for example, requires manual configurations to match the level of parallelism provided by the application. This implies having to define input data size and other execution details to reflect the characteristics of the system under test, which leads to

Á. Rocha et al. (Eds.): WorldCIST 2024, LNNS 989, pp. 242–251, 2024.
https://doi.org/10.1007/978-3-031-60227-6_22

the adoption of empirical fine-tuning strategies. This complexity in benchmark management calls for a more adaptive solution.

In this context, we introduce SA-HPL (Self-Adaptive High-Performance Linpack) for Multicore Systems, which is a self-adaptive benchmark designed to address these challenges without relying on complex fine-tuning techniques to optimize performance for specific hardware architectures. SA-HPL builds upon the widely recognized HPL benchmark, specifically tailored for assessing shared-memory parallel architectures frequently employed in scientific computing. It incorporates adaptive techniques and dynamic schedulers to seamlessly adapt to the underlying architecture's requirements.

The core concept behind SA-HPL is to dynamically adjust its parameters and task distribution according to the characteristics of the underlying architecture and the current workload. By introducing self-adaptive capabilities, SA-HPL optimizes resource allocation, enhances throughput, and maintains high levels of efficiency, even in the face of varying system configurations and workload demands. In essence, the purpose of SA-HPL is to simplify the benchmark's complexity for external users through autonomic computing.

To solve parallel linear systems of equations, SA-HPL harnesses the capabilities of self-adaptive scheduling policies, including dynamic and guided schedulers provided by OpenMP. These schedulers facilitate intelligent load balancing by dynamically adjusting task distribution based on the system's conditions and demands. SA-HPL divides the linear system of equations into smaller sub-problems, which are then assigned to parallel processes for independent solutions using the LU decomposition method from the Eigen library [5].

To evaluate the effectiveness of SA-HPL, we conducted extensive experiments across a range of hardware setups, including workstations and servers. Our results demonstrate significant performance improvements and enhanced efficiency. These findings offer users a basis for comparing MFLOPS (Million Floating Point Operations Per Second) throughput across different hardware configurations. The contributions of SA-HPL address the need for a self-adaptive benchmark that not only optimizes resource allocation but also improves throughput while maintaining high efficiency levels. Additionally, it simplifies manual configuration settings, reducing complexity.

2 Background

The autonomic computing initiative, launched by IBM [10], aims to address complexity in the software industry by adding self-managing properties to systems through the autonomous key elements in Fig. 1. These elements can adapt to changing conditions without constant human intervention, leading to reduced costs, improved performance, and increased safety.

A reference architecture called the MAPE-K loop (Monitor-Analyse-Plan-Execute with Knowledge) [10] was defined to structure those elements. It starts with monitoring, which collects and synthesizes data from sensors to build a model for analysis. The analysis activity assesses the situation and detects problems, sending them to the planner. The planner develops a plan to address the

244   C. Rista et al.

Fig. 1. Basic properties of autonomic computing to simplify system management.

identified issues. Next, the plan is put into action during the execution activity. The knowledge gained is used to continuously improve the system's performance. The ultimate goal is to simplify system management for administrators by reducing complexity and providing intuitive interfaces.

2.1 Linpack Benchmark

Linpack is a well-known open-source benchmark for scientific computing. It measures the floating-point arithmetic capabilities of a computer by using a dense matrix factorisation algorithm to solve a set of linear equations, a common problem in many scientific and engineering applications. In this paper, we leverage the Linpack as a foundational reference for our self-adaptive proposal, emphasizing that the it lacks self-adaptive management capabilities.

Its original implementation [3] was written in Fortran and included a set of BLAS routines for matrix operations. Since then, various implementations of Linpack have been developed in different programming languages, such as C, C++, and Python, and they also rely on BLAS for their core computational routines. Essentially, it measures the time taken to solve a set of linear equations of the form $Ax = b$, where A is an n-by-n matrix of coefficients, b is a vector of constants, and x is the solution vector [3]. The benchmark also allows for the calculation of throughput in terms of the number of MFLOPS that it can perform for future comparisons.

High-Performance Linpack (HPL) [3] is a widely used benchmark for measuring the performance of supercomputers. It is considered the standard implementation of the Linpack benchmark and includes optimizations specifically designed for high-performance computing. HPL requires MPI and either BLAS or VSIPL libraries. It provides a testing program to measure accuracy and performance, utilizing LU factorization with algorithms such as row partial pivoting and multiple lookahead depths. HPL is used to rank supercomputers on the TOP500[1].

HPL usually uses the LU decomposition algorithm to solve a set of linear equations, which involves transforming the coefficient matrix A into the product

---

[1] The TOP500 list (http://www.top500.org) shows the 500 most powerful commercially available computer systems.

of a lower triangular matrix L and an upper triangular matrix U using a series of elementary row operations.

HPL offers numerous input parameters (including BLAS and MPI routines) for users to tune the performance. These parameters pertain to problem sizes, machine configurations, and algorithm features. The two most important parameters are the problem size (N) and block size (NB). The HPL is essentially a volume-to-surface problem, where increasing the problem size can lead to higher performance scores (performed on MFLOPS).

Fine-tuning performance is possible, but it is limited by the parameters configuration, which is difficult and depends on specific knowledge and caution to avoid exceeding the physical memory size. As N increases, the execution time of HPL grows significantly. NB block size defines how data is distributed and the level of computational granularity. Thus, the optimal NB size is a trade-off between inter-node load balance and single-node performance.

2.2 Eigen Library

Eigen is an open-source C++ template library (it is licensed under the MPL2) for linear algebra and provides classes for matrices, vectors, numerical solvers, and factorisations. These flexible and comprehensive classes allow for high-performance and well-structured code representing high-level operations. It supports [5]: various matrix decompositions and geometry features; features such as non-linear optimisation, matrix functions, a polynomial solver, FFT, and much more.

It is important to emphasise that Eigen has no dependencies other than the C++ standard library. In short, it is a pure template library defined in the headers, using the standard C++98, and so should theoretically be compatible with any compliant compiler. Whenever using some non-standard feature, it is optional and can be disabled. Some traditional compilers supported are [5]: GNU C++, MSV, and Intel C++ Compiler.

Eigen is renowned for its performance, achieved through explicit vectorization using a range of SIMD instruction sets, including SSE 2/3/4, AVX, AVX2, FMA, AVX512, ARM NEON (32-bit and 64-bit), PowerPC AltiVec/VSX (32-bit and 64-bit), and ZVector (s390x/zEC13). From version 3.4 forward, Eigen explores the MIPS MSA instruction set with a graceful fallback to non-vectorized code [5]. Additionally, the library optimizes fixed-size matrices to avoid dynamic memory allocation and unrolls loops, further enhancing its performance.

The Eigen library is organised into a Core module and several additional modules, each providing a specific set of functionalities. To use a module, one has to include its corresponding header file in the code. For convenience, the library provides two header files - Dense and Eigen - allowing access to multiple modules simultaneously. Users can easily incorporate the necessary modules into their code by including these headers.

The interest of this paper, related to the Eigen library, is mainly on linear algebra solvers. More specifically, LU factorization that includes two solvers [5]: FullPivLU and PartialPivLU. The FullPivLU and PartialPivLU solvers are both used to efficiently and accurately solve systems of linear equations involving square matrices. The key difference lies in the pivoting strategy used during the LU decomposition, where FullPivLU allows both row and column permutations, while PartialPivLU allows only row permutations. These solvers are widely used in scientific computing and data analysis applications.

3   Related Work

Elasticity assessment and the development of benchmarks to measure the performance of adaptive systems are prosperous research trends. For example, [6] introduce BUNGEE, a flexible benchmark that allows users to define metrics and workload scenarios to optimise elasticity; [7] propose a performance model to evaluate elastic scaling strategies in the cloud; among others.

Other studies concentrate on specific domains. [12] develop a framework to assess the elasticity of graph processing systems, which exhibit dynamic workloads; [2] propose an Elastic Task Scheduling Scheme for coarse-grained reconfigurable architectures, which dynamically adjusts the allocation of processing elements to optimise resource utilisation and task execution time; [4] present an elastic parallel framework called Chronos for stream benchmark generation and simulation, suitable for data stream processing; and [11] introduce C-MART, a benchmark to assess the performance and cost-effectiveness of cloud computing services, which captures the trade-off between performance and cost, aiding in selecting the most suitable provider.

In summary, the literature offers various systems and architectures for evaluating the performance, scalability, cost-effectiveness, and efficiency of computing systems, but they often lack self-adaptive capabilities for dynamic reconfiguration. Nevertheless, by incorporating adaptive techniques and dynamic schedulers, these systems can seamlessly adapt to the underlying architecture's requirements. Self-adaptation is vital for simplifying the (typically complex) parameter configuration process and providing a user-friendly interface for desktops, workstations, or servers, all without human intervention. This paper aims to enable the benchmark to autonomously determine the optimal configuration for the degree of parallelism and the number of tasks, thereby reducing complexity.

4   SA-HPL for Multicore Systems

SA-HPL for Multicore Systems is a novel benchmarking approach that extends the capabilities of the HPL benchmark. It simplifies the computation of machine epsilon by setting it to a predefined value of 2.2e-16, which is adequate for IEEE 754 double precision. While our approach may not achieve the same level of precision as calculating the exact machine epsilon, it offers the advantage of convenient and efficient execution on shared-memory parallel architectures, including single and dual-socket multicore systems.

Furthermore, SA-HPL harnesses the capabilities of advanced CPU instruction sets by utilizing the *march=native* flag. This flag enables the compiler to leverage all instruction sets supported by the CPU, including optimization techniques provided by libraries like Eigen [5]. By integrating Eigen with SA-HPL, users can further enhance computational efficiency and take advantage of specialized linear algebra operations.

To efficiently and adaptively solve parallel linear systems of equations using OpenMP schedulers, including dynamic and guided modes [9] in C++. SA-HPL extends the capabilities of the traditional HPL benchmark and provides a viable option for performance evaluation without the complexity of configuring numerous input parameters, which typically demands domain-specific knowledge. SA-HPL adapt to different system configurations and workloads introducing self-adaptive techniques that optimize the computation process based on the characteristics of the underlying parallel architecture.

Breaking down problems into smaller components is a fundamental element when designing parallel scientific applications, such as LU factorization. This process involves dividing the main problem into smaller sub-problems that can be concurrently addressed by multiple processors, such as dual processors or cores. In the context of SA-HPL, we utilize the LU factorization method provided by the well-known Eigen library. This method is employed to solve linear systems of equations. Each thread operates independently to solve its designated sub-problem, utilizing self-adaptive scheduling techniques.

SA-HPL adjusts task scheduling in real-time based on system conditions and employs this approach to dynamically parallelize computational tasks among threads. In contrast to static scheduling, dynamic scheduling breaks down work into smaller tasks requested by threads, while guided scheduling starts with larger chunks, gradually reducing their sizes. While static scheduling excels in load balancing, it may not be ideal for irregular or dynamic workloads. Dynamic scheduling accommodates variable execution times, and guided scheduling reduces overhead while maintaining load balance, making it especially beneficial for irregular workloads [9].

In the case of SA-HPL, we solve a linear system of equations in the form $A * x = B$, where A is an N x N matrix, x is a column vector of size N, and B is a column vector of size N. The goal is to find the solution vector x that satisfies the equation. The core idea behind SA-HPL is to parallelize the linear system of equations across multiple threads and utilise self-adaptive scheduling (dynamic and guided) to balance the computational load. Each thread independently solves its assigned sub-problem using the LU factorizationmethod provided by the Eigen library [5], a popular linear algebra package.

SA-HPL incorporates self-adaptive schedulers that employ the MAPE-K reference model approach, which stands for Monitor-Analyse-Plan-Execute with Knowledge, as depicted in Fig. 2. This approach reduces complexity for system administrators and enables intelligent and adaptive decision-making by incorporating real-time monitoring and analysis of the system's state. Unlike static scheduling methods, SA-HPL does not rely on input configuration parameters for adaptation, making it a self-adaptive approach. Its goal is to find the optimal relationship between the degree of parallelism and the number of tasks.

To evaluate the performance of SA-HPL, metrics such as execution time and MFLOPS are employed. These metrics quantify the efficiency and computational performance of SA-HPL in solving linear systems of equations. The implementation uses a combination of OpenMP and the Eigen Library in C++. By dividing the problem into smaller subtasks executed by separate threads, parallelism is harnessed to potentially speed-up the overall execution time.

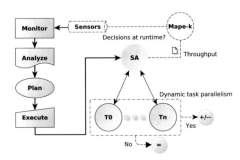

Fig. 2. Theoretical representation of SA-HPL using dynamic task parallelism to find the optimal relationship between the degree of parallelism and the number of tasks.

By breaking down the problem into smaller subtasks, each handled by a separate thread, you harness parallelism to potentially expedite the overall execution time. These parallel threads operate on distinct segments of the problem, and synchronization mechanisms, such as locking and unlocking to manage concurrent access to the shared vector, ensure accurate result consolidation. It's noteworthy that, in this particular case, we did not employ an OpenMP reduction mechanism to merge the results. This decision was driven by the fact that the Eigen library, utilized for solving the system of linear equations via LU factorization, lacks support for such reductions.

The chosen function for solving the linear system is the PartialPivLU() function from the Eigen Library [5]. This function performs LU factorization with partial pivoting, ensuring enhanced numerical stability and improved accuracy. SA-HPL also includes configurations to utilise vectorization support in the Eigen library, enabling the compiler to apply optimisations based on the CPU's supported instruction sets.

We enable vectorization support in the Eigen library by using compiler parameters. We set the *march=native* flag, which allows the compiler to enable all instruction sets supported by the CPU. Additionally, we employ other GNU C++ specific parameters such as *-std=c++1y*, setting the default C++ language standard to C++14, and *-O3*, indicating the highest level of optimisation for producing fast and optimised code. By adopting these settings, the compiler applies various transformations to improve performance, although this may also increase compilation time.

5   Experiments and Evaluations

Experiments were conducted on two hardware setups: a Dell OptiPlex 5080 workstation with an Intel Core i7-10700 CPU (16 cores, 8/8 HT), 16GB RAM, and a Dell PowerEdge R710 server with Intel Xeon E5620 CPUs (16 cores, 8/8 HT), 48GB RAM. Consistency was maintained with Ubuntu Server 20.04.6 LTS, GNU C++ 9.4, Eigen Library 3.3.7, and OpenMP 4.5. Each experiment was repeated 10 times with a 95% confidence interval, and efficiency and throughput were evaluated using metrics like MFLOPS. Two hardware configurations were tested under static and dynamic scheduling scenarios to evaluate the performance of our benchmark. These scenarios provide insights into the benchmark's throughput and efficiency under different scheduling strategies.

5.1   Baseline and Self-adaptive Scheduling

Static scheduling experiments establish a baseline to evaluate the system performance based on the number of threads used (1, 2, 4, 8, 16), being a factor/multiple of the size of our linear system of equations, which is $A * x = B$, where A is an 8000×8000 matrix, x is a column vector of size 8000, and B is a column vector of size 8000. This represents the best-case scenario of static scheduling for possible workload parallelization. However, we intentionally use 12 threads to create workload unbalance and introduce perturbations, deviating from optimal parallelization conditions.

In the context illustrated in Fig. 3, the scenario facilitates a comprehensive performance comparison between dynamic scheduling and the baseline, with a focus on measuring throughput and efficiency. When examining the MFLOPS performance trend in relation to the number of threads, it becomes crucial to highlight the identification of the optimal load balancing point, especially when compared to static scheduling. It is worth noting that the highest efficiency and throughput are achieved in all cases, regardless of the number of threads.

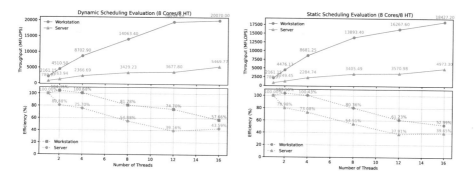

Fig. 3. Illustrates the scenario using **dynamic scheduling** to measure throughput and efficiency, enabling a comparison of performance between workstations and servers.

The behavior observed in the workstation is similar to that found in the server scenario, where dynamic scheduling outperformed static scheduling in all cases. However, in the workstation scenario, a significant drop in efficiency can be observed when using an unbalanced workload with 12 threads in both schedulers. Nevertheless, dynamic scheduling proved to be more effective again in terms of efficiency and throughput, achieving a percentage of 74.70% with 19534.50 MFLOPS, while static scheduling achieved a percentage of 61.23% with 16267.60 MFLOPS.

In Fig. 4, we observe the performance comparison between guided scheduling and the baseline. Guided scheduling consistently outperforms static scheduling in all server experiments, with the same focus on measuring throughput and efficiency. On the workstation, the behaviour was similar, but static scheduling was more efficient with 2 and 4 threads (super-linear), despite they were very close. In the sequence of the experiment, the guided scheduling outperformed static scheduling in all cases. This occurs because guided scheduling [9] reduces scheduling overhead and provides load balancing for workloads with irregularity.

Lastly, it is worth noting that, in certain cases, such as when using 2 or 4 threads (see baseline in Fig. 4), it is possible to observe super-linear efficiency specifically with the workstation. This phenomenon can be attributed to the workstation's substantial cache size, which effectively reduces memory access latency and enhances cache hit rates, ultimately resulting in improved performance beyond linear scaling. The phenomenon does not occur on the server due to it being an outdated generation server. It has already surpassed its peak performance capabilities, and its CPU and RAM, instruction set, and cache size are comparatively less robust when compared to modern workstations.

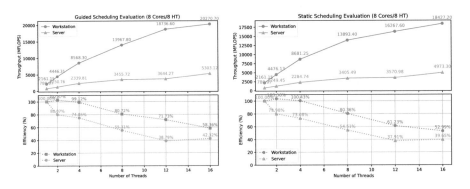

Fig. 4. Illustrates the scenario using **guided scheduling** to measure throughput and efficiency, enabling a comparison of performance between workstations and servers.

In summary, SA-HPL for Multicore systems significantly enhances performance by utilizing all CPU instruction sets, thereby leading to improved throughput and efficiency without the need for fine-tuning configurations.

6   Conclusion and Future Work

This paper introduces SA-HPL, a self-adaptive benchmark extending HPL for multicore system performance evaluation. SA-HPL employs novel schedulers and self-adaptive techniques for dynamically adjusting system configurations and workloads. It optimizes the solving of linear systems through dynamic task parallelization, achieving load balancing and maximizing throughput and efficiency. Scheduling decisions adapt at runtime based on current system conditions.

The SA-HPL is expected to address the gaps of the standard HPL by offering a flexible and adaptable framework that takes into account the characteristics of the linear system and the underlying parallel architecture. It provides researchers and practitioners with a convenient tool for assessing the performance of multicore systems without the complexity of configuring numerous input parameters.

In future work, we plan to further refine and optimise the SA-HPL benchmark by exploring additional self-adaptive techniques and considering other parallel architectures, such as cluster computing in distributed memory. Additionally, as part of our future work, we also plan to introduce a new module for predicting performance based on Generalized Stochastic Petri Nets (GSPN).

References

1. Bienia, C.: Parsec benchmark suite (2023). https://parsec.cs.princeton.edu/
2. Chen, L., et al.: An elastic task scheduling scheme on coarse-grained reconfigurable architectures. IEEE Trans. Parallel Distrib. Syst. **32**(12), 3066–3080 (2021)
3. Dongarra, J., Luszczek, P., Petitet, A.: HPL - a portable implementation of the high-performance Linpack benchmark (2023). http://www.netlib.org/benchmark/hpl/
4. Gu, L., Zhou, M., Zhang, Z., Shan, M.C., Zhou, A., Winslett, M.: Chronos: an elastic parallel framework for stream benchmark generation and simulation. In: 2015 IEEE 31st International Conference on Data Engineering, pp. 101–112 (2015)
5. Guennebaud, G., Jacob, B., et al.: Eigen: C++ templates for linear algebra (2010–2023). http://eigen.tuxfamily.org
6. Herbst, N.R., Kounev, S., Weber, A., Groenda, H.: An elasticity benchmark for self-adaptive iaaS cloud environments. In: IEEE/ACM International Symposium on Software Engineering for Adaptive and Self-Managing Systems, pp. 46–56 (2015)
7. Hwang, K., Bai, X., Shi, Y., Li, M., Chen, W.G., Wu, Y.: Cloud performance modeling with benchmark evaluation of elastic scaling strategies. IEEE Trans. Parallel Distrib. Syst. **27**(1), 130–143 (2016)
8. NPB: NAS benchmarks (2023). https://www.nas.nasa.gov/software/npb.html
9. OpenMP: The openMP API specification (2023). https://www.openmp.org
10. Petrovska, A., Hutzelmann, T., Kugele, S.: A theoretical framework for self-adaptive systems: specifications, formalisation, and architectural implications. In: Proceedings of the 38th ACM/SIGAPP Symposium on Applied Computing, pp. 1440–1449. SAC '23, ACM, New York, NY, USA (2023)
11. Turner, A., Fox, A., Payne, J., Kim, H.S.: C-mart: benchmarking the cloud. IEEE Trans. Parallel Distrib. Syst. **24**(6), 1256–1266 (2013)
12. Uta, A., Au, S., Ilyushkin, A., Iosup, A.: Elasticity in graph analytics? A benchmarking framework for elastic graph processing. In: 2018 IEEE International Conference on Cluster Computing (CLUSTER), pp. 381–391 (2018)

Computer Systems Analysis Focused on the Detection of Violence Against Women: Challenges in Data Science

Mariana-Carolyn Cruz-Mendoza[1,2] ⓘ, Roberto Ángel Meléndez-Armenta[1](✉) ⓘ,
María Cristina López-Méndez[1] ⓘ, and Narendra Velázquez-Carmona[2] ⓘ

[1] División de Estudios de Posgrado e Investigación, Tecnológico Nacional de México/Instituto Tecnológico Superior de Misantla, 93821 Misantla, Veracruz, México
{212T0509,ramelendeza,mclopezm}@itsm.edu.mx,
mariana.cm@vbravo.tecnm.mx

[2] División de Ingeniería en Sistemas Computacionales, Tecnológico Nacional de México/Valle de Bravo, Km. 31 de la Carretera Monumento-Valle de Bravo Ejido de San Antonio de la Laguna, Valle de Bravo, México
1201907042@vbravo.tecnm.mx

Abstract. Violence against women has increased. Throughout time, various strategies have been implemented to eradicate this public health problem. Researchers with diverse professional profiles have joined forces to propose solutions in the area of computer sciences. In this article, a systematic revision on the technologies implemented to eradicate violence against women was carried out, applying the PRISMA methodology. The country with the most research in the area of Artificial Intelligence is the United States of America. Research is focused on the predictive analysis of violence in different contexts, and studies related to Neural Networks and the treatment of medical reports data are the ones that are most similar to the analysis of partner violence. However, no evidence was found that predictions were made to prevent feminicide. This is why implementing data science in reports of violent incidents against women is imperative to restructure the protocol of action in governmental institutions and, with it, prevent the risk of feminicide.

Keywords: Data Science · Artificial Intelligence · Intimate Partner Violence · Feminicide

1 Introduction

Throughout the evolution of humanity, there has been an increase in violence against women. In developed countries like the USA, several strategies have been implemented with the aim of eradicating this public health problem. Researchers with diverse professional profiles have joined forces to propose solutions in the area of computer science. The ECLAC, in its 2020 report, indicated that two out of three women have been victims of violence due to gender reasons in various aspects of their lives, and one out of three has experienced physical, psychological, or sexual abuse from their partners. Additionally, at least 4,640 women were victims of feminicide in 2019 in Latin America and the Caribbean [1].

CONAVIM (National Commission to Prevent and Eradicate Violence against Women) leads initiatives against gender-based violence in Mexico. Their Integral Program aims to prevent, treat, sanction, and eradicate violence against women. BANAVIM (National Data and Information Bank on Violence Cases against Women) operates in each Mexican state, creating electronic files for women in violent situations, safeguarding personal information. BANAVIM tracks protection orders and identifies cases requiring urgent governmental measures [2].

1.1 Study Contribution

The PRISMA methodology is used to portray the elements of a systematic search of published research related to a study case. The PRISMA guidelines consist of a list of 27 elements for the systematic revision and a flowchart with 4 phases [5]. This study makes a valuable contribution by showcasing the creation of early warning systems for risk prediction, utilizing the capabilities of machine learning algorithms to analyze various data sources. These sources span interactions in social media networks, emergency hotline records, police databases, and healthcare information. By discerning patterns and risk factors associated with violence, these predictive models excel in identifying early indicators of potential harm. The integration of such models with existing support systems, including hotlines and counseling services, facilitates a proactive approach to intervention and support. Upholding ethical standards throughout implementation is imperative, addressing concerns related to data privacy, bias, and the potential misuse of predictive models. Furthermore, a commitment to continuous improvement, which involves incorporating feedback and adapting to evolving patterns, ensures the ongoing effectiveness and responsiveness of these data-driven approaches. Collaboration with experts in domestic violence, law enforcement, and social work remains critical for developing ethical and impactful solutions. To consolidate all the prominent evidence, a systematic search was carried out based on the following research questions.

2 Materials and Methods

The PRISMA methodology is used to portray the elements of a systematic search of published research related to a study case. The PRISMA guidelines consist of a list of 27 elements for the systematic revision and a flowchart with 4 phases [5]. To consolidate all the prominent evidence, a systematic search was carried out based on the following research questions.

2.1 Research Questions

RQ1: What data science techniques have been implemented to fight violence against women?; RQ2: Which variables are analyzed to determine the type of violence exercised?; RQ3: Which techniques are implemented in the process of information collection required for the analysis?; RQ4: Which techniques are more efficient in the classification of variables?

2.2 Selection Criteria

To determine the selection three phases were implemented: Phase 1: Select the indexed and high impact journals. Phase 2: Establish key words for the search criteria in the selected database engines. Phase 3: Revision by title, abstract, methods, and techniques.

In Table 1: the selected databases, criteria used for the selection of articles, and access dates are shown.

Table 1. Data Sources

Database	Selection Criteria	Date
ScienceDirect	Title, Summary, Keywords, Methods, References	2021-09-06
SpringerLink	Title, Summary, Keywords, Methods, References	2021-10-25
Google Scholar	Title, Summary	2021-12-15
IEEE	Title, Summary, References	2021-12-15

2.3 Inclusion Criteria

Intelligent systems represent a powerful tool wherein the solution to complex problems and their adaptability to the information environment enable us to achieve excellent results in various areas. Given the global imperative to address the issue of violence against women, a classification of the work conducted has been established into the following categories.

- Artificial Intelligence for the detection of partner violence.
  intelligence artificial AND "violence detection".
- Data science implementation in violence cases.
  "data science" AND "intimate partner violence".
- Algorithms, artificial intelligence, and data science to analyze gender violence cases.
  "Data science" AND "gender violence".

Once the analysis material is identified, articles that covered the following inclusion criteria were added.

- Articles that applied data science in violence against women cases.
- Articles that have been revised by peers.
- Articles written in the period of 2020–2023.

2.4 Exclusion Criteria

In the exclusion process, articles that address violence from a perspective different to computer sciences are not considered, in the same manner, articles presented in congresses, symposium abstracts, and thesis articles focused on violence from a social perspective.

3 Results

The analyzed articles were classified according to the techniques implemented to combat violence, with the most relevant being "Classification of violence variables," artificial intelligence, big data, and mobile applications. he research with the best results from each country was chosen, highlighting China, Brazil, Mexico, Argentina, France, South Africa, the United States, the United Kingdom, and Spain, using citation and download criteria.

3.1 Analysis of Relevant Findings

Ye, Wang, Ferdinando, Seppänen, and Alasaarela used motion sensors and AI to detect school violence. They applied Relief-F and Dempster-Shafe algorithms to nine activities, extracted 39 time and 12 frequency features. Improved Relief-F enhanced recognition accuracy to 93.6%, suggesting value in real-time monitoring to prevent gender-based violence [3].

Intimate partner violence (IPV) is a pervasive societal issue, particularly affecting women in South Africa. This machine learning study, using 2016 survey data, identifies the fear of one's husband as a key factor influencing IPV. Unlike traditional regression analysis, machine learning uncovers nuanced interrelationships, providing insights for intervention strategies by social workers, policymakers, and stakeholders [8].

In Mexico, an analysis of data banks recording violence against women revealed inconsistencies. Registration forms filled out at care institutions lacked importance or contained errors. The integrated information primarily focused on users from Xalapa and the central region, leaving gaps in the northern, southern areas, and 47 indigenous municipalities, highlighting a lack of incident knowledge [14].

González and Postay created an intelligent system rooted in classical conditioning theories, enhancing it with elements like building rules and historical data. Employing prototype methodology, they tested various agent objectives. While results were mostly satisfactory, analysis identified unexpected rules. To address this, they introduced a validation module categorizing rule levels based on expected situation frequency, distinguishing casual rules [15].

Guélorget, Gadek, Zaharia, and Grilheres applied active learning to assess opinions and violence in French newspapers. Analyzing major American media, they proposed measuring communication performance based on demographic data from platforms like Facebook. Emphasizing feature extraction and classification, they used precomputed word embeddings (FastText, ELMo, BERT) and recurrent or convolutional neural network architectures. With data from 25 French media outlets, they collected 39,611 news articles. Their violence detection model achieved a micro F1 score of 0.747, supporting versatile artificial intelligence development, aiding professionals in adopting AI for analysis and exploring research questions [16].

Paim et al. highlight the need for inclusive platforms to address diverse linguistic situations and abilities. Their study analyzes the instruments in Women's Institutes in Brazil, concluding that the applications to report violence are not inclusive. They classified women into six groups, emphasizing the lack of inclusion in technology. They suggest future focus groups specialized in the graphic interpretation of information [18].

Living in an unhealthy environment perpetuates global violence against women. This paper proposes a multi-modal approach, using satellite imagery to measure the Gender-Based Violence Index (GBVI). It integrates green canopy detection and atmospheric pollution levels to anticipate violence in neighborhoods. Employing computer vision for the Vegetation Index and IoT sensors for intoxicants, the results show promise. This approach could enable swift responses from entities such as the United Nations, World Health Organization, and governments to safeguard women's rights and bolster community security [19].

The Fig. 1 presents an information flowchart illustrating different phases of the PRISMA methodology.

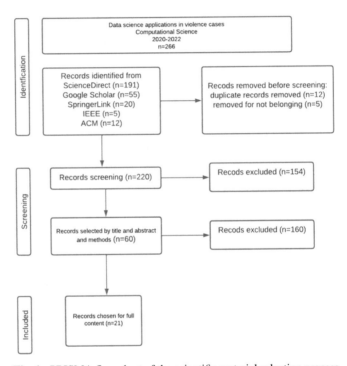

Fig. 1. PRISMA flow chart of the scientific material selection process.

It is important to mention that the use of the keywords "artificial intelligence" AND "violence detection," "data science" AND "intimate partner violence," and "data science" AND "gender violence" yielded enough information to determine that most of the articles are about corrective actions. However, there is little evidence that there are intelligent systems to prevent and eradicate violence. This is why there is a need to analyze the reports generated by government agencies with the aim of predicting possible femicides.

Based on the data presented in Fig. 2, one can argue that the number of publications developed from 2019 to 2022 is not considered a relevant number of publications. This shows the little importance that has been given to the problem of violence against women.

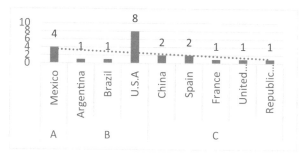

Fig. 2. Publications registered by country.

Regarding the technological tendencies used to combat violence and implement prevention activities, artificial intelligence is considered from different perspectives, such as machine learning and deep learning, which are the most widely used technologies. Please see Fig. 3.

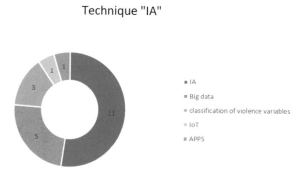

Fig. 3. Techniques used in the analysis of violence.

4 Discussion

This literature review provides an updated vision of how technology has been implemented to eradicate violence against women. From this revision, we can realize that the topic of partner violence has not been adequately addressed in the context of data science. In Mexico, academic production on this topic is almost inexistent. In response to the research questions, the following findings are observed:

RQ1: What data science techniques have been implemented to fight violence against women? Please see Table 2.

The tools implemented to combat violence against women have produced favorable results in the different environments where they were applied. However, in Mexico, valuable information that could be useful for carrying out predictive analysis is being wasted.

Table 2. Characteristics of the studies cited for RQ1

Number	Authors	Techniques	Year	Database
J1	[3] Ye, L., Wang, L. et al.	Improved algorithms Relief-F y Dempster-Shafe (DS)	2020	ELSEVIER
J2	[4] Fortune S., Mhlanga E. et al.	Analytical and simulation tools	2020	Magazine Society for Computer Simulation International
J3	[6] Zaman A., Kautz H., et al. Zaman A., Kautz H., et al.	API de Google, Google Takeout	2020	ScienceDirect
J4	[7] Díaz A.	K-Means, MLP, SVR, LS-SVM	2018	UPM
J5	[8] Amusa B	Machine Learning	2020	JIV

RQ2: Which variables are analyzed to determine the type of violence exercised? Please see Table 3.

Table 3. Characteristics of the studies cited for RQ2.

Number	Authors	Techniques	Year	Database
J1	[1] Lorenzano S.	Percentage of victims of femicide	2022	iMex
J2	[9] Hamida I.	Satellite images, Computer vision, IoT	2021	IEEE
J3	[10] Wang P.	Relief-F y Dempster-Shafe (DS)	2021	ELSEVIER
J4	[11] Hunt, d. X., Mark, T., et al.	Big Data y mHealth	2020	Frontiers in Artificial Intelligence
J5	[12] Blanco-Ruiz, M., Sainz-de-Baranda, et al.	Audiovisual stimuli with machine learning algorithm	2020	MDPI
J6	[13] Richard A.	Stochastic gradient augmentation, Genetic algorithm inspired by natural selection, Agglomerate clustering	2019	Criminology & Public Policy

The algorithms of machine learning and genetic algorithms are the ones used with better results in the data analysis.

RQ3: Which techniques are implemented in the process of information collection required for the analysis? Please see Table 4.

Table 4. Characteristics of the studies cited for RQ3

Number	Authors	Techniques	Year	Database
E-book	[14] Casados, E and Gómez, A.	Diagnosis of computer systems in Veracruz	2018	IVM
J1	[15] González D. y Postay J.	Production rules	2021	UCI
J2	[16] Guelorget P., Gadeka G., et al.	RNN, CNN	2018	ScienceDirect
J3	[17] Edward, L., Francisco, S-S.	Gradient Boosting Machine (GBM), Decision trees	2019	PLOS ONE
J4	[18] Paim P.,Sánchez L.et al.	inclusion of vulnerable groups	2020	Association for Computing Machinery

Researchers highlight the importance of database design and the quality of data in order to achieve favorable results when implementing techniques like decision trees and GBM.

RQ4: Which techniques are more efficient in the classification of variables? Please refer to Table 5.

The literature indicates that the efficiency of data analysis relies on the high availability of computer systems that gather information.

Table 5. Characteristics of the studies cited for RQ3.

Number	Authors	Techniques	Year	Database
J1	[20] Khurana B	Complaint Process	2020	BMJ Quality
J2	[21] Jung, S., Himmen M	app software ODARA	2022	APA
J3	[22] Havro, S., Freed, D. et al.	Establish behavior patterns of the aggressor	2019	Usenix
J4	[23] Withanage, N.	Gamma regression model (GR)	2021	Faculty of Applied Sciences

Nowadays, there are technological devices implemented for the care and monitoring of criminal incidents committed against women. These devices are generally activated through a type of signal at the moment the incident is happening. The main disadvantage of IoT devices is that they are not preventive but rather provide momentary or corrective data. Additionally, most articles use small and controlled samples for data analysis, leading to a lack of precision.

5 Conclusions

Thus far in the investigation, no evidence has surfaced to support the predictive analysis of reports made by victims of violence. Additionally, the presence of time delays in statistical data suggests that not all available information is being adequately considered, highlighting a potential gap in current analytical approaches.

While there has been some research on predicting certain violent behaviors, notably absent is its specific application within the context of couple relationships. This study gains paramount importance due to the alarming increase in feminicides observed in recent years. Among the noteworthy research findings, the spotlight is on various techniques, including the utilization of classification algorithms, support vector machines, longitudinal and lateral behavioral models, convolutional networks, fuzzy control systems, multilayer perceptron, and IoT devices. These findings underscore the potential of advanced technologies in addressing and mitigating the impact of gender-based violence.

In moving forward, it is imperative to bridge the gap identified in the current research landscape. Further investigations should focus on refining predictive models tailored to couple relationships, considering the unique dynamics and complexities inherent in such contexts. Additionally, efforts should be directed towards enhancing the real-time integration of statistical data, ensuring a more comprehensive and timely understanding of the factors contributing to violence against women. By continually advancing our technological and analytical approaches, we can strive towards more effective prevention and intervention strategies in the critical mission to eradicate violence against women. Due to the complexity of information management, the research used a small and controlled sampling, which suggests the need for more robust data and the creation of knowledge bases analyzed in real-time for advantageous use.

Acknowledgements. The first author acknowledgements the financial support from the National Council of Humanities, Sciences and Technologies (CONAHCYT, Scholarship 402743).

References

1. Lorenzano, S.: Estímulo Hasta que la justicia se siente entre nosotras Hasta que la justicia se siente entre nosotras (2022). https://doi.org/10.23692/iMex.21
2. CONAVIM (2022). Obtenido de https://www.gob.mx/conavim
3. Ye, L., Wang, L., Ferdinando, H., Seppanen, T., Alasaarela, E.: A video-based DT-SVM school violence detecting algorithm. Sensors (Basel, Switzerland) **20**(7), 2018 (2020). https://doi.org/10.3390/s20072018
4. Mhlanga, F.S., Perry, E.L., Kirchner, R.: Toward a predictive model ecosystem for interpersonal violence (WIP). Simul. Ser. **46**(10), 455–462 (2014)
5. Moher, D., Liberati, A., Tetzlaff, J., Altman, D.G., The PRISMA Group: preferred reporting items for systematic reviews and meta-analyses: the PRISMA statement. PLoS Med. **6**(7), e1000097 (2009). https://doi.org/10.1371/journal.pmed.1000097

6. Zaman, A., Kautz, H., Silenzio, V., Hoque, M.E., Nichols-Hadeed, C., Cerulli, C.: Discovering intimate partner violence from web search history. Smart Health **19**, 100161 (2021). https://doi.org/10.1016/j.smhl.2020.100161

7. Díaz Álvarez, A.: Modelado del comportamineto. Estructura de la población 2000, 2010 Y 2020 (2021), pp. 1–3 (2018). Descargado de http://censo2020.mx/

8. Amusa, L.B., Bengesai, A.V., Khan, H.A.: Predicting the vulnerability of women to intimate partner violence in south Africa: evidence from tree-based machine learning techniques. PubMed **37**(7–8) (2022)

9. Hamida, K., Iheb, A.: A multi-modal approach for gender-based violence detection. In: IEEE Xplore, pp. 144–149 (2020)

10. Wang, P., Wang, P., Fan, E.: Violence detection and face recognition based on deep learning. ACM Digit. **142**, 20–24 (2021)

11. Hunt, X., et al.: Artificial intelligence, big data, and Mhealth: the frontiers of the prevention of violence against children. Front. Artif. Intell. **3**, 1–16 (2020). https://doi.org/10.3389/frai.2020.543305

12. Blanco-Ruiz, M., Sainz-De-baranda, C., Gutiérrez-Martín, L., Romero-Perales, E., López-Ongil, C.: Emotion elicitation under audiovisual stimuli reception: should artificial intelligence consider the gender perspective? Int. J. Environ. Res. Publ. Health **17**(22), 1–22 (2020). https://doi.org/10.3390/ijerph17228534

13. Berk, R.A., Sorenson, S.B.: Algorithmic approach to forecasting rare violent events. Criminol. Public Policy **19**(1), 213–233 (2020). https://doi.org/10.1111/1745-9133.12476

14. Institucional, D.: Diagnóstico sobre la Violencia de G´enero contra las Mujeres en el Estado de Veracruz (2014)

15. González, D., Postay, J.D.: Aprendizaje autónomo en sistemas inteligentes. In: XIX Workshop de Investigadores En Ciencias de La Computación (WICC 2017, ITBA, Buenos Aires), pp. 35–39 (2017)

16. Guélorget, P., Gadek, G., Zaharia, T., Grilheres, B.: Active learning to measure opinion and violence in French newspapers. Procedia Comput. Sci. **192**, 202–211 (2021). https://doi.org/10.1016/j.procs.2021.08.021

17. Lannon, E., et al.: Predicting pain among female survivors of recent interpersonal violence: a proof-of-concept machine-learning approach. PloS One **16**(7), 0255277 (2021). https://doi.org/10.1371/journal.pone.0255277

18. Paim, P., Garcìa, L.S., Pereira, E.G.: No to violence against any woman! In: Proceedings of the 19th Brazilian Symposium on Human Factors in Computing Systems, pp. 1–6 (2020). https://doi.org/10.1145/3424953.3426645

19. Khatri, H., Abdellatif, I.: A multi-modal approach for gender-based violence detection. In: 2020 IEEE Cloud Summit, pp. 144–149. IEEE, October 2020

20. Khurana, B., Seltzer, S.E., Kohane, I.S., Boland, G.W.: Making the 'invisible' visible: transforming the detection of intimate partner violence. PubMed **29**, 241–244 (2020)

21. Jung, S., Himmen, M.K.: A field study on the police use of the Ontario domestic assault risk assessment (ODARA). APAPsycNet **9**, 204–217 (2022)

22. Havron, S., Freed, D., Chatterjee, R., McCoy, D., Dell, N., Ristenpart, T.: Clinical computer security for victims of intimate partner violence. In: Proceedings of the 28th USENIX Security Symposium, pp. 105–122 (2019)

23. Withanage, N., Wijekoon, S.: Identification of associated factors and prediction for the level of intimate partner violence against women in Sri Lanka. Int. J. Res. Appl. Sci. Biotechnol. **8**(5) (2021). https://doi.org/10.31033/ijrasb.8.5.2

Azure Kubernetes Services WebRouting AddOn

Andra-Isabela-Elena Ciobanu, Eugen Borcoci, Marius Constantin Vochin,
and Serban Georgica Obreja(✉)

National University of Science and Technology Politehnica Bucharest, Bucharest, Romania
`serban.obreja@upb.ro`

Abstract. In an era of cloud and data centers, compact and versatile data are important and needful. This paper proposes and develops an experiment as a demonstration of a web routing application inside Azure cloud, more specifically on a web routing AddOn feature on Azure Kubernetes Service. The goal of this add-on is to make it easier for any user to expose their application to the world in a secure manner while reducing some of the operational overhead that goes into that. This paper's contribution is useful for developers working in the domain of implementing Azure Kubernetes Services.

Keywords: Containers · Kubernetes · Azure · Cloud Computing · Ingress · Web Application · Managed Identity · DNS · SSL/TLS authentication

1 Introduction

In an IT world of fast revolution and technological speed, users expect applications to be available 24/7 and developers expect to deploy new versions of those applications several times a day. In here comes containerization part. Among many containerization solutions, Kubernetes is a portable, extensible, open-source platform for managing containerized workloads and services, that facilitates both declarative configuration and automation. Kubernetes is a production-ready, open-source platform designed with Google's accumulated experience in container orchestration, combined with best-of-breed ideas from the community [1].

This paper could address developers who work in cloud computing development, especially with containerized applications. Moreover, it is very suitable for web developers that wish to automate more on front-end website elements or back-end components. The scope of this paper is to demonstrate an easy solution to quickly set up an ingress controller, configure DNS and SSL/TLS authentication in the context of Azure Kubernetes Service (AKS) WebRouting. Because the paper investigates the functionalities of a tool that improves the developer experience with AKS environment for cloud services management, the performance of the services is expected not to be affected. That's why a performance evaluation as in [2] is not performed.

Web routing is the process of directing web requests from clients to the appropriate servers or services that can handle them. It involves matching the URL, path, method, headers, and other criteria of a request with the rules and policies defined by the web

© The Author(s), under exclusive license to Springer Nature Switzerland AG 2024
Á. Rocha et al. (Eds.): WorldCIST 2024, LNNS 989, pp. 262–271, 2024.
https://doi.org/10.1007/978-3-031-60227-6_24

developer or administrator. Web routing also enables features such as load balancing, traffic shaping, security, authentication, and caching.

The structure of the paper is presented below. Section 2 has the role to make the text more self-contained; they summarize the Kubernetes concepts, Azure Kubernetes services and NGINX Ingress Controller. Sections 3 and 4 contain the main contribution of the paper, i.e., the experimental setup and results. The benefits of the approach are discussed. Section 5 contains the conclusions.

2 Azure Kubernetes Services

There are so many things to be debated about Kubernetes concepts, clusters, and its administration parts. This paper will be focused on the main parts which are going to be used in a practical demonstration.

2.1 General Kubernetes Concepts

The smallest unit of computing that you can create and manage in Kubernetes is called a Pod. A Pod consists of one or more containers, with shared storage and network resources, and a specification for how to run the containers. A Pod models an application-specific "logical host": it contains one or more application containers which are relatively tightly coupled. A Pod always runs on a Node. A Node may be either a virtual or a physical machine, depending on the cluster [3].

These pods can be generated from a deployment. A deployment provides declarative updates for the Pods. The deployments control the way the pods will be generated based on the labels, selectors, and number of replicas. They are useful as the images can be adjusted along the way and scaled accordingly the needs from the cluster [4].

These objects and resources in Kubernetes are isolated into what is called a names-pace. A namespace provides a mechanism for isolating groups of resources within a single cluster. Names of resources need to be unique within a namespace, but not across namespaces [5].

In Kubernetes, a Service is a method for exposing a network application that is running as one or more Pods in your cluster. The exposure can be made inside the cluster or outside the cluster. Each Pod gets its own IP address [6].

The functional component Ingress is a Kubernetes resource which exposes HTTP and HTTPS routes from outside the cluster to services within the cluster. Ingress lets you configure an HTTP load balancer for applications running on Kubernetes, represented by one or more Services. Such a load balancer is necessary to deliver those applications to clients outside of the Kubernetes cluster [6].

2.2 Azure Kubernetes Services and NGINX Ingress Controller

This section shortly presents a selective view on some related work dedicated, for example, to companies that wish to develop their infrastructure. A development company expands in size, the infrastructure team encounters many growing pains managing their environment. A developer wishes for an easy solution to enable quick secure access that setup an ingress controller and configure DNS.

For the Ingress resource to work, the cluster must have an ingress controller running. Unlike other types of controllers that run as part of the kube-controller-manager binary, Ingress controllers are not started automatically with a cluster. Kubernetes as a project supports and maintains different cloud providers, as well as open-source projects, such as NGINX ingress controllers. The NGINX Ingress Controller is an application that runs in a cluster and configures an HTTP load balancer according to Ingress resources. NGINX Ingress Controller works with both NGINX and NGINX Plus and supports the standard Ingress features - content-based routing and TLS/SSL termination. Additionally, several NGINX and NGINX Plus features are available as extensions to the Ingress resource via annotations and the ConfigMap resource [7].

For this demo, the installation has been made with Helm on an Azure Kubernetes Services (AKS) cluster. The pre-requisites and all the steps can be found here [8]. Azure Kubernetes Service simplifies deploying a managed Kubernetes cluster by offloading the operational overhead to Azure. As a hosted Kubernetes service, Azure handles critical tasks, like health monitoring and maintenance. When you create an AKS cluster, a control plane is automatically created and configured. This control plane is provided at no cost as a managed Azure resource abstracted from the user. You only pay for and manage the nodes attached to the AKS cluster [9].

AKS provides a single-tenant control plane, with a dedicated API server, scheduler, etc. You define the number and size of the nodes, and the Azure platform configures the secure communication between the control plane and nodes. Interaction with the control plane occurs through Kubernetes APIs, such as `kubectl` or the Kubernetes dashboard. To run applications and supporting services, a Kubernetes node is needed that runs the Kubernetes node components and container runtime.

SSL/TLS termination is another common feature of Ingress. On large web applications accessed via HTTPS, the Ingress resource handles the TLS termination rather than the application itself. To provide automatic TLS certification generation and configuration, the Ingress resource can be configured to use certificate providers.

Customers of Azure Kubernetes Service (AKS) can use Azure's native Application Gateway L7 load-balancer to expose cloud services to the Internet, thanks to the Application Gateway Ingress Controller (AGIC), a Kubernetes application.

2.3   Other Azure Concepts

A common challenge for developers is the management of secrets, credentials, certificates, and keys used to secure communication between services. Managed identities eliminate the need for developers to manage these credentials.

While developers can securely store the secrets in Azure Key Vault, services need a way to access Azure Key Vault. Managed identities provide an automatically managed identity in Azure Active Directory (Azure AD) for applications to use when connecting to resources that support Azure AD authentication. Applications can use managed identities to obtain Azure AD tokens without having to manage any credentials [10].

One of Azure's key management options, Azure Key Vault, aids in the following issues' resolution [11]:

- Secrets Management: Tokens, passwords, certificates, API keys, and other secrets can be securely stored and access to them can be tightly controlled using Azure Key Vault.
- Azure Key Vault is a Key Management system: The encryption keys to encrypt data are simple to create and manage using Azure Key Vault.
- Certificate Management: Azure Key Vault makes it simple to provision, manage, and deploy both public and private Transport Layer Security/Secure Sockets Layer (TLS/SSL) certificates.

On an AKS cluster, the application routing add-on configures an ingress controller with SSL termination using certificates kept in Azure Key Vault. It has an optional integration with Open Service Mesh (OSM) enabling end-to-end mutual TLS (mTLS) encryption of inter-cluster communication. The add-on generates publicly reachable DNS names for endpoints on an Azure DNS zone when you deploy ingresses [12].

3 Experimental Setup and Benefits

The benefits of using AKS WebRouting AddOn are the following:

Managed NGINX Ingress Controller

The add-on will provide you with a highly available, managed Kubernetes Ingress based on open-source NGINX. This is similar to the HTTP application routing add-on but is intended for production use and will be fully supported by Azure.

Manages DNS Records

The add-on provides an optional integration with Azure DNS and uses an external-DNS controller to manage DNS records on your behalf. The older "HTTP application routing add-on" also provided this functionality; however, the difference here is that the new add-on can link to your custom Azure DNS resource.

As an example, let's say you have a custom domain called `myawesomesite.com` and delegated it to your Azure DNS resource's nameservers. When you deploy an Ingress with the hostname of `store.myawesomesite.com`, an A record will automatically be added to your Azure DNS resource to point the `store` subdomain to the Ingress' public IP.

Manages TLS Certificates

The add-on provides another optional integration with Azure Key Vault using the azure-keyvault-secrets-provider AKS add-on. When combined with the Web Application Routing add-on, the ingress controller will be able to load TLS certificates from Azure Key Vault and save it as a Kubernetes TLS secret with secret rotation.

As an example, if you have a TLS certificate for `store.myawesomesite.com`, simply upload the `.pfx` file to Azure Key Vault and bind the certificate to Ingress using the `kubernetes.azure.com/tls-cert-keyvault-uri` annotation.

Integrates with Open Service Mesh (OSM)

Finally, if you have security requirements such as needing end-to-end encryption between your Ingress and your backend pods, there is out-of-the-box support for Open Service Mesh (OSM). OSM is an open-source light-weight service mesh to help secure intra-cluster communications.

In most cases, TLS termination is done at the ingress, and traffic flows un-encrypted to the backend pods. With OSM, you can configure it to proxy connections between Ingress and IngressBackends using mutual TLS (mTLS). This ensures traffic is encrypted all the way down. OSM can be manually installed and configured for NGINX. However, if you combine the OSM add-on with the Web Application Routing add-on, configuration for NGINX is automatically added for you.

To implement an easy solution to enable quick secure access that setup an ingress controller, configure DNS, and lastly SSL/TLS authentication, the following approaches are available:

1. HTTP Application Routing addon by AKS which might solve a lot of developer team's issues, but it is limited, and it does not provide secure SSL communication and is not recommended to be used in any production workload.
2. The developer tries to deploy a nginx-controller or AGIC on AKS but does not know how to configure ingress yaml file using TLS and how to configure DNS on azure. This is s a challenge for new users.
3. A more manageable solution that has been used on this paper would be *Web application Routing*.

The HTTP Application Routing addon and the AKS Web Application Routing addon are two different ways to configure an Ingress controller in an AKS cluster. Some of the main differences between the two addons are:

- The HTTP Application Routing addon uses the NGINX Ingress controller and automatically creates DNS records for your web applications using the External-DNS project. The AKS Web Application Routing addon uses the Application Gateway Ingress controller and integrates with Azure Key Vault to store and manage SSL certificates.
- The HTTP Application Routing addon is fully managed by AKS and does not require any additional configuration or resources. The AKS Web Application Routing addon is more flexible and customizable but requires you to create and manage an Application Gateway instance and a Key Vault instance [11].

The HTTP Application Routing addon is generally available and free of charge, but it does not support advanced features such as SSL termination, path-based routing, or health probes. The AKS Web Application Routing addon incurs costs for the Application Gateway and Key Vault resources, but it supports these features. It will automate the pain points of initial deployment of an ingress controller and SSL/TLS authentication.

To access the application from the web (NGIX, AGIC, Http Routing) we need many configurations and changes inside NGINX, which should be applied for the ingress YAML file. Using this feature, it works out of the box to integrate this AddOn. It will reduce complexity to develop web applications.

For example, with the WebRouting AddOn it is not necessary anymore to have a separate ingress YAML file where to store the secrets up to date. This is going to be automatically integrated and adjusted along the development and setup. There will be no need to work on ingress or add DNS zone manually. Everything will be automatically. The add-on provides a managed ingress controller based on NGINX and integrates out of the box with Open Service Mesh (OSM) to secure intra-cluster communications using mutual TLS and use OSM to encrypt the traffic between the ingress and the application. There will be secured traffic internally and externally.

4 Web Application with WebRouting AddOn

This section presents the steps to set up a cloud Service using the WebRouting AddOn and its functional evaluation. The benefits of this approach are also highlighted.

4.1 User Experience

The end user should be able to check the "Web Application Ingress" on a new cluster deployment. After deployment they should be able to access their cluster via their specified URL: https://mywebapp.com. The only change from end-user on the service yaml file to add annotation as below:

```
- kubernetes.azure.com/ingress-host: myapp.contoso.com
- kubernetes.azure.com/tls-cert-keyvault-uri:
- https://<MY-
  KEYVAULT>.vault.azure.net/certificates/<KEYVAULT-
  CERTIFICATE-NAME>/<KEYVAULT-CERTIFICATE-REVISION>
```

The user only adds this annotation on the service, mentions the hostname and use this URI only. Once he uses this annotation on the service, a secret and a file using TLS will be created automatically and `A Record` will be updated on DNS, without any change on the user side.

4.2 Web Application Routing Deployment

This subsection describes the steps to prepare and install the environment for the web application routing deployment.

1. **BYOC + BYOD should be ready**

```
- az extension add --name aks-preview
```

2. **Import the certificate to keyvault**

```
- az keyvault certificate import --vault-name psammancert -n
  psamman -f aks-ingress-tls.pfx
```

3. **Deploy web app routing**

a)
```
az aks create --resource-group aks --name akss --enable-
addons       azure-keyvault-secrets-provider,       open-
service-mesh,web_application_routing --generate-ssh-keys
--network-plugin                 azure                  -
-dns-zone-resource-id /subscriptions/2bd82879-5354-47a1-
a3a930041cc8e121/resourceGroups/dns/providers
/Microsoft.Network/dnszones/psamman.com
```

b) Buy your own certificate from cx side and import this certificate from KeyVault to use it.

c) After that, create AKS cluster using KeyVault and OSM to encrypt the traffic and deploy web app routing.
d) Also, download binary of OSM to configure the namespace Web application routing Deployment.

4. **Configure OSM latest binaries are available OSM GitHub releases page**
- wget https://github.com/openservicemesh/osm/releases/download/v1.2.0/osm-v1.2.0-linux-amd64.tar.gz
- `tar -xf osm-v1.2.0-linux-amd64.tar.gz`
- `sudo mv linux-amd64/osm /usr/local/bin/osm`

5. **Create the application namespace**
- `kubectl create namespace hello-web-app-routing`
- `osm namespace add hello-web-app-routing`

Any pod from this namespace will use a sidecat container to encrypt the traffic internally between the namespaces.

4.3   Grant Permission for Web Application Routing

We need to grant permission for KeyVault and Azure DNS zone. Managed Service Identity (MSI) for web application routing will be created.

- object ID to grant GET permission for web application routing to retrieve certificates from azure Key Vault;
- client ID to grant permission role assignment for DNS to write the A record automatically.

  – *az keyvault set-policy --name psammancert --object-id 62ca18ec-24a8-456b-8ba4-87618a60eaeb --secret-permissions get --certificate-permissions get*
  – *az role assignment create --role "Contributor" --assignee 95c3e6d0-91e1-431c-aedb-edf8558442b1 --scope /subscriptions/2bd82879–5354-47a1-a3a9-30041cc8e121/resourceGroups/dns/providers/Microsoft.Network/dnszones /psamman.com*

App Routing will use the controller and to have access to the KeyVault to recreate the certificate it will use the managed identity. Is very important to take the object ID for the certificate to the KeyVault and the ClientID to create a role assignment to the Resource ID of the DNS.

4.4   Demo Start

After importing the certificate to Azure KeyVault (Fig. 1), AKS cluster creation is done, and a separate namespace will be deployed to configure OSM and policy.

Then, the Client ID from the Web App MSI will be taken, to make sure that the Web App Controller Deployment will have access to the DNS. After that, the certificate will be taken and the URI which will bring the annotation will be given. So, the new user or developer will not create a new Ingress Yaml file or secret. Everything will be created automatically, and the certificate will be retrieved automatically using the KeyVault.

Fig. 1. Certificate in Azure KeyVault

The pods from Fig. 2 are created automatically using the addon once the creation of the AKS cluster is done and OSM is used to encrypt the traffic between the NGINX and application.

NAMESPACE	NAME	READY	STATUS	RESTARTS	AGE
app-routing-system	external-dns-7b688b969f-gjb2j	1/1	Running	2 (6h18m ago)	10d
app-routing-system	nginx-57cfb7dd9d-22kp6	1/1	Running	0	6h18m
app-routing-system	nginx-57cfb7dd9d-g6cv8	1/1	Running	2 (6h18m ago)	10d
kube-system	aks-secrets-store-csi-driver-pwbqp	3/3	Running	10 (6h18m ago)	10d
kube-system	aks-secrets-store-csi-driver-tcd5m	3/3	Running	9 (6h18m ago)	10d
kube-system	aks-secrets-store-csi-driver-tq7gj	3/3	Running	8 (6h18m ago)	10d
kube-system	aks-secrets-store-provider-azure-2kmtb	1/1	Running	2 (6h18m ago)	10d
kube-system	aks-secrets-store-provider-azure-54bqv	1/1	Running	2 (6h18m ago)	10d
kube-system	aks-secrets-store-provider-azure-ftf7h	1/1	Running	2 (6h18m ago)	10d
kube-system	azure-ip-masq-agent-82xml	1/1	Running	2 (6h18m ago)	10d
kube-system	azure-ip-masq-agent-ggd9x	1/1	Running	2 (6h18m ago)	10d
kube-system	azure-ip-masq-agent-rxrnl	1/1	Running	2 (6h18m ago)	10d
kube-system	cloud-node-manager-6x1bs	1/1	Running	2 (6h18m ago)	10d
kube-system	cloud-node-manager-g7kk7	1/1	Running	2 (6h18m ago)	10d
kube-system	cloud-node-manager-plwcs	1/1	Running	2 (6h18m ago)	10d
kube-system	coredns-autoscaler-7d56cd888-w6m8j	1/1	Running	1 (6h18m ago)	9d
kube-system	coredns-dc97c5f55-4kld7	1/1	Running	1 (6h18m ago)	9d
kube-system	coredns-dc97c5f55-7sz58	1/1	Running	2 (6h18m ago)	10d

Fig. 2. Pods Creation

The ingress is created and uses the hostname that it was specified (in Fig. 3). And it worked and it used the certificate to make a tunnel- secret between the client and NGINX. And inside NGINX, it will encrypt the traffic using OSM between the NGINX and the Application. So, everything is secure.

From the client side, he must add the annotation and URI and that's it. AKS automatically created the A record in DNS zone (see Fig. 4):

Fig. 3. Ingress URL

Fig. 4. DNS Zone Record

5 Conclusions

This paper provided a pragmatic experiment howing a clear picture of a feature that Azure Kubernetes services can use, that is WebRouting AddOn. The demo described in the previous sections constitutes an effective tool, possible to be used by interested developers in their activities to implement services in the Azure Kubernetes complex framework.

The contribution and the biggest advantages of the presented approach are that it can be provisioned via infrastructure as code templates (eg, Bicep, ARM, Terraform) all in one shot; there is no need to separately deploy the ingress controller via Helm. And, it also provides a quick and secure solutions for developers which might need it.

Acknowledgment. The research has received funding from the NO Grants 2014–2021, under Project contract no. 42/2021, RO-NO-2019-0499 –"A Massive MIMO Enabled IoT Platform with Networking Slicing for Beyond 5G IoV/V2X and Maritime Services" - SOLID-B5G.

References

1. Kubernetes basics Webpage. https://kubernetes.io/docs/tutorials/kubernetes-basics/. Accessed 29 Dec 2023
2. Sethy, K.K., Singh, D., Biswal, A.K., Sahoo, S.: Serverless implementation of data wizard application using azure kubernetes service and docker. In: 1st IEEE International Conference on Industrial Electronics: Developments & Applications (ICIDeA), Bhubaneswar, India, pp. 214–219 (2022). https://doi.org/10.1109/ICIDeA53933.2022.9970103
3. Kubernetes Pods Webpage. https://kubernetes.io/docs/concepts/workloads/pods/. Accessed 29 Dec 2023
4. Kubernetes controllers Webpage. https://kubernetes.io/docs/concepts/workloads/controllers/deployment/. Accessed 29 Dec 2023
5. Kubernetes namespaces Webpage. https://kubernetes.io/docs/concepts/overview/working-with-objects/namespaces/. Accessed 29 Dec 2023
6. Kubernetes ingress Webpage. https://kubernetes.io/docs/concepts/services-networking/ingress/. Accessed 29 Dec 2023
7. NGINX ingress controller Webpage. https://docs.nginx.com/nginx-ingress-controller/intro/overview/. Accessed 29 Dec 2023
8. Microsoft Azure AKS ingress basics Webpage. https://learn.microsoft.com/en-us/azure/aks/ingress-basic?tabs=azure-cli. Accessed 29 Dec 2023
9. Microsoft Azure AKS intro Webpage. https://learn.microsoft.com/en-us/azure/aks/intro-kubernetes. Accessed 29 Dec 2023
10. Microsoft Azure managed identities Webpage. https://learn.microsoft.com/en-us/azure/active-directory/managed-identities-azure-resources/overview. Accessed 29 Dec 2023
11. Microsoft Azure key-vault Webpage. https://learn.microsoft.com/en-us/azure/key-vault/general/overview. Accessed 29 Dec 2023
12. Microsoft Azure app routing Webpage. https://learn.microsoft.com/en-us/azure/aks/app-routing?tabs=without-o. Accessed 29 Dec 2023

Clustering and Optimization Algorithms to Enable Reliable 6G Mobile Molecular Communications

Borja Bordel$^{(\boxtimes)}$, Ramón Alcarria, and Tomás Robles

Universidad Politécnica de Madrid, Madrid, Spain
{borja.bordel,ramon.alcarria,tomas.robles}@upm.es

Abstract. Molecular communications are envisioned to transform medicine and environmental sciences, but currently only small, isolated networks have been deployed. It is essential, then, to develop new mechanisms to enable information flow from remote users to nanometric biological machines through next-generation networks, such as 6G mobile technologies. 6G mobile networks are characterized by an extreme Quality-of-Service, but in the context of molecular communications, two requirements turn critical: ultra-massive and extremely reliable communications. Molecular communications are simplex, so there is typically no channel to transmit acknowledgment messages. Furthermore, several nanometric receptors concentrated in just some square micrometers are an ultra-massive device density for 6G base stations. In this paper, we propose a computational algorithm to make reliable and massive 6G mobile molecular communications feasible. The proposed algorithm employs clustering to create a real-time map with the positions of the biological nanometric machines, and particle swarm optimization to track the identity of the different machines while slowly moving. To handle ultra-massive density, clustering operates by defining which magnetic particles used as communication interface belong to the same biological machine. While the optimization mechanism considers the current and previous clustering results and a probabilistic model to determine the identity of each cell. Reliability is achieved by an acknowledgment message generated by a 6G transceiver when the biological machines reach the expected destination. Simulation tools are employed to validate the proposed solution. Results show that the identification error is less than 15%, and the reliability achieves a probability of up to 90%.

Keywords: optimization algorithms · clustering · 6G communications · molecular communications · identity tracking · probabilistic models

1 Introduction

Molecular communications [1] are nanometric information exchanges, performed through biological machines (such as cells) transporting information molecules from one point to the other, separated by some micrometers. Biological machines are expected to offer an interface for connecting to external devices, usually based on artificially introduced magnetic particles in biological cells [2]. Then, external transceivers could excite these particles to force the cell to move towards the desired destination, or to capture or expulse information molecules. Internal sensors, transmitters, and receivers within the cells are usually employed to handle these information molecules. Figure 1 shows how all these elements relate through an internal controller, supported by the internal chemical signaling system in the biological cell.

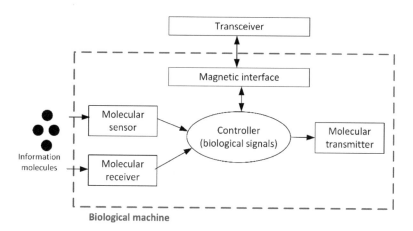

Fig. 1. General model for a biological nanomachine

This approach is envisioned to transform medicine and environmental sciences, but even though the proposed biological machine model is quite general and flexible, currently only small networks including some biological machines have been deployed under laboratory conditions [3]. The key open challenge is the adaptation between protocols and technologies operating in kilometric scales (Internet, mobile communications, satellite devices...) [4], and techniques whose purpose is to handle micrometric biological devices. These adaptation mechanisms are fully dependent on the network technology to be employed. However, in the context of future health, medical, and environmental applications, next-generation mobile technologies (6G) [5] are the most promising and commonly reported enabling solution in the scientific state of the art. Then, to allow real extreme-to-extreme communication channels, it is essential to develop new mechanisms to enable the information flows from remote users to nanometric biological machines through next-generation networks, such as 6G mobile technologies.

Among all the innovations expected from 6G technologies, the most critical and differential characteristic is extreme Quality of Service (QoS) [6]. Concepts such as Ultra-Massive Machine-type communications (umMTC) [7], enhanced ultra-broadband mobile communications (euMBBC) [8] or extremely ultra-reliable low-latency communications (eURLLC) [9] refer to networks where information must navigate with delays below 0.1 ms with a 99.99999% probability, and bitrates above one terabit per second, in scenarios with a device density above ten million device per square kilometer.

Although 6G molecular communications should preserve all the expected QoS indicators from next-generation mobile networks, in the context of molecular communications two requirements turn critical and the most complicated to be achieved: ultra-massive and extremely ultra-reliable communications. On the one hand, molecular communications are simplex, so no channel to transmit acknowledgment messages is typically available. However, a duplex channel is essential to build reliable communications and protocols, as acknowledgments and retransmissions are the only known way currently to ensure network reliability [10]. On the other hand, several nanometric receptors (biological machines) concentrated in just some square micrometers is not only an ultra-massive device density for 6G base stations, but an open problem as existing radio signaling techniques are mostly supported by identification schemes [11] which are not native nor available in biological cells.

In this paper, we propose a computational algorithm to address these challenges and make reliable and massive 6G mobile molecular communications feasible.

The proposed algorithm employs affinity propagation clustering to create a real-time map with the positions of the biological nanometric machines, and particle swarm optimization to track the identity of the different machines while slowly moving. To handle ultra-massive density, clustering operates by defining which magnetic particles used as communication interface belong to the same biological machine. While the optimization mechanism considers the current clustering result, the previous results and a probabilistic Bayesian model to determine the identity of each cell in the clustered map. Reliability is achieved by an acknowledgment message generated by the 6G transceiver when the biological machines reach the expected destination.

The rest of the paper is organized as follows. Section 2 describes the state of the art on mobile molecular communications, including techniques to ensure reliability and massive device density; Sect. 3 describes the proposed solution, including the clustering algorithm, the optimization scheme, and the probabilistic model; Sect. 4 presents an experimental validation; and Sect. 5 concludes the paper.

2   State of the Art on Mobile Molecular Communications

Because mobile molecular communications are a very innovative paradigm, most of the current works address basic communication challenges and mechanisms such as clock synchronization [12]. Solutions based on physical transmission signals [13] and molecular phase-locked loop (PLL) [14] may be found with respect to this key issue. But coding technologies based on the position of nanomachines [15], information theory approaches to calculate the maximum capacity in molecular channels [16], interference mitigation strategies [17], error correction and detection techniques [18] or molecular stochastic [19] and time-variant [20] channel models have also been developed. Although all these proposals address relevant open challenges in molecular communications, all of them are focused on the nanometric scale. And there is no evidence on how these innovations could interoperate with common network and communication technologies.

Some general or initial ideas on how the coexistence between the nanometric and the kilometric scale could be achieved have been described [2]. But always from a very abstract point of view without reference of a particular implementation. On the contrary, some authors are investigating mathematical models to describe an extreme-to-extreme mobile molecular communication system employing transmission systems [21] or graph theory [22]. Although these approaches can describe integration between molecular and mobile communications, there is no description on how to apply these models to the real problem of coexistence.

In the last years, some works investigating on the real coexistence between standard communications and molecular communications have been reported. Although some authors propose to exploit the integration of deep learning technologies in 6G networks to create intelligent models and tools that monitor the nanoscale [23], other researchers suggest the use of simulation techniques to create a virtual representation of biological machines and their movements [24]. In contrast, proposals to adapt common mobile networks to a bioinspired architecture [25] have also been communicated.

All these previous contributions are promising and report positive results, but (again) there is no evidence on how reliability, co-existence, or massive communications are actually achieved. Although all the proposed technologies have the potential to be used for that objective. Our paper fills this gap by addressing the provision of reliable massive mobile molecular communications with a specific and tangible solution.

3   Achieving Reliable Massive Mobile Molecular Communications

The proposed enabling scheme for reliable massive 6G mobile molecular communications operates as a supervisory control system (see Fig. 2).

In our scenario, the number of biological nanomachines B is unknown (cell may get destroyed, absorbed, etc.). But a set P of M magnetic particles p_i (1) can be monitored very precisely. Each biological machine b_j has absorbed an unknown number of magnetic particles $n(b_j)$ making up a partition π_j of set P. All partitions must be π_j disjoint (2) [2].

$$P = \{p_i \, i = 1, \ldots, M\} \tag{1}$$

$$P = \bigcup_{j=1}^{B} \pi_j$$

$$M = \sum_{j=1}^{B} n(b_j) \tag{2}$$

The first problem to be addressed is to determine which magnetic particles are absorbed by the same biological cell and then they represent the same machine. To do that, we employ an affinity propagation clustering algorithm, so we can calculate the particle representatives (exemplars). But to track the communication flows (and make them reliable) it is essential to determine the individual movement of each biological machine. Thus, a permanent identity must be associated to each machine (exemplar particle). Therefore, an association function is needed to link the permanent identities to the biological machines. A Particle Swarm Optimization (PSO) algorithm is deployed to perform this association. The PSO scheme operates in such a way that the most probable identity for each exemplar magnetic particle is obtained. The optimization function is a Bayesian probabilistic model, based on the expected cell movement. Finally, a supervisory engine in the transceiver controls the molecular flows and generates an acknowledgment when the target destination is achieved.

The next subsections describe every algorithm in detail.

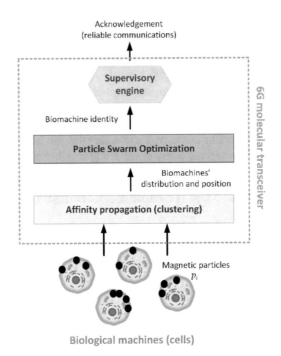

Fig. 2. Global architecture

3.1 Evaluation of the Distribution and Position of Biomachines

Using a very sensible magnetic interface, the tridimensional position (3) of every magnetic particle p_i is collected. This position is expressed as Euclidean coordinates. Through this position, the similarity between any two magnetic particles p_i and p_k is calculated using function $S(\cdot)$, based on exponential laws and where D_{cell} is the standard expected diameter for the employed biological machines (4). This similarity function represents a measure of how probable is that two given magnetic particles are embedded into the same cell.

$$p_i = (x_i, y_i, z_i) \tag{3}$$

$$S(p_i, p_k) = 1 - e^{\frac{3 \cdot \|p_i - p_k\|}{4 \cdot D_{cell}}}$$

$$= 1 - exp\left\{\frac{3 \cdot \sqrt{(x_i - x_k)^2 + (y_i - y_k)^2 + (z_i - z_k)^2}}{4 \cdot D_{cell}}\right\} \tag{4}$$

Using the similarity function, the set P of magnetic particles is divided into partitions π_j (clusters). Every cluster π_j is understood as a different biological machine b_j, whose global position is inherited from the position of the exemplar particle $e^*(\pi_j)$ for cluster π_j. The number of defined clusters B, as well as the exemplar particle for each cluster π_j, , are proposed to maximize (5) the accumulated affinity function $\Phi(\cdot)$.

$$\Phi = \sum_{i=1}^{M} \sum_{\forall j} S(p_i, e^*(\pi_j)) \tag{5}$$

To calculate the exemplar particles, two different recursive functions are employed: the responsibility function $\mathcal{R}(\cdot)$ and the availability function $\mathcal{A}(\cdot)$.

The responsibility function $\mathcal{R}(\cdot)$ represents how suitable is a potential exemplar particle $e(\pi_j)$ to be the actual representative for particle p_i (6), considering all possible options. This is achieved using the similarity function, and subtracting the maximum value detected for similarity (so the different represents how much similar is the current candidate to exemplar particle compared to previously visited candidates).

$$\mathcal{R}(p_i, e(\pi_j)) = S(p_i, e(\pi_j)) - \max_{e(\pi_k) \neq e(\pi_j)} \{\mathcal{A}(p_i, e(\pi_k)) - S(p_i, e(\pi_k))\} \tag{6}$$

Particles having the same representative make up a cluster are embedded into the same biological machine.

On the other hand, the availability function $\mathcal{A}(\cdot)$ represents how suitable is a potential exemplar particle $e(\pi_j)$ to be the actual representative for cluster π_j (7). The value is accumulative and does not consider any possible alternative. As similarity is a negative magnitude in our proposal, maximum and minimum functions must be employed to ensure that no positive value is assigned to availability.

$$\mathcal{A}(p_i, e(\pi_j)) = min\left\{0, \left(\mathcal{R}(e(\pi_j), e(\pi_j)) + \sum_{\forall p_k \neq p_i, e(\pi_j)} \mathcal{R}(p_j, e(\pi_j))\right)\right\} \tag{7}$$

Finally, the individual affinity function $\alpha(\cdot)$ represents how suitable is particle p_i to be the exemplar for partible p_k (8). For those particles p_i for which the self-affinity is positive (9), they are transformed into potential exemplar particles. And clusters π_j are build by selecting each particle p_i the candidate for exemplar particle $e(\pi_j)$ for which the individual affinity function is maximum. Algorithm 1 shows the final affinity propagation process and exemplar calculation. Final exemplar particles $e^*(\pi_j)$ are obtained when potential exemplar candidates do not change for some iterations.

$$\alpha(p_i, p_k) = \mathcal{A}(p_i, p_k) + \mathcal{R}(p_i, p_k) \tag{8}$$

$$\alpha(p_i, p_i) = \mathcal{A}(p_i, p_i) + \mathcal{R}(p_i, p_i) > 0 \tag{9}$$

Algorithm 1. Affinity propagation algorithm

Input: Magnetic particles $\{p_i \ i = 1, \dots, M\}$, convergence threshold c_{th}
Output: Representative particles E^*
$\mathcal{R}(p_i, e(\pi_j)) = \mathcal{A}(p_i, e(\pi_j)) = 0 \ \forall i, j$
Define E^* and E^*_{old} as empty sets
Define convergence parameter $c = 0$
while $c < c_{th}$ **then**
    $E^*_{old} = E^*$
    Remove all elements in E^*
    Update $\mathcal{R}(p_i, e(\pi_j)) \quad \forall i, j$
    Update $\mathcal{A}(p_i, e(\pi_j)) \quad \forall i, j$
    for $\forall p_i$ **do**
        if $\alpha(p_i, p_i) > 0$ **then**
            Add p_i to E^*
        end if
    end for
    if $E^*_{old} = E^*$ **then**
        $c = c + 1$
    else
        $c = 0$
    end if
end while

3.2 Identity Management and Tracking: Reliable Communications

The affinity propagation algorithm generates a set of exemplar particles E^* every T seconds. Given a set of identities Id (10), it is required to associate every exemplar particle $e^*(\pi_j)$ (which represent a biological machine b_j) with a permanent identity I_r. But this association must be coherent in the long term and current identities must be compatible with previous identity associations. In addition, the number of biological machines within a given scenario is expected to remain constant. So new identities need to be minimized, especially after a few time steps. Then, the optimal (most probable) list of permanent identities $\vec{\Lambda}$ at time $t = T_o$ must minimize a fitness function $\mathcal{F}(\cdot)$ (11), where $\rho(e^*(\pi_j), I_r; T_0, T_q)$ is the probability of exemplar particle $e^*(\pi_j)$ to be the biological machine identified as I_r at $t = T_0$, given the list of permanent identities $\vec{\Lambda}$ at time $t = T_q$.

This probability can be obtained using a Bayesian model (12), where a maximum number of different identities R_{max} is considered to make the model computable. And being $e^*(\pi_j; T_q)$ the representative for the π_j cluster at $t = T_q$, and $\rho(e^*(\pi_j; T_q), I_r)$ the probability of representative $e^*(\pi_j; T_q)$ to have the identity I_r. Probability $\rho(e^*(\pi_j; T_q), I_r)$ is obtained from previous calculations and the average operator (13), and conditional probability $\rho(e^*(\pi_j; T_0), I_r|e^*(\pi_j; T_q), I_r)$ is obtained through the movement model for biological machines (14). In this model, the probability of machine I_r to move from $e^*(\pi_j; T_q)$ position to $e^*(\pi_j; T_0)$ position in $T_0 - T_q$ seconds is analyzed. The uniform rectilinear motion model is employed, where v_{cell} is the expected movement speed for a biological machine. As a large number of independent random factors are affecting the biological movement, a Gaussian distribution describes the final probability, where σ_{cell} is a configuration parameter.

Finally, the number of new identities is obtained using the cardinality operator and the union and intersection of current and past lists of permanent identities $\vec{\Lambda}$.

$$Id = \{I_r r = 1, \ldots, \infty\} \tag{10}$$

$$\mathcal{F}\left(\vec{\Lambda}[T_0]\right) = \frac{card\left\{\vec{\Lambda}[T_0] \cap \left(\bigcup_{q=1}^{-\infty} \vec{\Lambda}[T_q]\right)\right\}}{\sum_{\forall j} \sum_{q=1}^{-\infty} \rho\left(e^*(\pi_j), I_r; T_0, T_q\right)} \tag{11}$$

$$\rho\left(e^*(\pi_j), I_r; T_0, T_q\right)$$
$$= \sum_{r=1}^{R_{max}} \rho\left(e^*(\pi_j; T_q), I_r\right) \cdot \rho\left(e^*(\pi_j; T_0), I_r|e^*(\pi_j; T_q), I_r\right) \tag{12}$$

$$\rho\left(e^*(\pi_j; T_q), I_r\right) = \frac{1}{m_{max}} \sum_{m=q-1}^{m_{max}} \rho\left(e^*(\pi_j), I_r; T_q, T_m\right) \tag{13}$$

$$\rho\left(e^*(\pi_j; T_0), I_r|e^*(\pi_j; T_q), I_r\right)$$
$$= \frac{1}{\sigma_{cell} \cdot \sqrt{2\pi}} \cdot e^{-\frac{\left(\|e^*(\pi_j; T_0) - e^*(\pi_j; T_q)\| - (T_0 - T_q) \cdot v_{cell}\right)^2}{2\sigma^2}} \tag{14}$$

The optimum list of permanent identities $\vec{\Lambda}$ is then calculated using a Particle Swarm Optimization (PSO) algorithm (15). The PSO algorithm will minimize the fitness function through a maximum of U_{max} iterations. In this algorithm, λ_j^u represent the identity for exemplar particle $e^*(\pi_j)$ in the u-th (current) iteration. λ_j^{best} is the most suitable identity proposed for representative $e^*(\pi_j)$ according to function $\mathcal{F}(\cdot)$, and $\lambda_{gloobal}^{best}$ is the most suitable identity ever proposed. Parameters $\mu_{(0,1)}^{1,2,3}$ are random values produced through a uniform distribution in the interval $[0, 1]$. And parameters ω_{min}, ω_{max} and $\beta_{1,2}$

are configuration values.

$$\lambda_j \in Id$$
$$\lambda_j^u = \lambda_j^{u-1} + v_j^u$$
$$v_j^u = \begin{cases} \left\lfloor \xi_j^u \right\rfloor & if \ \mu_{(0,1)}^3 < \frac{1}{2} \\ \left\lceil \xi_j^u \right\rceil & otherwise \end{cases} \tag{15}$$
$$\xi_j^u = \beta_1 \cdot \mu_{(0,1)}^1 \cdot \left(\lambda_j^{best} - \lambda_j^{u-1}\right) + \beta_2 \cdot \mu_{(0,1)}^2 \cdot \left(\lambda_{global}^{best} - \lambda_j^{u-1}\right) + \omega_u \cdot v_j^{u-1}$$
$$\omega_u = \omega_{max} - \frac{u \cdot (\omega_{max} - \omega_{min})}{U_{max}}$$

Finally, when the PSO algorithm determines that machine I_r has achieved position $e^*(\pi_j)$, defined as target position in the supervisory control engine, the engine generates an acknowledgment message which can be employed as support for reliable communications.

4 Experimental Validation: Simulation and Results

In order to analyze the performance and behavior of the proposed solution, an experimental validation was designed and carried out, employing simulation tools and scenarios. Simulation scenarios were built using MATLAB 2022b software, which was deployed on a Linux server (Ubuntu 22.04 LTS) with the following hardware characteristics: Dell R540 Rack 2U, 96 GB RAM, two processors Intel Xeon Silver 4114 2.2G, HD 2TB SATA 7,2K rpm.

In the simulation scenarios, a variable number of biological machines were considered, whose movement was defined according to previous models [26] that describe cellular flows. Embedded in this random number of cells, a random number of magnetic particles were integrated. These particles were free to move within each cell, according to previously reported simulation models [27]. The scenario has a size of 0.5 square millimeters. Within this area, a target point was randomly defined. And all cells were initially placed in the furthest possible position.

Simulation had no temporal limit, and they finished when the target position was achieved. Every simulation was repeated twelve times to ensure that spurious variables or effects are influencing our results. The final experimental results are calculated as the average value of all partial simulations.

Two different variables were monitored and evaluated. First, the identification error, understood as the number of times a biomachine is associated with the wrong permanent identity. This is calculated by comparing the output of the identity management algorithm to the actual identities (described in the simulation scenario). Besides, global reliability is also measured. This is understood as the ratio between the number of simulations and the number of time when the proposed system detected the cells at the target position correctly. This is calculated by comparing the output of the proposed algorithm, and the real position of biomachines indicated by the simulation engine. The experiment was repeated for different numbers of biological machines.

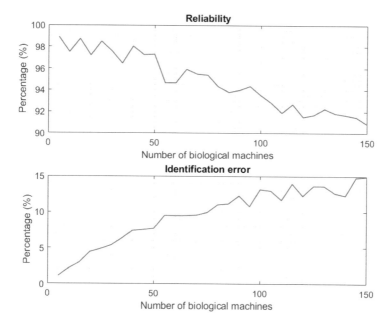

Fig. 3. Experimental results

Figure 3 shows the results obtained. As can be seen, the identification error and the reliability evolve in opposite directions. Although the identification error increases following an exponential-like law as the number of machines goes up; the reliability reduces under similar conditions. Although scalability could be improved, results are successful and promising. The identification error is always less than 15%, while reliability never drops below 90%.

5 Conclusions and Future Works

In this paper, we propose a computational algorithm to make reliable and massive 6G mobile molecular communications feasible. The proposed algorithm employs affinity propagation clustering to create a real-time map with the positions of the biological nanometric machines and particle swarm optimization to track the identity of the different machines while slowly moving. To handle ultra-massive density, clustering operates by defining which magnetic particles used as communication interface belong to the same biological machine. The optimization mechanism considers the current and previous clustering results and a probabilistic model to determine the identity of each cell. Reliability is achieved by an acknowledgment message generated by a 6G transceiver when the biological machines reach the expected destination.

Simulation tools are employed to validate the proposed solution. Results show the identification error is less than 15%, and that reliability achieves a probability of up to 90%.

Future works will consider real molecular communication environments in laboratory scenarios, in order to evaluate the performance of the proposed solution in real biological nanomachines.

Acknowledgments. The research leading to these results has received funding from the Ministry of Science, Innovation and Universities through the COGNOS project (PID2019-105484RB-I00).

References

1. Farsad, N., Yilmaz, H.B., Eckford, A., Chae, C.B., Guo, W.: A comprehensive survey of recent advancements in molecular communication. IEEE Commun. Surv. Tutor. **18**(3), 1887–1919 (2016)
2. Nakano, T., Okaie, Y., Kobayashi, S., Hara, T., Hiraoka, Y., Haraguchi, T.: Methods and applications of mobile molecular communication. Proc. IEEE **107**(7), 1442–1456 (2019)
3. Furubayashi, T., Sakatani, Y., Nakano, T., Eckford, A., Ichihashi, N.: Design and wet-laboratory implementation of reliable end-to-end molecular communication. Wireless Netw. **24**, 1809–1819 (2018)
4. Robles, T., Bordel, B., Alcarria, R., de Andrés, D.M.: Mobile wireless sensor networks: modeling and analysis of three-dimensional scenarios and neighbor discovery in mobile data collection. Ad Hoc Sens. Wirel. Networks **35**(1–2), 67–104 (2017)
5. Bordel, B., Alcarria, R., Robles, T.: An optimization algorithm for the efficient distribution of resources in 6G verticals. In: Rocha, A., Adeli, H., Dzemyda, G., Moreira, F. (eds.) WorldCIST 2022. LNCS, vol. 468, pp. 103–114. Springer, Cham (2022). https://doi.org/10.1007/978-3-031-04826-5_11
6. Bordel, B., Alcarria, R., Robles, T., Sanchez-de-Rivera, D.: Service management in virtualization-based architectures for 5G systems with network slicing. Integr. Comput.-Aided Eng. **27**(1), 77–99 (2020)
7. Bordel, B., Alcarria, R., Chung, J., Kettimuthu, R.: Predictor-corrector models for lightweight massive machine-type communications in Industry 4.0. Integr. Comput.-Aid. Eng. (Preprint) **30**, 1–25 (2023)
8. Bordel, B., Alcarria, R., Chung, J., Kettimuthu, R., Robles, T.: Evaluation and modeling of microprocessors' numerical precision impact on 5G enhanced mobile broadband communications. In: Rocha, Á., Ferrás, C., López-López, P.C., Guarda, T. (eds.) ICITS 2021. AISC, vol. 1330, pp. 267–279. Springer, Cham (2021). https://doi.org/10.1007/978-3-030-68285-9_26
9. Bordel, B., Alcarria, R., Chung, J., Kettimuthu, R., Robles, T., Armuelles, I.: Towards fully secure 5G ultra-low latency communications: a cost-security functions analysis. Comput. Mater. Continua **75**(1), 855–880 (2023)
10. Kafi, M.A., Othman, J.B., Badache, N.: A survey on reliability protocols in wireless sensor networks. ACM Comput. Surv. (CSUR) **50**(2), 1–47 (2017)
11. Bordel, B., Alcarria, R., Robles, T., Iglesias, M.S.: Data authentication and anonymization in IoT scenarios and future 5G networks using chaotic digital watermarking. IEEE Access **9**, 22378–22398 (2021)
12. Huang, L., Lin, L., Liu, F., Yan, H.: Clock synchronization for mobile molecular communication systems. IEEE Trans. Nanobiosci. **20**(4), 406–415 (2020)
13. Luo, Z., Lin, L., Ma, M.: Offset estimation for clock synchronization in mobile molecular communication system. In: 2016 IEEE Wireless Communications and Networking Conference, pp. 1–6. IEEE, April 2016

14. Lo, C., Liang, Y.J., Chen, K.C.: A phase locked loop for molecular communications and computations. IEEE J. Sel. Areas Commun. **32**(12), 2381–2391 (2014)
15. Qiu, S., Asyhari, T., Guo, W.: Mobile molecular communications: positional-distance codes. In: 2016 IEEE 17th International Workshop on Signal Processing Advances in Wireless Communications (SPAWC), pp. 1–5. IEEE, July 2016
16. Lin, L., Wu, Q., Liu, F., Yan, H.: Mutual information and maximum achievable rate for mobile molecular communication systems. IEEE Trans. Nanobiosci. **17**(4), 507–517 (2018)
17. Chang, G., Lin, L., Yan, H.: Adaptive detection and ISI mitigation for mobile molecular communication. IEEE Trans. Nanobiosci. **17**(1), 21–35 (2017)
18. Haselmayr, W., Aejaz, S.M.H., Asyhari, A.T., Springer, A., Guo, W.: Transposition errors in diffusion-based mobile molecular communication. IEEE Commun. Lett. **21**(9), 1973–1976 (2017)
19. Ahmadzadeh, A., Jamali, V., Schober, R.: Stochastic channel modeling for diffusive mobile molecular communication systems. IEEE Trans. Commun. **66**(12), 6205–6220 (2018)
20. Ahmadzadeh, A., Jamali, V., Noel, A., Schober, R.: Diffusive mobile molecular communications over time-variant channels. IEEE Commun. Lett. **21**(6), 1265–1268 (2017)
21. Yu, W., Liu, F., Yan, H., Lin, L.: Evaluation of non-coherent signal detection techniques for mobile molecular communication. IEEE Trans. Nanobiosci. **22**(2), 356–364 (2022)
22. Iwasaki, S., Nakano, T.: Graph-based modeling of mobile molecular communication systems. IEEE Commun. Lett. **22**(2), 376–379 (2017)
23. Isik, I., Er, M.B., Isik, E.: Analysis and classification of the mobile molecular communication systems with deep learning. J. Ambient. Intell. Humaniz. Comput. **13**(5), 2903–2919 (2022)
24. Shrivastava, A.K., Das, D., Mahapatra, R.: Particle-based simulation of the differential detectors for mobile molecular communication. IEEE Commun. Lett. **25**(9), 3008–3012 (2021)
25. Zhai, H., Yang, L., Nakano, T., Liu, Q., Yang, K.: Bio-inspired design and implementation of mobile molecular communication systems at the macroscale. In: 2018 IEEE Global Communications Conference (GLOBECOM), pp. 1–6. IEEE, December 2018
26. Freund, J.B., Shapiro, B.: Transport of particles by magnetic forces and cellular blood flow in a model microvessel. Phys. Fluids **24**(5) (2012)
27. Plouffe, B.D., Murthy, S.K., Lewis, L.H.: Fundamentals and application of magnetic particles in cell isolation and enrichment: a review. Rep. Prog. Phys. **78**(1), 016601 (2014)

Security Assessment of an Internet of Things Device

Daiana Alexandra Cîmpean$^{(\boxtimes)}$, Marius-Constantin Vochin,
Răzvan-Eusebiu Crăciunescu, Ana-Maria-Claudia Drăgulinescu,
and Laurențiu Boicescu

National University of Science and Technology Politehnica Bucharest (UNSTPB),
060042 Bucharest, Romania
`daiana.cimpean@stud.etti.upb.ro`

Abstract. The rapid advance of Internet of Things (IoT) and its immersion in every domain brought into attention, besides many of its technical, social, and economic advantages, a panoply of security vulnerabilities and attack vectors that threaten IoT interconnected devices. IoT devices face many security issues such as weak authentication, insufficient encryption, deficient device management, insecure interfaces, inadequate physical security, lack of standardization, privacy concerns, insecure networks, resource constraints, and non-compliance with security standards. This highlights the pressing need for comprehensive security measures that currently seem insufficient to address the evolving landscape of threats in the dynamic IoT ecosystem. This paper conducts a security evaluation of a physical IoT device through the penetration testing methodology. It then focuses on a known vulnerability from the Common Vulnerabilities and Exposures (CVE) database. It presents the execution of a brute force attack to uncover credentials and the device's buffer overflow vulnerability to cause a denial of service (DoS) on the device's server. Employing a hands-on approach, the research emphasizes the practical execution of these exploitation scenarios providing a step-by-step guide on how they were performed. Lastly, it delves into the development of a proof of concept (PoC) application created to automate the process of firmware analysis and running the buffer overflow exploit for this particular use case.

Keywords: Exploit · Privacy · Brute-force · Firmware analysis · Penetration Testing

1 Introduction

1.1 Current Insights in Internet of Things Security

Nowadays, the realm of human-machine interaction has materialized as a feasible and practical concept due to the significant advancements in technology and the widespread accessibility of the Internet. Consequently, the Internet of Things (IoT) has emerged as a cutting-edge and state-of-the-art technology, capturing substantial attention, and garnering considerable interest within the realm of research [1].

© The Author(s), under exclusive license to Springer Nature Switzerland AG 2024
Á. Rocha et al. (Eds.): WorldCIST 2024, LNNS 989, pp. 284–294, 2024.
https://doi.org/10.1007/978-3-031-60227-6_26

The rapid growth of IoT technology and smart devices has revolutionized industries and daily life, but it has also brought intricate security challenges. The vast network of IoT devices lacks inherent security [2], needing real-time visibility and intelligent policy enforcement to counter risks and vulnerabilities. There are several significant challenges associated with IoT security, the main one being that IoT devices were often not designed with security as a primary consideration [3]. Also, in many cases, it is not possible to install security software directly on the IoT devices themselves [4]. Thus, robust security measures are essential for identifying, monitoring, mitigating potential threats, and ensuring protection against costly and hazardous breaches.

The state-of-the-art IoT research reflects a notable concentration on diverse facets like applications, architecture, protocols, and standards. When discussing how this technology is employed, it becomes obvious that the IoT serves diverse purposes, from personal use (e.g., smart house appliances, wearable devices, etc.), to community needs (e.g., equipment used in healthcare or in the industrial sector). Despite its rapid growth, the IoT faces challenges related to limited computing power and resources, prompting significant research efforts to address security concerns [5]. The state-of-the-art in IoT architecture reflects a comprehensive and layered approach, encompassing physical, virtual, connectivity, and user-specific aspects, to design sophisticated and interconnected systems for enhanced functionality and user experience [6]. It is characterized by four primary stages (Device Layer for physical identification, Gateway & Aggregation Layer for data collection, Event Processing Layer for behavior control, and Application Layer for versatile interface and analysis), but can contain many other layers (e.g., for virtualization, interoperability, communication) depending on the use case it was designed for [7].

However, there is a conspicuous lack of exploration into threats and attacks within the IoT landscape, particularly concerning architecture and protocols. A meticulous analysis of the SCOPUS database, performed in the paper referenced at [8], bridges this gap, showcasing a rapid rise in researchers' interest in these security aspects. While initially underexplored, threats and attacks analysis in IoT architecture and protocols has gained significant momentum, motivating the present work to review and address these gaps. Moreover, there is an issue of interoperability in IoT, emphasizing the need for a holistic understanding of security standards and goals to overcome challenges [8].

1.2 Paper Objectives and Structure

This paper aims to evaluate the security of an indoor surveillance camera from the D-Link DCS series, focusing on a known vulnerability listed in the CVE database. The investigation covers a brute-force attack for acquiring login credentials, reverse engineering of the device's firmware, and the development of a proof-of-concept application automating security assessments. The goal is to highlight the security vulnerabilities in IoT devices and provide practical insights into exploiting and addressing these issues. The paper answers the following research questions:

- How secure IoT devices are and what are their vulnerabilities if there are any?
- What is the impact of outdated software and bad practices on IoT security?

   Therefore, the specific objectives of this paper are:

- Conducting an assessment for an IoT device by identifying its vulnerabilities;
- Analyzing the security implications of said vulnerabilities;
- Assessing the impact of an outdated and vulnerable software and the influence of bad practices on IoT security;
- Based on the findings, offer practical recommendations to improve the security of the device under test.

In the following sections, we delve into distinct aspects of the study, focusing on conducting the security assessment. Section 2 outlines the methodology employed for executing a brute-force attack to acquire the login credentials for the vulnerable IoT device and describes the reverse engineering process applied to the device's firmware, uncovering information on possible security flaws. Section 3 details the implementation of a web application meant to automate both the firmware analysis and running the exploit. Section 4 presents the findings regarding the IoT device under investigation and the future prospects for how the application could be further improved.

2 Vulnerability Assessment of the IoT Device

2.1 Research Methodology

While it is certain that the pervasive development of the Internet of Things is accompanied by a notable deficiency in security measures, this paper aims to offer insight on just how secure an IoT device really is. Thus, this paper investigates the security landscape surrounding an indoor surveillance camera from the D-Link DCS series, renowned for a CVE-listed vulnerability triggering a server buffer overflow [9]. The study delves into the exploitation of the D-Link DCS5009L camera, a versatile device supporting Ethernet/Fast Ethernet and wireless networks for adaptable connectivity [11].

In a simulated yet realistic scenario, an adversary possessing information about a potentially vulnerable device uses a tool such as NMAP (Network Mapper) to perform a network scan to discover operational devices, including the one in question. When the IoT device is up (and discovered running a vulnerable firmware version), the attacker proceeds to exploit its vulnerability. Upon confirming the operational status of the IoT device, the attacker strategically exploits the stack-based buffer overflow within the alphapd web server (documented by several databases with disclosed information about security issues in accordance with the Common Vulnerabilities and Exposures system) [9]. 'alphapd' refers to the web server component within the D-Link DCS series of Wi-Fi cameras. It plays a crucial role in managing wireless functionalities and web-based interactions for the camera. Next, the steps taken to carry out the security assessment consisted of obtaining the credentials through a brute force attack, analyzing the firmware, and finally running the buffer overflow exploit as depicted in Fig. 1.

It is worth noting that network security scanners such as NMAP are normally not limited just to identifying accessible devices and services but can also obtain more in-depth information regarding the current version of firmware, operating systems or servers running on the devices under scrutiny. This information is valuable to both penetration testers and attackers, as it would point out to vulnerable devices.

Fig. 1. Exploitation scenario workflow

2.2 Uncovering the Credentials Through a Brute-Force Attack

In the process of obtaining login credentials for the exploit, a brute force attack was employed using the Burp Suite tool, as depicted in Fig. 2.

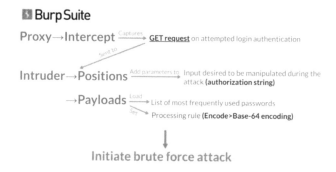

Fig. 2. Burp Suite brute-force attack systematic approach.

Upon trying to authenticate into the management interface of the camera, Burp Suite captures the HTTP (Hypertext Transfer Protocol) GET request using the Intercept mode. The HTTP header captured contains an authorization string that serves as means of user authentication. Since this string can potentially reveal the credentials, it is sent to the intruder module in Burp Suite where it is flagged for manipulation during the brute force attack. Next, Burp Suite requires the user to upload a list of passwords among which the tool will try to find matches. Base64 Encoding should be also applied to ensure compatibility with the target system and URL-encoding was intentionally disabled to avoid interference with special characters during the attack. After these steps are completed, Burp Suite starts the brute-force attack.

The brute-force attack is a time-consuming yet methodical process in which every password from the word list is systematically tested against the target system. This exhaustive approach significantly improves the chances of discovering the correct password. Upon successfully identifying the password, the intruder tool displays an HTTP 200 status code (indicating a successful request). All other previous attempts returned an HTTP 401 status, for an unauthorized request (an authentication failure). **3**Further investigation provides more detailed information, including the authorization string containing the credentials. With the obtained login information, the adversary can now access and control the IP camera's management interface.

This section's focus has been on the process of uncovering passwords rather than on the quantization of how long it would take to complete it. Thus, the device used a vulnerable, simple password. Normally, a complex device would use at least use a simple CAPTCHA (Completely Automated Public Turing test to tell Computers and Humans Apart) to identify and stop automated password-cracking attempts, as well as delays (or progressively increasing delays) in case of unsuccessful authentication attempts, to considerably increase the password cracking duration. However, due to their limitations, IoT devices tend to be inherently vulnerable to brute-force password-cracking attempts, as they rarely implement the above mechanisms.

One of the most recent studies regarding password cracking difficulty, the so-called "Hive Systems Password Table" [12], attempts to provide the means of comparing the amount of time it would take to crack a password, depending on its complexity. The study shows that, for example, simple ten-character long alpha-numeric passwords can be cracked in less than a day, whilst the effort increases to two weeks if the password contains special characters. Increasing the size of a password to 15–20 (or more) characters exponentially increases the complexity of brute-force attacks to the point where they become infeasible with today's computing systems.

2.3   Reverse Engineering the Firmware

This section delves into firmware analysis and reverse engineering using Binwalk, focusing on the examination of the camera's firmware, specifically version 1.08.11, through a detailed investigation. The primary objective is to extract the file system and scrutinize its contents—a crucial step in identifying potential vulnerabilities within the IoT ecosystem. Leveraging Binwalk, a tool equipped with distinctive pattern recognition, the binary file of the firmware was analyzed. Thus, key elements such as the use of Universal Boot Loader (U-Boot) and lzma-compressed data were identified.

The section containing the LZMA compressed data was then duplicated and extracted and the data obtained was analyzed again using Binwalk, revealing that the firmware uses an outdated Linux kernel version, as seen in Fig. 3. It showcases that the firmware uses version 2.6.21 which dates back to the 25th of April 2007, while the most recent version is 6.6.9 released on the 1st of January 2024 [13].

Since the kernel also contains a section of LZMA compressed data, the extraction process was repeated and the file obtained was an ASCII cpio archive used by the manufacturers, including D-Link, to store the file system. This archive can be extracted and mounted in a dedicated folder, from which the filesystem can be further scrutinized.

After this phase, the unpacked file system becomes accessible, granting the adversary the capacity to explore directories, manipulate files, execute programs, and conduct various actions within the file system. For example, the 'etc_ro' directory houses a web interface, potentially useful for identifying web application vulnerabilities. Additionally, the 'sbin' directory contains human-readable scripts. This level of access to the firmware's system files, whether through the terminal or directly from directories, provides a substantial advantage.

```
┌──(kali㉿kali)-[~/Downloads]
└─$ binwalk kernel

DECIMAL       HEXADECIMAL     DESCRIPTION

3276876       0x32004C        Linux kernel version 2.6.21
3317264       0x329E10        SHA256 hash constants, little endian
3323216       0x32B550        AES Inverse S-Box
3323984       0x32B850        AES S-Box
3364384       0x335620        Unix path: /usr/gnemul/irix/
3366308       0x335DA4        Unix path: /usr/lib/libc.so.1
3423052       0x343B4C        Copyright string: "Copyright (c) 2011 Alpha Networks Inc."
3432948       0x3461F4        Unix path: /var/run/udhcpc.pid
3508480       0x358900        Unix path: /usr/bin/killall
3511744       0x3595C0        Unix path: /etc/Wireless/RT2860STA/e2p.bin
3520464       0x35B7D0        Unix path: /etc/Wireless/RT2860STA/RT2860STA.dat
3520596       0x35B854        Unix path: /etc/Wireless/RT2860/RT2860.dat
3570923       0x367CEB        Neighborly text, "neighbor %.2x%.2x%.2x%.2x%.2x%.2x%.2x%.2x%.2x%.2x lost on port %d(%s)(%s
)"
3679248       0x382410        CRC32 polynomial table, little endian
3682752       0x3831C0        AES S-Box
3874816       0x3B2000        LZMA compressed data, properties: 0x5D, dictionary size: 1048576 bytes, uncompressed siz
e: 16888320 bytes
```

Fig. 3. Binwalk analysis for the kernel file

Because the camera's firmware has a vulnerability that causes a buffer overflow in its server, an adversary would focus not only on disrupting the service (causing a Denial of service), but also use more capabilities and push further.

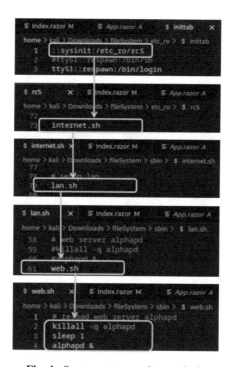

Fig. 4. System startup scripts analysis

An initial analysis of the script /etc_ro/inittab, depicted in Fig. 4 reveals that, upon startup, the server's processes are terminated, followed by a one-second pause and the server's restart with processes running in the background.

This unrestricted accessibility to critical system operations raises security concerns, potentially leading to service disruption, privilege escalation, and information exposure. Publicly available firmware can be scrutinized by attackers to discover vulnerabilities. However, if an attacker were able to access an IoT device to copy its firmware image, he would be able to extract even more valuable information, including passwords, any type of credentials or even private encryption keys if they are not properly encrypted (as shown in the above case). Thus, reverse engineering such vulnerable binary firmware images can open new attack vectors for adversaries. Physical protection of IoT devices, alongside proper management, and update policies, is of paramount importance to protect against this type of attack. Furthermore, an adversary with this type of access can also replace information in the reverse-engineered firmware, and then update the firmware on the IoT devices to obtain a backdoor to the IoT network.

3 Automation Proof of Concept Application

3.1 Security Assessment Application

This research culminated in the development of a proof-of-concept application designed to automate the security assessment of a vulnerable IoT device. The chosen software architecture is the client-server model, ensuring efficient interaction between the client, responsible for making requests, and the server, responding accordingly. This architecture supports scalability and clear separation of concerns.

Fig. 5. Security Assessment Web Application

The application automates the commands used in firmware analysis, providing an intuitive stage-based mechanism for users, as seen in Fig. 5. In its second tab, the application provides information on the buffer overflow exploit for the D-Link DCS-5009L camera. The UI (User Interface) offers general information about the exploit, lists vulnerable models, and can run the buffer overflow exploit upon entering the credentials.

The code uses a Python script created based on the model provided in [14]. The script checks for credentials, constructs URLs, and sends a GET request to initiate the exploit. Successful execution renders the camera's management interface unresponsive.

3.2 Validation of the Proposed Investigation

The validation of the proposed investigation is paramount to affirming the credibility and effectiveness of the research methodology employed in this study and is ensured as follows:

- The network scanning methodology using NMAP ensures a thorough and accurate discovery of operational devices. Our staged approach, starting with a simulated yet realistic scenario, reflects practical considerations. Employing NMAP as the tool of choice aligns with industry standards for network reconnaissance.
- The brute force attack approach, executed through Burp Suite, ensures a reliable credential acquisition process. The systematic workflow, detailed in Fig. 2, demonstrates a well-defined and replicable methodology. Moreover, the HTTP status code indicators provide transparency—HTTP 200 for success and HTTP 401 for failure.
- The firmware analysis, conducted using Binwalk, substantiates the effectiveness of our approach in extracting critical information from the IoT device under scrutiny. By employing Binwalk's pattern recognition capabilities, we successfully identified key elements such as the Universal Boot Loader (U-Boot) and lzma-compressed data within the firmware binary. The duplication and extraction of the LZMA compressed data unveiled an outdated Linux kernel version, a significant security risk. The subsequent unpacking of the firmware file system provided extensive access, allowing us to explore directories, manipulate files, and execute programs within the system. This validated our methodology's ability to uncover vulnerabilities in firmware, demonstrating the robustness of our approach in firmware analysis and potential threat identification.
- The successful development and implementation of a proof-of-concept application affirms the robustness and usability of our approach. The Python script's effective utilization for exploit execution aligns with existing models, adding credibility to our investigation. The visual representation in Fig. 5 serves as tangible evidence, illustrating the application's structure and functionality.

4   Conclusions

4.1   Paper Contributions

The Internet of Things (IoT) has revolutionized connectivity but introduced security challenges. This project focused on assessing the security of a D-Link Camera due to a known vulnerability affecting DCS series devices. The first step of the security assessment was a systematic brute force attack on the IP camera system's password-based authentication, emphasizing the significance of strong, unique passwords and advanced authentication methods in IoT security. Additionally, the use of Base64 encoding revealed its vulnerabilities, highlighting the necessity for stronger data protection measures like encryption and hashing. There are several take-aways from this work for other researchers in the field:

i) Outdated Linux kernel version vulnerability. The reverse engineering process allowed a deeper understanding of the firmware's inner workings and revealed exploitable features since several specific areas could serve as access points for potential attacks, including the firmware binary file and the system's architecture. Thus, one of the key revelations from this exercise was the identification of the outdated Linux kernel version used by the firmware, a significant security risk as vulnerabilities associated with this version have been publicly disclosed.

ii) Successful unpacking of the firmware file system. Another crucial finding was the successful unpacking of the firmware file system which could provide extensive access to an adversary. It allows a potential attacker to navigate directories, manipulate files, and execute programs within the file system. This level of access could lead to the identification of further vulnerabilities and manipulation of the system for nefarious purposes.

iii) Restrictions on system startup scripts. The analysis also delved into critical system operations. Unpacking the firmware exposed open accessibility to these operations, a significant security fault due to its potential misuse. The absence of restrictions on system startup scripts, for example, can lead to service disruption, privilege escalation, and unintended information disclosure.

iv) Possibility of running arbitrary code. The buffer overflow vulnerability within the camera's firmware server is of particular concern as well since it can be further exploited to run arbitrary code on the system, allowing attackers to modify the system's functions, steal sensitive information, or potentially take complete control of the system.

Overall, the application developed focused on two crucial parts of the security assessment - the firmware analysis and the exploit. In terms of architecture, the client-server model was employed due to the clear separation of concerns and efficient distribution of resources it offers. The application's capability to automate firmware analysis and exploitation processes signified how software can drastically simplify complex and repetitive tasks. By doing so, the application saved time and minimized the risk of human error.

4.2 Future Directions of Development

So far, the research is at a proof-of-concept stage and there are several possible changes that could be further implemented to enhance the security assessment application. As user interaction is an important aspect of any application, the UI could use some improvements to become more intuitive and allow customization options. An assessment engine, responsible for automated processing and real-time monitoring, could ensure a dynamic and proactive approach to security evaluations. It could be used in conjunction with a reporting module to give users detailed insights through comprehensive reports and visualization tools. An integration module could be introduced to seamlessly incorporate established security frameworks, incident response protocols, and continuous updates, fostering compatibility and resilience against emerging threats. Key actions performed while interacting with the application should be logged, as well as any possible errors that may take place.

Documenting the activity within the application should ensure that users only perform authorized activities and could facilitate faster remediation of any errors that might occur and may disrupt the workflow. Finally, the scalability module would ensure efficient resource utilization, incorporating parallel processing, distributed architecture, and scalable database structures. Leveraging auto-scaling capabilities and load balancing algorithms, it dynamically adjusts resources based on demand, preventing bottlenecks, and optimizing performance.

Acknowledgment. The research leading to these results has been partially supported by the NO Grants 2014-2021, under Project contract no. 42/2021, RO-NO-2019-0499 - "A Massive MIMO Enabled IoT Platform with Networking Slicing for Beyond 5G IoV/V2X and Maritime Services" - SOLID-B5G.

References

1. Singh, J., Singh, G., Negi, S.: Evaluating security principals and technologies to overcome security threats in IoT world. In: 2023 2nd International Conference on Applied Artificial Intelligence and Computing (2023). https://doi.org/10.1109/ICAAIC56838.2023.10141083
2. Tariq, U., Ahmed, I., Bashir, A.K., Shaukat, K.: A critical cybersecurity analysis and future research directions for the internet of things: a comprehensive review. Sensors (Basel) (2023). https://doi.org/10.3390/s23084117
3. Schiller, E., Aidoo, A., Fuhrer, J., Stahl, J., Ziorjen, M., Stiller, B.: Landscape of IoT security. Comput. Sci. Rev. **44** (2022). https://doi.org/10.1016/j.cosrev.2022.100467
4. Zenarmor: Cyber security solutions on IoT security (2023). https://www.zenarmor.com/docs/what-is-iot-security. Accessed 07 Nov 2023
5. Srhir, A., Mazri, T., Benbrahim, M.: Security in the IoT: state of the art, issues, solutions, and challenges. Int. J. Adv. Comput. Sci. Appl. **14**(5) (2023). https://doi.org/10.14569/IJACSA.2023.0140507
6. Weigong, L.V., Meng, F., Zhang, C., Yuefei, L.V., Cao, N., Jiang, J.: A general architecture of IoT System. In: IEEE International Conference on Computational Science and Engineering and IEEE International Conference on Embedded and Ubiquitous Computing, Guangzhou, China, pp. 659–664 (2017). https://doi.org/10.1109/CSE-EUC.2017.124
7. Bouaouad, A.-E., Cherradi, A., Assoul S., Souissi, N.: The key layers of IoT architecture. In: 5th International Conference on Cloud Computing and Artificial Intelligence: Technologies and Applications (CloudTech) (2020). https://doi.org/10.1109/CloudTech49835.2020.9365919
8. Krishna, A., Priyadarshini, R.R., Jha, A.V., Appasani, B., Srinivasulu A., Bizon, N.: State-of-the-art review on IoT threats and attacks: taxonomy, challenges and solutions. Efficiency and Sustainability of the Distributed Renewable Hybrid Power Systems Based on the Energy Internet, Blockchain Technology and Smart Contracts-Volume II (2021). https://doi.org/10.3390/su13169463
9. National Institute of Standards and Technology (NIST): National Vulnerability Database. https://nvd.nist.gov/vuln/detail/CVE-2017-17020. Accessed 15 Nov 2023
10. D-Link: DCS-5009L Pan & Tilt WiFi Camera. https://eu.dlink.com/uk/en/products/dcs-5009l-pan-tilt-wifi-camera. Accessed 29 Oct 2023
11. Wikipedia Contributors: Wikipedia List of HTTP Status Codes. https://en.wikipedia.org/wiki/List_of_HTTP_status_codes. Accessed 25 Nov 2023

12. Neskey, C.: Hive Systems Cybersecurity Solutions. https://www.hivesystems.io/password. Accessed 1 Nov 2023
13. Linux Kernel End-of-Life. https://endoflife.date/linux. Accessed 5 Jan 2024
14. Tacnetsol: D-Link exploit Python scripts. https://github.com/tacnetsol/CVE-2019-10999. Accessed 15 Nov 2023.https://nvd.nist.gov/vuln/detail/CVE-2017-17020

Applying DevOps Practices for Machine Learning: Case Study Predicting Academic Performance

Priscila Valdiviezo-Diaz[✉] and Daniel Guamán

Department of Computer Science, Universidad Técnica Particular de Loja,
San Cayetano Alto, Loja 1101608, Ecuador
{pmvaldiviezo,daguaman}@utpl.edu.ec

Abstract. This paper presents the application of DevOps practices in the development and implementation of a machine learning model to predict the academic performance of students enrolled in a Higher Education Institution in Ecuador. The study encompasses three key phases: 1) Dataset preparation to build and validate the model, 2) Experimental evaluation of machine learning algorithms for academic performance prediction, and 3) Implementation of DevOps practices to design and build a software application using Microservices architecture, that supports the prediction model and visualize the results to take decisions. The dataset employed for the model and software application incorporates socio-demographic data and academic records of undergraduate students within the Information Technology career. Logistic Regression, Random Forest, and Neural Network algorithms are applied directly to the student dataset. Compared with other algorithms, experimental results present favorable outcomes with Random Forest, particularly in precision, sensitivity, and f-score metrics.

Keywords: academic performance · software application · microservices · machine learning · prediction

1 Introduction

The prediction of academic performance in Higher Education is a topic of great interest for educational institutions. Currently, research in this field has advanced significantly thanks to the use of data mining and machine learning techniques, which have made it possible to identify hidden factors and patterns that influence academic performance [2,9].

Predicting academic performance holds significant value in identifying students at risk of dropping out within an educational institution. According to [1], one effective approach to evaluating student success involves predicting their performance using their prior academic grades. This foresight enables educators to implement proactive measures and offer targeted support to students requiring assistance.

In the context of distance learning systems, Karalar et al. [6] consider the characteristics of students' academic activity to identify those who may be at risk of academic failure. Nahar et al. [11] presents academic results and the behavior of some engineering students are analyzed with the aim of predicting student performance. In [5] explores the potential of machine learning algorithms to reliably estimate students' academic performance.

The application of Machine Learning (ML) techniques in predicting academic performance in higher education has attracted considerable attention in recent years due to its potential to provide valuable and personalized information to students and educators. Predictive modeling has been a focal point for enhancing learning outcomes. Researchers have applied various machine learning algorithms to predict academic performance, leveraging features such as student demographics, previous academic records, and socio-economic factors. However, effective implementation of ML algorithms involves not only deep understanding of the models but also efficient management of workflows and continuous collaboration between development and operations teams. This is where DevOps practices emerge as a comprehensive solution, optimizing the development, deployment, and maintenance of machine learning models.

The novelty of this research lies in the fusion of two crucial areas: Machine Learning and DevOps. Although both disciplines have proven their value individually, their integration to improve the prediction of academic performance in higher education settings is an underexplored area. The successful implementation of ML models is not only based on the quality of the algorithm but also on the agility and efficiency of the processes related to its development, implementation, and maintenance. This is where the DevOps methodology plays a crucial role.

Therefore, the main objective of this research is to apply DevOps practices in the development and implementation of a machine learning model for a case study on the prediction of academic performance in higher education. The idea is to predict whether a student will approve or fail a subject based on the student's available data (prior academic performance and enrollment data).

This effort encompasses a meticulous selection and processing of relevant variables, followed by developing and deploying a predictive model.

2   Related Work

The integration of DevOps practices and machine learning applications has caught the attention in the literature. Recent studies have explored the synergies between these domains, emphasizing the potential for rapid models deployment. For example, in [13] proposes the practice of applying the culture of DevOps with machine learning to build and deploy a model based on neural network rapidly.

Kuma et al. [7] develop a framework for effectively managing DevOps practices. The authors conduct an empirical study using the publicly available HELENA2 dataset to identify the best practices for effectively implementing DevOps. They use prediction algorithms such as Support Vector Machine

(SVM), Artificial Neural Network (ANN), and Random Forest (RF) to develop a prediction model for DevOps implementation. Reference [3] proposes a model-based DevOps process for incremental design and continuous testing of Database Cost Models. The design approach that authors adopt is based on a hybrid model that is a combination of an analytical model and a machine learning model.

From the literature review, we observe few studies have specifically addressed the combination of DevOps practices and machine learning workflows in the educational domain. However, some studies have focused on developing prototypes or applications that support the identification of students at risk, for example, Nuankaew et al. [12] present research to identify the risk of dropout in tertiary students through a prototype for dropout predictions of higher education students. The authors use CRISP-DM to develop predictive models and the Software Development Life Cycle (SDLC) for application development. Nagy et al. [10] present a web application to identify at-risk students based on their grades. The authors use machine learning algorithms including gradient-boosted decision trees (XGBoost), for this purpose. The development of this software application was executed using the Dash web framework.

The approaches presented in related studies have demonstrated their effectiveness in predicting students' academic performance. However, the integration of DevOps practices in the educational domain remains a relatively unexplored area. Hence, this study seeks to implement a machine learning model good enough in terms of precision in order to predict academic performance by applying DevOps practices.

3   Materials and Methods

This section shows the process used to predict the academic performance of students enrolled in a Higher Education Institution in Ecuador. First, we expose the phases used to prepare the dataset that we will use in the machine learning algorithms. Then we explain the machine learning algorithms used for predicting academic performance. Finally, we present the process to design and build a software application that allows us to analyze and visualize the results obtained.

3.1   Dataset Preparation

The dataset used for experimenting with machine learning algorithms comprises information gathered from students enrolled in the Information Technology career at a Higher Education Institution. The academic data cover the range from April–August 2020, until October 2021–February 2022. In total, a dataset with 27.440 records from 2001 students and 48 attributes related to socio-demographic and academic characteristics are obtained (see Table 1).

The information sourced and extracted from the educational institution's electronic academic system underwent a thorough anonymization process. The process of creating the dataset started by filtering the information of students enrolled in the Information Technology career. Subsequently, data cleaning and

Table 1. Variables Description

Category	Variable name
Socio-demographic	Age
	Sex
	Province
	City
	Disability
	Payment methods
	Scholarship percentage
Academics	Period_admission
	Courses
	Partial grades for each component:
	Autonomous learning (aab1, aab2),
	Experimental practical (apeb1, apeb2),
	Teacher-guided Learning (acdb1, acdb2),
	Final grades of each two-month term
	Status
	Approval rate
	Number of failures

dataset preparation for analysis were realized, including handling missing values and removing outliers. In summary, as part of the preparation of the dataset, the following was carried out:

- The records lacking sufficient information for analysis and variables with over 70% missing values were excluded from consideration.
- Only records of students who received at least one grade in the subjects were taken into account.
- In cases where a subject was taken in multiple cycles, the grade for the most recent academic term was used.
- If a student has not yet taken any of the subjects, a blank value is assigned to that specific subject.
- The variables exhibiting an imbalance in the amount of data were excluded, for example, number_failures, percentage_of_disabilities, disability, and number_disabilities.
  As can be seen in Fig. 1, these variables tended to have a single value, which could affect the performance of the algorithm.

It is important to highlight that tests were also conducted by balancing the variables in the dataset, employing the RandomOverSampler technique [15], which utilizes the nearest neighbor algorithm to balance the classes. Nevertheless, maintaining the balanced variables yielded evaluation metric results similar to those obtained by removing them from the dataset.

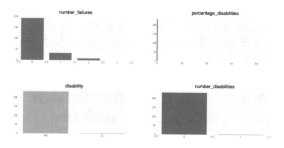

Fig. 1. Unbalanced variables

3.2 Academic Performance Prediction

Numerous techniques have been suggested to address binary classification tasks. In our study, experiments were conducted with three supervised machine learning algorithms for predicting academic performance: Logistic regression, Random Forest, and Neural Networks. We selected these algorithms based on their distinct strengths and capabilities in handling predictive tasks for academic performance.

Furthermore, machine learning algorithm evaluation metrics were employed to assess the quality of the classification such as precision, sensitivity, specificity, and f-score. As a result of the evaluation, the Random Forest algorithm presented the most promising results in the mentioned metrics.

3.3 Process to Design and Build the Software Application

To design, build, and validate the software application for the predictive model of academic performance, some software frameworks, practices, and methodologies are used, among which the following stand out:

1. The DevOps (Development and Operations) framework, encompassing its various phases and the corresponding tools employed at each stage, as well as the automated continuous integration and deployment (CI/CD) process are used to build the software application.
2. The scope of the software application is to help interpret the academic performance data, for that reason we have to identify some business capabilities that can be automated into the DevOps pipeline. To determine the level of granularity of the software components, a decomposition technique called Capability Map is applied. The Capability Map offers a comprehensive overview of the functionalities addressed by the software application.
3. The logical and physical architecture of the software application is documented using the C4 model [16]. The software architecture design allows to maximize the benefits of the Random Forest model applied in the classification of student performance.

4. Finally, through communication flow diagram, the interaction between the client, dataset, and software application is represented, outlining the entire process required to classify a new student within the software application, as a part of the Case Study.

4 DevOps Practices to Build the Software Application for Academic Performance Prediction

In order to evaluate and visualize the results of the prediction of academic performance to enable decision-makers and academic advisors to predict a student's approval status easily and quickly, it is propose to build a software application where the architectural design, services and data model are used. In addition, we propose to use a DevOps approach for continuous integration and deployment issues.

Prior to the design and construction of the software application, several machine learning models were tested. As a result, Random Forest was selected as the base algorithm due to its superior performance on the metrics used. The results of data processing and analysis through this algorithm are displayed in the software application.

The software application was developed in Python using the DevOps framework [8]. This framework allows the production of a good quality and reliable solution.

Figure 2 illustrates a set of automated processes (Pipeline), providing a breakdown of each implemented phase and the corresponding tools utilized for each phase. Pipeline DevOps integrates several tools that automate and control software development, these vary according to the software to be developed and the technology used. Some tools, depending on the technology, do not integrate perfectly with others, so before applying the pipeline, this must be taken into account.

The DevOps framework lifecycle consists of eight phases. Each phase builds on the others and in this way guides the effectiveness and efficiency of the development of the software application for the prediction of students' academic performance.

In the Plan phase, all the features and capabilities of the application were added. In the Code phase, all aspects of coding and best practice code writing were included. In the Build, Test, Release, and Deploy phases, the automated continuous integration and deployment (CI/CD) process was used, which allows all of these phases to be integrated. In the Operate phase, all the maintenance and infrastructure of the application were managed. Finally, in the Monitor phase, all supervision of each version of the implemented application was carried out.

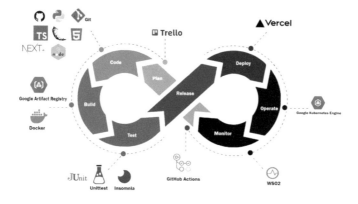

Fig. 2. DevOps used for software application

4.1   Capability Map

The capability map, also referred to as the value chain, serves as a model to comprehensively outline the activities required for the creation of the software application aimed at predicting students' academic performance.

Figure 3 provides a detailed representation of the capabilities map constructed for the development of the software application.

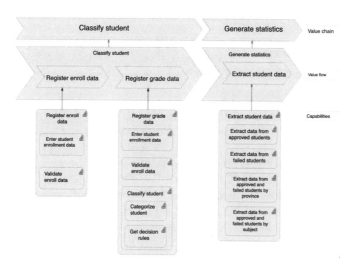

Fig. 3. Capability map for software application

The capabilities map comprises three distinct levels. The first level, labeled as "Value Chain," details all those activities that add value to the application

302 P. Valdiviezo-Diaz and D. Guamán

and create a competitive advantage. At the second level called "Value Flow" those support activities that help the activities of the value chain to be more efficient are considered. Finally, the last level called "Capabilities" are all the functionalities that were included in the development of the software application.

4.2 Software Application Architecture

The proposed architecture is a microservices-oriented architecture, which seeks to fully utilize the benefits of the student academic performance classification model.

A microservices-oriented architecture allows the application to be structured as a collection of services that are highly maintainable, testable, loosely coupled, and implemented independently [14] and above all organized according to the analyzed capabilities map. Each microservice contains a service that must be configured correctly, the registry of these services contains some configurations of interest, such as the real locations of the servers, and the ports on which they are running. To identify the microservices, the dependency, and coupling between each capability within the value flow were analyzed.

The microservices selected for the development of the software application are explained as follows:

- Microservice for classifying students' academic performance: Responsible for receiving both the student's enrollment data, as well as the values of the grades obtained in a specific component. From this, it sends the enrollment and grade values to the machine learning model so that it evaluates and predicts the student's approval status.
- Microservice for generating statistics: Essentially responsible for retrieving the stored data of all "approved" and "failed" students who have been classified so far, based on the filters that the end user selects in the software application.

Figure 4 presents software application and microservices-oriented architecture where each of the components and the interactions between each of them are visualized.

According to the architecture, end users connect to the application through a web client through the HTTP or Web Sockets protocol. These connections are attended to by the API manager (API Management) who is in charge of orchestrating the interconnection of the APIs, that is, it processes the request received and communicates with the Cluster that contains each of the microservices of the application and determines which microservice to connect to serve the request. Once the microservice receives the request, it communicates with the PostgreSQL or Redis database through read and write operations. In addition, communication through the WebSockets protocol allows real-time updating when classifying a new student. Each of the microservices is packaged in artifacts through Docker that are managed by Google Kubernetes Engine.

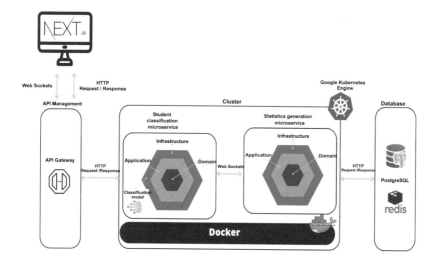

Fig. 4. Software application and microservices architecture

4.3 Interaction Flow with the Software Application

The software application features a main screen enabling the user to access a dashboard displaying statistical graphs representing both "approved" and "failed" students who have been classified through machine learning algorithms. Each of these graphs has filters that permit the user to make selections, thereby allowing for a more specific view of student information tailored to individual user requirements.

In addition, the software application allows the user to input demographic information and student scores, and then predict the student's approval status. These results are visualized thanks to the model implemented with machine learning algorithms.

The process of predicting the student's approval status begins when the user chooses the option 'Classify New Student'. This option presents a Frontend for the user to input the enrollment details of the student. Subsequently, a validation of the entered data occurs. If everything is correct, a new Frontend is presented, allowing the user to input the academic data, including the grades obtained by the student in the learning components of a specific subject.

After registering all the requested values, the application validates, processes, and forwards the data to the machine learning model for the prediction process. Finally, the prediction result is presented, offering detailed information on the estimated approval status, along with the decision rules used to determine whether a student will 'approve' or 'fail' a subject.

5   Results

In this section, we provide details on the experimental setup and the results of evaluating the machine learning algorithms using quality metrics that enable the selection of the best prediction model.

5.1   Experiment Setup

Several experiments were conducted with the Random Forest (RF) algorithm, Neural Networks (NN), and Logistic Regression (LR) using different values for the employed hyperparameters. The optimal configuration that allowed for the maximization of algorithm performance was:

- RF: number of trees in the forest n_estimators $= 100$, maximum tree depth max_depth $= 5$.
- Neural network: hidden_layer $= 4$, hidden_layer_sizes $= 120$, max_iter $= 1300$, activation $=$ "relu".
- Logistic   regression: Intersection with the decision function fit_intercept $=$ True, max_iter $= 100$, multi_class $=$ "auto".

For the experiments, the Monte Carlo cross-validation technique [4] was used to compute the classification accuracy. The trained models were used to predict the Status class with each algorithm selected. Testing and training sets percentages are same for the two tested datasets: testing set $= 20\%$, training set $= 80\%$

5.2   Experimental Results

Table 2 contains the metric values obtained for each tested machine learning algorithm.

Table 2. Metrics Results

Metric	RF	LR	NN
Precisión	0.940	0.880	0.925
Sensitivity	0.900	0.820	0.770
Specificity	0.896	0.822	0.771
F-Score	0.920	0.850	0.820

Based on the Table 2, the Random Forest algorithm exhibits superior performance across all utilized metrics compared to the other algorithms. The Neural Network is the second algorithm to present good precision; however, the results obtained in the other metrics are lower than Logistic Regression and Random

Forest. Therefore, from the purpose of prediction, we can conclude that the Random Forest model displays a very high level of precision in predicting approved cases. This means a degree of precision of 94% in the predictions made. The model generated with the Random Forest was chosen to be implemented in the software application.

The Random Forest model inferred 13 decision rules for predicting the approval status of a student in a specific subject. These rules showed the variables identified as significant are of academic nature, corresponding to the components of autonomous learning, teacher interaction, and practical experimentation, as well as the pass rate and the weighted variables for each component.

6   Conclusions

This article introduces a case study focused on predicting academic performance in a Higher Education Institution, employing DevOps practices to build a software application that contains some tasks that require executing a prediction model to classify the student.

We evaluated three distinct machine learning algorithms: Logistic Regression, Random Forest, and Neural Network using the students' dataset. The results indicated that Random Forest outperformed the other algorithms across all evaluation metrics. We find that this algorithm gives relatively accurate predictions and are able to provide more reliable prediction. Furthermore, this algorithm determined that academic variables were more significant in predicting students' academic performance. Subsequently, the model generated with the Random Forest was integrated into a software application developed in Python, employing DevOps practices.

The prediction model can serve as an early warning mechanism to identify at-risk students and enhance the performance of those with lower academic achievement. Additionally, the development of a software application will assist teachers in accessing information regarding the academic performance of their students.

As a future work, there is an intention to expand the prediction model with data on students' interaction with the virtual learning platform. Additionally, there are plans to extend the software application by implementing new functionalities that will enable the bulk upload of enrollment and academic data for new students in the Information Technology career.

The results presented in this article seek not only to show the efficiency of ML models for predicting academic performance in higher education but also to contribute to the advancement of research in applying DevOps practices in the educational context, providing valuable insights for the academic and research community.

References

1. Abdul Bujang, S.D., et al.: Imbalanced classification methods for student grade prediction: a systematic literature review. IEEE Access **11**, 1970–1989 (2023). https://doi.org/10.1109/ACCESS.2022.3225404
2. Alkayed, M., Almasalha, F., Hijjawi, M., Qutqut, M.H.: Factors analysis affecting academic achievement of undergraduate student: a study on faculty of information technology students at applied science private university (2023)
3. Chikhaoui, A., Chadli, A., Ouared, A.: A model-based DevOps process for development of mathematical database cost models. Autom. Softw. Eng. **30**(2), 23 (2023)
4. Chlis, N.: Machine learning methods for genomic signature extraction. Thesis doctoral, Technical University of Crete school of Electronic & Computer Engineering Digital signal & Image Processing Lab (2015). https://doi.org/10.13140/RG.2.1.1515.6324
5. Huang, C., Zhou, J., Chen, J., Yang, J., Clawson, K., Peng, Y.: A feature weighted support vector machine and artificial neural network algorithm for academic course performance prediction. Neural Comput. Appl. **35**(16), 11517–11529 (2023). https://doi.org/10.1007/s00521-021-05962-3
6. Karalar, H., Kapucu, C., Gürüler, H.: Predicting students at risk of academic failure using ensemble model during pandemic in a distance learning system. Int. J. Educ. Technol. High. Educ. **18**(1), 63 (2021). https://doi.org/10.1186/s41239-021-00300-y
7. Kumar, A., Nadeem, M., Shameem, M.: Machine learning based predictive modeling to effectively implement DevOps practices in software organizations. Autom. Softw. Eng. **30**(2), 21 (2023)
8. Leite, L., Rocha, C., Kon, F., Milojicic, D., Meirelles, P.: A survey of DevOps concepts and challenges. ACM Comput. Surv. **52**(6) (2019). https://doi.org/10.1145/3359981
9. Mohd, N., Abd, N., Sahran, S.: Identification of student behavioral patterns in higher education using k-means clustering and support vector machine. Appl. Sci. (Switzerland) **13**(5) (2023). https://doi.org/10.3390/app13053267
10. Nagy, M., Molontay, R., Szab, M.: A web application for predicting academic performance and identifying the contributing factors, pp. 1794 – 1806 (2019)
11. Nahar, K., Shova, B., Ria, T., Rashid, H., Islam, A.: Mining educational data to predict students performance: a comparative study of data mining techniques. Educ. Inf. Technol. **26**, 6051–6067 (2021)
12. Nuankaew, P., Nasa-ngium, P., Nuankaew, W.S.: Application for identifying students achievement prediction model in tertiary education: learning strategies for lifelong learning. Int. J. Interact. Mobile Technol. **15**(22), 22–43 (2021)
13. Parihar, A.S., Gupta, U., Srivastava, U., Yadav, V., Trivedi, V.K.: Automated machine learning deployment using open-source CI/CD tool. In: Khanna, A., Polkowski, Z., Castillo, O. (eds.) Proceedings of Data Analytics and Management. LNNS, vol. 572, pp. 209–222. Springer, Singapore (2023). https://doi.org/10.1007/978-981-19-7615-5_19
14. Sorgalla, J., Wizenty, P., Rademacher, F., Sachweh, S., Zündorf, A.: Applying model-driven engineering to stimulate the adoption of DevOps processes in small and medium-sized development organizations: the case for microservice architecture. SN Comput. Sci. **2**(6), 459 (2021)

15. Vanarase, R.: Building farsighted intrusion discovery employing ML algorithms (2018). https://doi.org/10.1109/ICCUBEA.2018.8697692
16. Vázquez-Ingelmo, A., García-Holgado, A., García-Peñalvo, F.J.: C4 model in a software engineering subject to ease the comprehension of UML and the software. In: 2020 IEEE Global Engineering Education Conference (EDUCON), pp. 919–924 (2020)

Author Index

Á. Rocha et al. (Eds.): WorldCIST 2024, LNNS 989, pp. 309–310, 2024.
https://doi.org/10.1007/978-3-031-60227-6

Printed in the United States
by Baker & Taylor Publisher Services